Failure Characteristics Analysis and Fault Diagnosis for Liquid Rocket Engines

Wei Zhang

Failure Characteristics Analysis and Fault Diagnosis for Liquid Rocket Engines

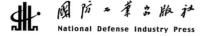

National Defense Industry Press

Springer

Wei Zhang
Xi'an Research Institute of High-Tech
Xi'an
China

ISBN 978-3-662-56994-8   ISBN 978-3-662-49254-3   (eBook)
DOI 10.1007/978-3-662-49254-3

Jointly published with National Defense Industry Press

Printed on acid-free paper

This Springer imprint is published by Springer Nature
The registered company is Springer-Verlag GmbH Berlin Heidelberg

Preface

Being composed of different components working with different programs, the liquid rocket engine (LRE) is a complex system that is widely used in the aerospace field. It can provide thrust for a vehicle outside of the atmosphere. However, the LRE is always working in execrable conditions with high temperatures and heavy corrosion. Therefore, it is necessary to monitor the condition of an LRE to enhance the reliability and safety of the thrust system and understand its working performance; it would also be useful to establish an advanced condition monitoring and fault diagnosis system for the LRE. Involving several disciplines and technologies, the core of LRE health monitoring is the technologies of fault detection and fault diagnosis. In recent decades, a special theory of fault detection and diagnosis was developed using digital computer technology, the theory of automatic control, signal processing, artificial intelligence, reliability theory, and engineering to provide a theoretical foundation for the development of fault detection and diagnosis in the LRE.

In this book, the mode, model, and characteristics of the LRE's fault are investigated. Then, the corresponding fault diagnosis methods, including methods based on the model, signal processing, and artificial intelligence are studied. Developed under the guidance of modern control theory and the modern optimization method, the model-based method carries out fault diagnosis by analyzing the residual obtained by the Luenberger observer, equivalent space equation, Kalman filter, parameter estimation, and identification based on the specialized criterion or threshold. It has attracted much attention because the model-based method can be closely combined with the control system and is a precondition of condition monitoring, fault tolerance control, system modification, and reconstruction.

By referring to the production of the residual, the model-based method can be divided into several methods, including the condition estimation method, parameter estimation method, and the equivalent space method. Comparatively, the method based on signal processing does not rely on an accurate model and has good suitability. Although most parts of signal processing-based methods are proposed on the assumption that the system is linear, they are easily expanded to nonlinear systems. The methods involving Kullback information criterion, wavelet transform, time series analysis, and information fusion are included in the signal processing-based methods. With the development of the system, the system is becoming more and more complex and it is very difficult to establish an accurate model. In this instance, an artificial intelligence-based method, which does not rely on an accurate model, can be used for fault diagnosis. For example, the artificial neural net (ANN), expert system, pattern recognition, revolution algorithm, fuzzy logic method, and a combination of these methods are included in the intelligence-based method.

An LRE is a complex system involving high temperatures, high pressure, heavy corrosion, and strong power release. It is very difficult to establish an accurate model for the whole system, and the model-based method may be restricted in the engineering application. Comparatively, the signal processing-based method and the artificial intelligence-based method are advanced in the processing of the complex character of the object. In this instance, by combining the character and developing trends of the LRE fault diagnosis, wavelet transform, artificial intelligence, fuzzy logic, statistic learning theory, and gray theory are applied in the fault diagnosis of the LRE. This book concentrates on the subjects of monitoring technology of a healthy LRE, including its failure analysis, fault diagnosis, and fault prediction; several fault diagnosis methods have been investigated by combining the fault cases of LRE. We hope to have achieved the following characteristics in the writing of this book:

Advanced. The latest investigation achievements are summarized in this book and corresponding content is advanced and novel.
Applicable. As it is focused on the application of the fault character analysis and diagnosis technology in the LRE, this book can be used as the reference by researchers and engineers.
Readable. Although the theory is very complex, the content of this book, including the basic mechanisms, synthesis methods, and application cases, are logically arranged for the sake of understanding.

The content of this book is the culmination of research by our group. Many graduate students, including Tian Gan, Xu Zhi-gao, Yang Zheng-wei, Li Ming, Lin Xiang-jin, Gao Zheng-ming, Liu Chong-yang, Tian Lu, Xu Hai-bo, and Ming An-bo, worked on different parts of this book. All authors of references are appreciated because they provide rich material for the content in this book. We also thank the publishers for their help in the publishing process.

Because of restrictions in the knowledge of the authors, errors may appear in this book. All comments and corrections are cordially accepted!

January 2016   Wei Zhang

Contents

Prologue

Astronautical technology is a type of synthesis engineering technology that aims to explore, develop, and use celestial bodies in outer space; it is an integrative sign of modern technology development. Moreover, military astronautical technology—an important part of the national defense industry—is very important for improvement in military information levels and national safety. However, the structure of an astronautical vehicle is very large and the environment of space flights is very complex. The launch and reclamation of the vehicle is very difficult. Furthermore, the demand for astronautical technology that carries people is increasing, while maintaining the safety and reliability of these vehicle faces great challenges.

Generally, a liquid rocket engine (LRE)—a kind of heat hydrodynamic system—is the main power source for space shuttles; its running condition is critical for the safety and reliability of the vehicle. Therefore, the investigation of the fault diagnosis and prediction of an LRE is very important for the improvement of astronautical activities. Using achievements in the areas of propulsion theory, signal processing, pattern recognition, artificial intelligence, sensor technology, and computer technology, the aim of the investigation of the fault diagnosis and prediction of an LRE is to enhance running reliability and flight safety. Recently, investigations on the fault diagnosis and prediction of LREs have achieved great improvements in China. The theoretical system of fault diagnosis and prediction was founded on the background of LREs and successfully applied in a ground test of space vehicle and military LREs. However, a systematical technical monograph on fault character and fault diagnosis technology has not yet been written in the area of military LREs. I am very pleased to see that this monograph on the basic theory and methods of the fault character, diagnosis, and prediction of military LREs summarizes the basis of the investigation and engineering practice of the past decades. This book is very significant for the improvement of military LRE fault diagnosis and engineering applications throughout the world.

Based on the LRE model with fault and characteristic analysis, the achievements on the fault characteristics and diagnosis technology of military LREs have been introduced. In particular, theoretical methods, including the artificial neural net, wavelet analysis, intelligence computing, fuzzy theory, statistic learning theory, time series analysis, and the gray system model, are applied on the fault diagnosis and prediction of LREs and provide technological support for the novel design, scientific maintenance, and safe operation of weapons.

This book will help in the expansion and dissemination of fault diagnosis technology of military LREs and space vehicles. It will play an important role in the future of engineering.

Jinji Gao
Member of the China Engineering Academy
Vice President of the Equipment Management Association of China

Chapter 1
Introduction

Nowadays, the space and missile technologies have become the metrics of the state scientific and technological development level in the world and led the development of physics, astronomy, mechanics, materials science, electronics, thermal, optical, chemical, and automation technology, computer technology, remote sensing technology, and other disciplines and technology. A great many of crossover and interdisciplinary are produced, and the developments of a national machinery industry, electronic industry, metallurgical industry and chemical industry, and other basic industries are also promoted. As an important manifestation of the comprehensive national strength, the space and missile technology have great influence on the political, economic, military, scientific, and cultural status of a country. In this instance, various countries have taken a great number of attentions on the development of the astronomy technology and space technology by treating them as important areas for the future development. Soon afterward, some developing countries such as China, India, Brazil, Pakistan, and South Korea also began to develop independent missile and astronautics technology [1, 2].

1.1 Necessity for the Fault Diagnosis and Condition Monitoring of Liquid Rocket Engine

Bringing the power source by itself, the liquid rocket engine (LRE) is not restricted by the operation environment and is the major power device working outside the atmosphere. Furthermore, owing to the advantages of high performance, powerful adaptability, high reliability, and low costs, the LRE has been widely used in the astronautics areas.

It is known that the LRE is composed of many different components and comprehensive relations exist between each other. With the increasing of the payload, values of the running parameters are also increased and the working

© Springer-Verlag Berlin Heidelberg and National Defense Industry Press 2016
W. Zhang, *Failure Characteristics Analysis and Fault Diagnosis
for Liquid Rocket Engines*, DOI 10.1007/978-3-662-49254-3_1

environment of the component is become more and more complicated. For example, the temperature and pressure are higher, the ablation and erosion of the fluid medium are worse, and the amplitude and scope of the vibration are bigger. Therefore, the required reliability and safety of the LRE is become more and more higher than any time before and corresponding investigation on the condition monitoring has been attracted more and more attentions in the astronautics areas in pace with the fast developments of the corresponding science and technology [2].

However, as the most widely used engine in the space launches, the LRE is the most frequent and easiest part in the spacecraft owing to the poor working environment. According to statistics, 60–70 % of lunching failures are caused by the faults in the LRE. During 1990–1994, there were 15 times of lunching failures and 14 times of them were caused by the fault in LRE. In addition, no matter manned or not, aerospace vehicles will cause significant economic losses once any breakdowns occur. For the manned spacecraft, the safety of the astronaut is much more important than the economic consideration. In this instance, improving the reliability of the LRE is the key to enhance the reliability and safety of space activities by the consideration of both personnel safety and economic benefits.

Apparently, researches on the health monitoring system (HMS) require additional costs and the payload will be decreased by the development and usage of HMS. At the same time, the system complexity may be increased and the reliability may be affected. However, combining the above analysis, it can be seen that the HMS is overall significant for the safety, reliability, availability, and economy of the spacecraft system and the fault diagnosis on the LRE has become one of the key issues at present [3–11, 180].

Synthesized by the computer technology, automatic control theory, mathematical statistics, artificial intelligence, signal processing, sensor technology, and engine technology, the condition monitoring technology of LRE is primarily composed of the system or subsystem fault mode analysis methods, fault detection and diagnostic technology, fault prognostic technology, and the abnormal condition control technology, and is very important for the enhancement of the operability, reliability and security, the improvement of the fighting capability, and the reduction of the launch costs. Generally, applications of the condition monitoring of LRE include the following aspects.

- Improve the safety of manned flight vehicle

In January 1986, the United States even the whole world were shocked by the crash of Challenger shuttle. It is known that the loss and bad influence caused by the crash were very large. Since then, many attentions have been taken on the reliability of the craft engine and the development of the LRE. Corresponding condition monitoring technology was also effectively pushed by this disaster.

- Enhance the safety of rocket and missile at the off stage

In the off stage, loads on LRE change very fast with huge amplitudes and many faults may be happened. In March 1992, a launch disaster of CZ-2E rocket was avoided by the turn off signal sent by the emergency shutdown system on the rocket when an abnormal signal was detected. It is indicated that a fast condition monitoring system is needed in the off stage.

- Establish the fault detection and diagnostic system for the ground tests of LRE

The fault detection and diagnostic system for the ground test can not only improve the safety and reliability of the test program, but also being the test stand of the existed fault detection and diagnostic system. Moreover, the existed fault diagnosis methods and the sensor technologies can also be evaluated.

- Establish a condition monitoring system for the missile at the storage stage

For the storage of missile, there has been no efficient method for the evaluation of the safety, reliability, good rates, and reusability owing to the poor storage environment. In this case, a condition monitoring system used in the storage stage is needed.

- Establish the fault detection and diagnostic system for the launching stage of LRE

The reliability in the launching stage is one of the key elements related to the successful launch and corresponding fault detection and diagnosis need to provide real-time decision owing to the fast condition variation.

- Develop the fault monitoring and diagnostic system of LRE on the rocket or missile

Operation conditions of the LRE are very important for the improvement of the shooting accuracy. A fault monitoring and diagnostic system can regularly assistant the daily work of LRE. Once any fault occurs, appropriate control measures can be taken in a timely manner and the rocket can be protected as most as possible.

- Develop the health analysis and evaluation system for the LRE

Rooting on the synthesis analysis of the data obtained in ground test or flight, the health analysis and evaluation system can provide basis for the comprehensive evaluation of expert. In the design and manufacture stage, the fault detection and diagnostic system can be used for the fault analysis and location determination, which is helpful for decreasing the design costs and shortening the development cycle. In the real application, it can maintain the normal running of the LRE and provide basis for the emergency determination.

From 1980s, investigations on the HMS of Space Shuttle Main Engine (SSME) have made great improvement and corresponding HMS is the most advanced at present. In this instance, various HMSs have to be developed for the safety improvement of different LREs.

1.2   History and Development of LRE Fault Diagnostics Technology

Fault detection and diagnosis technology were not generated with the birth of the LRE. Only some redundancy and fault tolerance technology were used to prevent the occurrence and development of the fault in the original LRE. In the early 50s of twentieth century, the condition monitoring of LRE, including the integrity tests and parameter monitoring, was performed by the manual observation in the ground test and the emergency off button was ready to be pushed down by the leader at any time. As the increasing of the propulsion force, transducers were applied on the condition monitoring of the LRE and corresponding monitoring and control could be performed in the control center by researchers. In the last 1960s, digital recording system designed for the test of LRE were developed and parts of red line alarm technologies performed by the manual observation were replaced by the automatic red line alarm technologies. However, such system has not been still applied on the spacecraft. In the early of 1970s, a novel digital processing and control system were designed for the SSME and the application of computer technology on the LRE came true. In the early 1980s, the ground tests of LRE have already been performed completely by the computer controlling and the conditions of LRE were monitored or controlled by lots of signal channels with different colors. In the medium 1980s, the system for anomaly and failure detection (SAFD) was established by the enhancement of the red line alarm capability. Since the late 1980s, investigations on fault mode, fault detection, diagnostic algorithm, fault control, and special sensor technology were performed and several fault diagnosis systems, such as SAFD, flight accelerometer safety cutoff system (FASCOS), engine data interpretation system (EDIS), PTDS, and APDS, were applied or tested on SSME.

Compared with America aiming to SSME, former USSR focused on the LREs of РЛ-170, РЛ-120, and РЛ-0120 to develop the technical diagnostic system. Emphasized on the investigation on mechanism and mathematic method, the technical diagnostic system establishing static and dynamic model first could simulate LRE faults and study the fault condition identification methods [12].

In China, investigations of the health monitoring technology were begun at late 1980s. Establishing the static and dynamic model of LRE [13], Chen and Zhang introduced the artificial neural net (ANN) [14], cluster analysis [15], fuzzy theory [16], and qualitative model reasoning [17, 18] to build a framework for the fault detection and diagnosis of LRE. Improvements were obtained in recent years. For example, qualitative bond graph model [19], support vector machine [20], rough set theory [21], plume spectrum [22], etc. were gradually applied on the fault detection and diagnosis of LRE. For some subsystems of LRE, corresponding diagnostic systems have become quite mature [23, 24].

All above works have built a solid foundation for the development of HMS. Summarized the development of fault detection and diagnosis methods, primary achievements were concentrated on the following several aspects to solve problems of the fault detection, feature selection and extraction, and fault mode recognition [12, 25, 26].

1. Signal processing-based methods, including threshold detection algorithm, vibration analysis methods, etc. As one of the most concise and common threshold detection methods, the red line alarm was applied on the SSME with the advantages of simple, clear logical structure. However, the red line alarm was not able to detect the incipient fault and was easy to make mistakes or miss faults. What worse is that the red line alarm cannot even identify faults of the LRE or those of sensors. In this instance, some adaptive threshold algorithms (ATA) were proposed and improved. For example, the systems of abnormal and fault detection (SAFD), ATA [27, 28], adaptive correlation algorithm (ACA) [29], adaptive weighted sum square algorithm (AWSSA), envelop algorithm (EA) [30], and adaptive correlation safety band were developed. Calculating the thresholds of parameters by three steps, SAFD can detect the LRE fault earlier than the red line alarm method in the ground test of SSME in early of 1990s. Although the reliability has been improved, SAFD was just suitable for the stable condition program and was not sensitive to some faults. Comparatively, ATA, ACA, and AWSSA have improved the suitability no matter they were still limited in the fault detection of stable conditions.

Vibrations and fluid pressure pulsations were important parameters for the condition indication of LRE. Generally, vibration signal can be analyzed in time domain, and frequency domain of the time–frequency domain. Many statistic indexes, such as peak, average amplitude, RMS amplitude, root amplitude, skewness, kurtosis [31], etc., were used in time domain, while FFT, PSD, high-order spectrum analysis, cepstrum analysis, and maximum entropy spectrum analysis were performed in the frequency domain. Wavelet analysis and WD distribution were the most widely used time–frequency analysis methods.

FASCOS is a red line alarm system based on the RMS detection in SSME. Several channels of vibrations on the turbopump were collected and corresponding thresholds were determined by the probability distribution. Although used in several times of ground test of SSME, FASCOS has not been applied in flight test owing to the lower reliability of the equipments than that of the turbopump.

In 2001, real-time vibration monitoring system (RTVMS) was developed to apply on the vibration analysis of the turbopump in SSME with much higher safety and reliability. Involving 14 DSPs and 32 A/D converters, RTVMS can deal with 32-channel vibration signal at the same time and 10 channel vibrations, including two channel vibrations on the high-pressure hydrogen turbopump and a channel vibration of hyperbaric oxygen turbopump, can be independently alerted by red line alarm analysis. The RTVMS was the major vibration monitoring system of SSME by experience 150 times of statistic tests with none of the system faults.

2. Model-based method, including time series analysis, modal analysis, state space method, and fault tree method [31]. As the most widely used time series method, ARMA (Autoregressive Moving Average) [33] model can evaluate the current value of signal by the usage of the previous value based on a time series model. Two ARMA models have been developed using the signals of oxidizer

flow, fuel flow, pressure before oxidizer injection, chamber pressure, and turbopump speed on the liquid missile. Strategy of the fault detection was performed by calculating the confidence intervals of the autocorrelation function of error signal. Comparatively, the gray model (GM), using the data to determine parameters of the differential equation, was applied to detect the leakage and blocking faults in LRE [34]. The diagnosis method based on the gray relational analysis can be used for the mode classification. When some faults occur, the dynamic characteristic features of the LRE may be different from the normal condition. For example, the structure stiffness may reduce and the damping may increase if cracks were appeared in LRE. In this instance, modal analysis can be used for the structural fault detection for the reason that the modal shapes are sensitive to the structural parameters. Comparisons of the modal shapes between the normal and abnormal conditions can make the fault recognized. Furthermore, a quantitative identification can be performed by modal assurance criterion (MAC).

Developments of the chaos and fractal dynamic theory have provided novel technologies for the fault diagnosis of the large complex nonlinear systems. Phase space can be reconstructed by the time series of the LRE parameters and singular spectrum can indicate the fault feature. Combining the fuzzy cluster method, the singular spectrum is able to classify different faults. On the other hand, the fractal dimension is efficient in the feature extraction and can be used for the fault diagnosis, although calculation of the fractal dimension is quite complicated.

Placing the failure at the top and unfolding the related faults layer by layer based on the structure and function of the object, the fault tree is a kind of digraphs which can reflect the propagation relationship of the faults and holds the advantages of high speed detection, easily modifying, and good consistency. Moreover, the fault diagnosis can be achieved in any area once the fault tree was constructed. However, construction of the fault tree is very difficult because that the mode and mechanism analysis, basis of fault tree analysis, are very complicated. The diagnostic result is relied on the complication of the fault tree and the unknown faults cannot be identified.

3. Artificial intelligence-based method, including ANN, pattern recognition, expert system, and information fusion. Not relying on an accurate mathematic model, the artificial intelligence-based methods can deduce the condition of the system by the knowledge learned from the collected data [37]. It is very convenient because that the condition of the system can be obtained even when the internal structure of the system is unknown.

With capabilities of parallel processing, high adaption, powerful self-learning, and good fault tolerance, the ANN has gotten a great development in the fault detection and diagnosis of LRE in the past several decades. The primary applied ANNs include adaptive resonance theory-based ANN [38], back propagation (BP) net ANN [39], winner-take-all-based ANN, multiple layer ANN [31, 40, 41], and dynamic ANN. Belonging to unsupervised learning net, the ART-based ANN includes models of ART-1, ART-2, and ART-3. Using the power spectral density to calculate the weight of input vectors, ART-2 model can detect the abnormal vibration and nozzle ablation failure in LRE. Comparatively, the feature collection, extraction, and cluster were combined in the BP-ANN, although lower amounts of calculation were required. Therefore, the BP-ANN was beneficial for the real-time fault diagnosis. In this instance, combining both WTA and BP-ANN, a compound ANN-based fault detection system can be used to evaluate the fault size. Using the collected data as the input vectors, the dynamic ANN can detect faults by the error signal of the condition identification. Faults separation was further performed by the autocorrelation function of the error signal even though a LRE model was not involved.

Pattern recognition [42] was introduced in the fault diagnosis of LRE by comparing condition parameters with a database constructed on the basis of test data and experience. The optimal statistical clustering analysis, proposed by the combination of evolution strategies and genetic algorithm, was applied on the clustering of 560 simulated samplings. Fourteen classes were involved and each sampling contained 68 features. Comparatively, the fuzzy Kohonen clustering networks (FKCN) [43] were used in the cases with higher degrees of noise, since a sliding window was involved in the beginning of the signal processing. If the priori probability density of random noise can be calculated by the test data, the FKCN can be further improved to the real-time fault diagnosis.

Expert system is an intelligent computer program which is able to solve complex problems by the usage of knowledge and reasoning model developed by the knowledge and experience of top experts [31]. Lots of expert systems have been developed on the fault diagnosis of LREs. Pre-launch expert system (PLES) developed by Rocketdyne was used for the pre-launch data analysis of LRE. Expert system for fault diagnosis (ESFD) proposed by Tennessee University can reason faults using the normal and abnormal mode extracted in the test data of SSME. Furthermore, the knowledge database of ESFD can be expanded by the combination of the structural knowledge to enhance the fault identification capability. In addition, LeRS, MSFC, and Aerojet jointly developed an expert system, named automated post-test data review system for liquid rocket engine (APTDRS) to assist the determination of the fault detection. For the first-stage engine of Titan rocket, Aerojet developed the Titan Health Assessment Expert System (THAES) to deal with the test data analysis and condition identification. Data analysis work needing 6 h manually can be completed within only one minute by THAES. Developed by the combination of Alabama University and MSFC, EDIS was the first practical expert system used for the ground assessment of SSME. However, EDIS involved very complicated reasoning program and was inefficient, because the system model of the whole engine and many subsystems were integrated together. Comparatively, investigations on the expert system in china were aimed at the main subsystem of LRE by the construction of fault database and basic diagnostic strategies. In summary, the knowledge acquisition, system self-learning, and real-time application of expert system need further improvements.

1.3   Development Trend of the LRE Fault Diagnosis

From above analysis, it can be seen that investigations of fault diagnosis system of LRE are concentrated on the typical LRE of America, Russia, France, and China and corresponding fault diagnosis methods are improved with the time pass by. However, only those designed for the SSME are mature for applications. Therefore, the development of the LRE fault diagnosis system needs to be improved in the following several aspects.

1. Establish fault databases for different LREs and construct corresponding failure development gallery for different faults [26].

   As shown in Fig. 1.1, the mechanism of the fault is clearly illustrated in the failure development gallery, sketch of the occurrence, and development of the fault. The relationship of the faults is very visible and is convenient for the evaluation of the location of fault. Therefore, the failure development gallery plays an important guiding role in the design and manufacture of the LRE.

2. Extract similar fault information from different LREs and establish a universal diagnostic platform based on a unified model.

   Generally speaking, structures of LREs are different from each other if they are marked by different types and the detailed faults may be different. However, a unified model may be proposed based on the commonality of different types of LREs because the design philosophies of different LREs are almost similar. The synthesized model can integrate common features of different types of LREs and corresponding faults are shared by all different types of LREs. In this instance, investigations on the synthesized model are suitable for many different types of LREs and much work can be saved. Apparently, corresponding fault diagnosis system aiming to the synthesized model is more universal and portable. Many attentions can be saved and focused on the problems which are more serious.

3. Introduce new theory and propose novel method to solve difficult problems in the fault diagnosis.
4. Develop an accurate model which can simulate the whole working program of the LRE and introduce different types of fault by changing corresponding parameters. In this case, the fault mechanism can be further understood and the control capability of the fault can be improved.

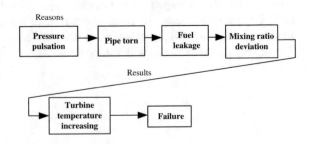

Fig. 1.1 Failure development gallery

5. Develop a health monitoring and fault diagnosis system with good performance
   and high efficiency.

Currently, kernels of fault diagnostic systems on different LREs are generally developed on different fault diagnosis algorithms and the transplantation of fault diagnostic systems on different LREs cannot work with high efficiency. Therefore, the integration of different fault diagnostic algorithms needs to be developed in a diagnostic system and the probability of misdiagnosis and missed diagnosis are inclined to be decreased as little as possible.

On the other hand, the combination of the condition simulation and fault diagnostic system is needed. The collected parameters on the truth LRE can be input into the simulation system and the real-time work in the interior of LRE can be seen on the simulated system. In this case, the understanding on the LRE condition can be further enhanced and comparisons between the normal and real-time conditions on the simulated system will indicate the fault immediately. Both occurrence moment and location of the fault can be visualized by the simulate system. In this instance, the health monitoring and fault diagnosis system with higher efficiency, more powerful function, and more intuitive visualization is the developing trend with the development of LRE. Knowledge of multiple subjects and joint efforts of researchers around the whole world are needed.

Chapter 2
Failure Pattern and Corresponding Mechanism Analysis of LRE

2.1 Introduction

Liquid propellant rocket (LPR) has been studied, produced, stored, and used in China for over 50 years, the performance and fault analysis of the engine, core component of LPR, is involved throughout the process of study, production, storage, usage, and even retirement. The reason is mainly shown in the following aspects. First, the mission of space flight and the rocket weapon needs the LRE holding high reliability. It is known that each launch of the space vehicle costs more than hundred millions dollars. However, about 60 % fault of the carrier rocket is caused by the LRE failure. Second, fault analysis is the development basis of the design, production, and improvement of new products. By finding the weak point of the product and inheriting the mature technology, lot of cost can be saved and the development period can be shortened in the development of the new product. Third, the fault analysis can provide an important guarantee for the storage of rocket weapon. Since, the rocket is a long-term storage and one-time use object in poor environment, little failure information can be collected. Therefore, the fault analysis must be carried out to meet the increasingly demand of the storage, usage and launching decision, usage, to provide a basis for launching decision-making and optimize storage environment to the rockets, extending service life and provide the basis for weapon system, for weapon system maintenance, repair, management to provide a basis, and method guidance.

In this chapter, the mechanism of the LRE fault is studied based on statistical analysis of the rocket engine components and system failure modes. The standard models and performance characteristics of the common faults are established to facilitate the burden of fault analysis, detection, and diagnosis.

© Springer-Verlag Berlin Heidelberg and National Defense Industry Press 2016
W. Zhang, *Failure Characteristics Analysis and Fault Diagnosis
for Liquid Rocket Engines*, DOI 10.1007/978-3-662-49254-3_2

2.2 Structure of LRE

LPR is composed of the propellant charging system, engine system, and propellant transportation system [44–46]. The engine system, choosing the double component gas generator cycle, is mainly comprised of the thrust chamber, turbine pump, gas generator, valve, cavitation pipe, heat exchanger, etc. In the thrust chamber, the propellant is burnt to produce high temperature and high pressure gas and transformed to the nozzle to produce the thrust. The turbo pump including gas turbine and centrifugal pump supercharges the fuel and oxidant. The gas generator using the fuel and oxidant as same as the thrust chamber is used to produce the working gas to turbine with the lower oxygen coefficient than the thrust chamber. In the supply lines of fuel and oxidant pumps, the pump valves are installed in order to separate the propellant tank from engine and the engine can be stored safely when the tank is filled with propellant. In the supply line of the propellant, the main valve and throttle valve are installed. Furthermore, to ensure the stability of the system, cavitation venturi tube, one-way valve, and the auxiliary valve are installed in the supply line behind the pump. The starting material of the turbine is provided by the explosive actuator. There are two heat exchangers. The first one, pressurizing oxidizer tank by the vaporization of the oxidant is located in the turbine exhaust pipe, the other one is placed in the supply line between the fuel pump and the thrust chamber partition and is used to pressurize the fuel tank by part of cool fuel-rich gas.

The propellant transportation system is comprised of propellant tank, relief valve, safety valve, sensors, pipe lines, and other components and is mainly used for charging, discharging, and storing the required propellant by the demand.

Propellant charging system, used to charge the rocket tank, releases propellant and stores the propellant with long time and is comprised of the hydraulic system which is comprised of gas path system, liquid road system, circuit system, and measurement control system.

2.3 Failure Pattern Analysis of the LRE

Development of liquid rocket fault monitoring and diagnosis system is beginning with the deep understanding of various failure modes. However, the failure mode which is varied with the type of engine brings great difficulty and inconvenience to the study. Foreign country (especially the United States and the Soviet Union) does the amount of research on failure mode, such as the United States announced "the list of SSME Failure mode, effect analysis and key project" in 1987 after analyzing the failure record during the past 70 years on the SSME. Finally, the fourteen critical faults with SSME were determined which covered five failure mode with occurrence frequency.

In the 80s later, the Soviet Union analyzed the fault of various liquid rocket to propose that faults were divided into three categories according to the time of failure: The first failure which occurs is shorter (about 0.05 s) than the completion of the work; The second occurs for 0.04–0.05 s mainly in the oxidant pump and can be tested and diagnosed by the algorithm based on the parameters such as pressure, temperature, etc.; The third occurs about tens seconds which is mainly in the bad sealing condition of combustion agent and gas tank, which can be tested and diagnosed by the algorithm based on parameter information. But, practice shows that it cannot completely prevent the appearance of the failure especially for the first failure during the liquid rocket works especially during flight, such as turbine and gas channel wear, turbine blade after working for a long time. Therefore, it is necessary to study the regional and failure mode of the fault.

In our propulsion technology field, the analyses of failure mode and effect are widely used in the engine fault analysis, which are based on test the statistical data, from the management of production and quality, test procedure control, design improvement, etc. These analyses and studies are very beneficial for the engine fault detection and diagnosis. Due to the previous failure analysis focused on quality management and design improvement, few qualitative and quantitative analysis to the variation of the engine parameters under failure is done. Even to the monitor of the main parameters of test or flying (such as oxidant injection pressure, turbine pump speed, and engine thrust, etc.), people usually use deviation zone test with strong empirical deviation inspection which often causes large false alarm and leakage alarm to error shutdown and drain off. Therefore, it is imperative to strengthen and improve the study of engine failure mode.

For the rocket engine in the storage and service condition, the main contents of fault analysis are composed of the investigations on the working condition of the failure occurrence, the manufacturing and service history before the failure, mechanical analysis of the failure engine, analysis and evaluation of the failure materials, analysis of abnormal situations, failure mechanism statistics and decision suggestion, etc. It notes that the mechanics analysis is a generalized conception, which contains the stress and strength analysis of the engine when the fault occurs. The stress is composed of all factors that may cause the failure and the strength is the ability to prevent any failures.

According to statistical analysis of the previous run-in, storage, usage, and other conditions, the failures occurred in the start-up procedure of the rocket engine are mostly caused by starting device including starter and starter valves, etc. In the stable working stage, the obstruction, leakage, and turbine system failure, etc., may occur. In the shutdown process, the failure may appear at the main valve and shutoff valve. From the failure phenomenon, the liquid path failure is mainly divided into three categories: the obstruction, the leakage, and the efficiency drop of pumps. According to the location of the fault, there are the thrust chamber fault, turbine fault, pump fault, gas generator fault, pipeline, and pressure vessel fault, etc. All faults may cause the performance degradation, parameter inconsistent and component damages. Comparatively, there are three types of gas path faults, such as the blocking, the leakage, and function failure. All these faults may occur at the high pressure vessel, the automatic device, the valve and the pipeline fault, etc. For the circuits, measurement and control system, failures are mainly composed of the short circuit, open circuit, sensor failure, or drift and actuator failure, etc.

2.4 Failure Mechanism Analysis of the LRE

2.4.1 Thrust Chamber and Gas Generator

By converting the chemical energy of liquid propellant to the jet kinetic energy and generating the thrust, the thrust chamber is composed of the liquid propellant injector, the combustion chamber, and the nozzle. Liquid propellants are injected in the combustion chamber with specified flow rate and mixing ratio at first. Then by atomization, evaporation, mixing, and combustion process, high temperature and pressure gas is produced and accelerated in the nozzle to be supersonic flow. The thrust is the anti-force of the supersonic flow performed on the thrust chamber.

Generally, the turbine power is supplied by the gas generator in the pump LRE. Similar to the structure of the thrust chamber, the gas generator works in the condition with lower mix ratio than the thrust chamber and provides power for the turbine by the gas. The power cycle of the engine will be influenced and expanded to the whole engine system if the gas turbine failure occurs.

The common failures of thrust chamber and gas generator include the injector clogging and ablation, combustion chamber tearing and cracking, unstable combustion, and nozzle ablation, etc., and all of these failures may occur in the process of study, production, and usage.

1. Injector clogging

Injector is a special device in which propellant component is introduced into the combustion chamber and atomization, which is easy to nozzle blockage because of the small diameter of the nozzle and the special structure form when the processing debris, foreign body, and floccule in propellant flow through the nozzle. Nozzle blockage will change the mixing ratio of the combustion chamber which can cause very high local temperature, the combustion chamber can still work when the degree is not enough to change the main system propellant flow that the venture-tube still works and the overall mix of the combustion chamber is not changed.

When the degree has caused the change of the main system propellant flow, the overall mixing ratio and flow rate of the combustion chamber have been changed which can cause the drop of the engine specific impulse, pressure drop of combustion chamber, and engine thrust reduction.

2. Injector ablation

Ablation of the injector nozzle will occur when unreasonable cooling design, decreased cool effect due to blockage of cooling nozzle, backflow of high temperature gas in combustion chamber, and high frequency unstable combustion in the combustion chamber happens. Ablation of the nozzle will cause the nozzle flow characteristics, decrease atomization effect, reduce and fluctuation of combustion temperature, pressure and thrust drop.

3. Ablation of thrust chamber throat

The insufficient cooling flow, blockage of cooling channel, the uniform mixture ratio of the propellant and asymmetric local mixing lead to combustion lag, which is easy to cause ablation of thrust chamber throat. Ablation of throat will decrease the rate of engine nozzle to area and increase throat to radius ratio, which can lead the increase of the flow rate of the nozzle, the pressure drop of the combustion chamber, the engine's specific impulse, and thrust loss.

4. High frequency instability combustion of thrust chamber

The phenomenon is called unstable combustion when the working parameters of the combustion chamber are disturbed and the vibration of large amplitude of the pressure takes place. When unstable combustion occurs, strong vibration of the engine, mechanical damage of thrust chamber and other parts, and injector face, nozzle, ablation of combustion chamber interior, and melt will cause engine failure. It is called high frequency instability combustion when the vibration frequency is more than 1000 Hz, which is the most easy to occur, the most difficult to suppress and must be solved, and occurs with the character of high volatility, high pressure fluctuations, sometimes up to 50–100 % of stable pressure, and vibration acceleration of tens to one hundred g, even to the flame appears strong flash and "sharp whistle".

The length of combustion chamber can be set to L_k, and the radius is R_k with closed two ends, in which it is filled with stationary and uniform gas and the natural frequency is:

$$f = \frac{a}{2}\sqrt{\left(\frac{q}{L_k}\right)^2 + \left(\frac{\alpha_{m,n}}{R_k}\right)^2} \qquad (2.1)$$

Formula:

a Sound velocity of gas at steady state

q, m, n The integer, respectively, for the longitudinal, radial, and tangential vibration mode of the resonance frequency

$\alpha_{m,n}$ Coefficient.

During the working process of the combustion chamber, the flow fluctuation of the transportation system, start or stop of the engine, the instability and inconsistency of nozzle, and other parts of the mechanical vibration transfer will cause the disturbance of gas parameters (pressure, temperature); when the frequency of the parametric disturbance is consistent with the vibration frequency of the combustion chamber, all gas is regularly shocked with the inherent acoustic frequency in the combustion chamber, if the source of the disturbance is maintained or continuously excited, the high frequency instability combustion can be maintained and increased.

High frequency instability combustion has been effectively suppressed through several decades of research and experience from continuous summary [7], the effective measure to prevent the high frequency unstable combustion is to prevent the existence of the initial disturbance source for the rocket in the service.

5. Low frequency unstable combustion of gas generator

It is called low frequency unstable combustion, when the frequency of the unstable combustion chamber is lower than 300 Hz. Low frequency unstable combustion is the interaction between the pressure in the combustion chamber and propellant flow of conveying system, usually occurred in the engine start, stop or turn, which is prone to occur under the condition of low thrust or lower nozzle pressure drop relatively. Low frequency unstable combustion rarely occurs in modern engines with high pressure in the combustion chamber, but occurs in gas generator with low mixed and bad combustion organization or small thrust chamber. When low frequency unstable combustion occurs, it will lead to the transmission line and the other connecting components which are damaged, kinds of automatic devices cannot work properly, and it is possible to stimulate high frequency instability.

6. Abnormal cooling process of cooling jacket

Thrust chamber and gas generator are working in the high temperature, pressure, and speed environment, and with huge heat transfer, so it is necessary to effectively cool to thrust chamber and gas generator to guarantee the normal work.

It usually chooses combined cooling method, but regenerative cooling is the most basic and effective way which utilizes one propellant component (or two propellant components) as the cooling fluid, flows in certain velocity through the cooling channel between the inner and the outer walls of thrust chamber (or gas generator) before it enters the combustion chamber to take away the heat that transfers through gas to the inner wall. The cooling passage between the inner and outer wall is called cooling jacket, which is made in special processing and with narrow channel gap, and is easy to block or fluid channeling caused by sealing off after high temperature which will cause the reduced flow rate, the decreased cooling, and the abnormal cooling process. When it is in very serious conditions, it will result the ablation of injector face nozzle, thrust chamber inner wall and throat, the reduced thrust, and even to the failure of engine.

7. Mechanical damage

Mechanical damage is manifested in the cracks, tear, cracking, and fracture of the body or part. The reason is that the unreasonable structure design, the poor manufacturing process, the material defect, the big work load, and the other faults cause the violent vibration. The failure easily lead to leakage of propellant and even explosion, the accident will be disastrous. It mainly occurs in the engine development, rarely in the product after stereotypes.

2.4.2 Turbo Pump

1. Dynamic model of a turbo pump rotor system

Figure 2.1 shows Turbine pump system of liquid rocket.

Combustion agent pump, oxidant pump, turbine, inducer of two pump, and external force gear are placed on the same shaft which is supporting by two ball bearings. Seal is used for the isolation of each working medium. The whole system is firmly fixed on the engine. The shaft and the rotary table can be dispersed into the mass block and the rotor engine with no mass, so the dynamic model of the rotor system of the turbine pump can be established, as shown in Fig. 2.2.

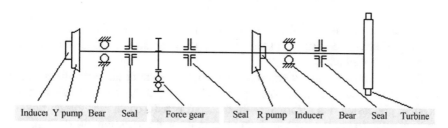

Inducer Y pump Bear   Seal   Force gear   Seal R pump Inducer   Bear   Seal   Turbine

Fig. 2.1 Schematic diagram of turbo pump system

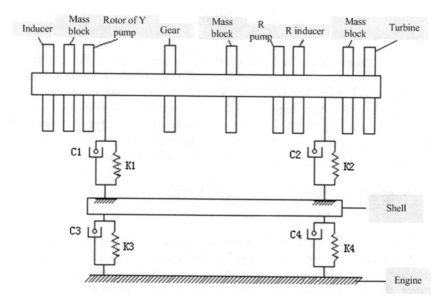

Fig. 2.2 Rotor system model of turbine pump

There are 4° of freedom per mass on the rotor (shift x,y and angle of rotation $\theta x, \theta y$), 6° of freedom on shell foundation (shift x,y,z and angle of rotation $\theta x, \theta y, \theta z$), and generally 2° of freedom on bearings, so the differential equation of the system is:

$$M\ddot{X} + \Omega J\dot{X} + C\dot{X} + KX = R \tag{2.2}$$

Formula: Mass matrix of the system:
J Matrix associated with the moment of the system, the elements of which are the polar moment of inertia of the mass block
Ω Rotational angular velocity of the rotor
C System damping matrix
K Stiffness matrix of the system
R The external force array acting on the system.

The natural frequency and vibration mode of each order, unbalance response and stability of the system can be obtained by appropriate simplification of using various analysis methods.

For the turbine pump system, there is the gyro torque because the disks which are not in the middle of the two support, which is equal to elastic moment, in the positive precession, the deformation of the shaft is reduced and the elastic rigidity and critical speed of the shaft are improved, in the anti-precession, the deformation of the shaft is increased, the rigidity and critical speed of the shaft are reduced, in the case of small deformation and rotation speed, the gyro torque is small, and cannot be considered, here (2.2) is:

$$M\ddot{X} + C\dot{X} + KX = R \tag{2.3}$$

By formula (2.2), the stability, natural frequency, and vibration mode of the system will change when the system exists disturbance force or the variation of force R array, mass matrix M, damping matrix C, stiffness matrix K, etc. Some kind of symptom will appear when the parameter of the system changes or stimulated by the outside from the above formula, on the contrary, the system's change (fault) can be deduced according to the signs that is the basis of fault diagnosis of which is based on the symptoms. The relationship between the excitation force (external, internal), the symptom, and the system dynamic characteristic is shown as follows:

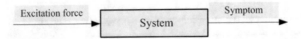

When the system is affected by some kind of stimulation (external or internal) which can cause the response, we can find out the incentive source and detect the fault by the response and analysis of the incentive. Response will change as the system change, which is in the form of self-excited vibration that is mainly the fractional frequency or fractional frequency hysteresis vibration, and with nonlinearity. During the analysis of fault mechanism for each fault, the first thing is to analyze the incentive and system and establish a simplified system model to find out the relationship between signs and failure.

2. Mechanism and characteristics of common failures of turbine pump system

The turbine pump is prone to failure, because it works with short start-up and shut down, the quick varied conditions, the turbine works in high temperature, high pressure, high speed, and pump also works with high pressure, high speed, flammable, explosive, highly toxic, strong corrosion in the propellant, which is shown in vibration mode, so that fault monitoring and diagnosis for the turbine pump is based on vibration parameters. It is necessary to the turbine pump vibration monitoring of whether the ground test or the flight. In order to monitor and analyze the fault, the mechanism and characteristics of the common failure of turbine pump should be analyzed and summarized. The mechanism and characteristics of common failures of turbine pump system are as follows:

1. Rotor unbalance

Rotor unbalance is divided into two kinds: rotor mass eccentricity and defect of rotor components. Mass eccentricity is caused by the manufacturing error, the assembly error, and the uneven quality of the rotor. Defect of rotor components refers that the parts such as (impeller, blade) damage locally, shed, and fragments fly off in operation to cause the imbalance, especially for turbine blades, induced blade wheel. In the turbo pump system, if the quality of the mass is imbalance in not considering the influence of the gyro moment, the external force excitation is caused by the mass eccentricity, formula (2.2) will be:

$$M\ddot{X} + C\dot{X} + KX = me\omega^2 \cos \omega t \tag{2.4}$$

Formula:
M   Eccentric mass
e   Eccentricity
ω   Rotor angular frequency.

The vibration of the system will be forced vibration caused by mass eccentricity, the frequency of which is the same as rotor angular frequency, the M will change when the m mass parts defect and the vibration mode, natural frequency of the system will change, which is not big. The features of the rotor unbalance are shown that: regular time domain wave form and the frequency domain vibration energy are concentrated in the fundamental frequency with stable phase; the peak value and phase inversion occurs at the critical value, orbit of shaft center is elliptical, synchronous forward precession. If the forced vibration frequency of the mass eccentricity is equal or close to the natural frequency of one component in the system, resonance occurs. Because the eccentricity of rotor is inevitable, resonance also occurs in the normal state with small amplitude. But, the component defect in the turbine pump system occurs with sudden nature, it is the basis for division between the component defect and eccentricity. The following parts are often damaged in the turbine pump system:

(1) Blade fracture in rotor

Blade crack, machining marks, and material defects are the fatigue source of rotor blade. When the rotor is rotating at high speed, the blade is subjected to alternating load to result in fatigue damage as to emerge broken marks. Due to the small gap of the turbine cavity, blade fracture is very easy to block rotor and lead rotor blade defect which makes the turbine fail to work, which will lead the failure of engine instantaneous.

(2) Crack of pump-induced wheel

The fault such as the deep machining marks, stress concentration, and poor anti-fatigue performance exists, which make the inducer to produce fatigue fracture in the repeated alternating loads.

(3) Abscission of Turbine blade shroud

Abscission of Turbine blade shroud is caused by the friction that thermal deformation of turbine covers and the little gap between the turbine cover and the turbine guard, which will cause aggravated vibration of turbo pump and the unbalance of turbine rotor at work.

2. Rotor bow

The bending of the rotor in turbo pump system is caused by the unreasonable design of the shaft structure, the large manufacturing error, the uneven quality of the material and the long storage of the material. For the rotor installed in the rocket and turbo pump in rocket, the main cause is that the long-term improper storage leads the plastic deformation of rotor which results in the shaft bending. Bending emerges the same rotational excitation force as the mass eccentricity, so the performance of the rotor bending fault and the mass unbalance caused by mass eccentricity has the same feature. However, the vibration of axial happens when the rotor bends, and the vibration frequency is consistent with the rotation speed, sometimes accompanied with two frequency multiplication.

3. Rotor rubbing

Rotor rubbing happens when the shaft contacts with the fixed part, which is caused by the influence of the imbalance and the constant reduction of the gap between the static and the movement during work. Rotor rubbing is divided into two cases, one is radial collision caused by the contact between the outer rotor (the outer edge of the impeller) and static friction, the other is axial rubbing which is caused by the contact between the rotor in the axial and stator component. For turbine pump, it is common that the radial rubbing between the outer edge of impeller and stator component, the radial rubbing between rotary shaft and the sealing parts, rubbing between shaft and seal rubbing, which cause fracture failure and propellant channeling cavity sometimes failure or explosion.

When rubbing between the rotor and stator component happens, constantly changing stress is produced in the rotating shaft and accompanied with very complicated vibration phenomenon serious to failure. The simplified model of collision and friction is shown in Fig. 2.3, the gap between the outer edge of the impeller and stator component is set as δ, so the positive rubbing F_N and tangential rubbing F_T can be expressed as follow:

$$F_N = \begin{cases} 0 & (e < \delta) \\ (e - \delta)K_c & (e \geq \delta) \end{cases}$$

$$F_T = fF_N$$

Formula:

f Friction coefficient between the rotor and stator component

K_c Integer, respectively, for the longitudinal, radial, and tangential vibration mode of the resonance frequency

$e = (x^2 + y^2)^{1/2}$ Radial displacement of the rotor.

The stiffness of the rotor increases as the rotor and stator component contact instantly decreases when rotor is divorced from the contact after bounce of stator component and with lateral free vibration. Therefore, the rotor stiffness changes between the contact and noncontact, the frequency of which is the rotor vortex frequency which will produce a unique frequency of complex vibration response. The characteristics of rubbing are: cutting head cosine waveform, larger frequency vibration, vibration performance in full frequency, especially larger in the multiplier and divider vibration component, multiples of 1/2 components in typical spectrum, approximate trapezoid of axis orbit.

Fig. 2.3 The simplified model of collision and friction

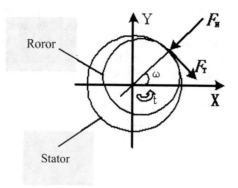

4. Crack of Shaft

The turbine pump works in high temperature, pressure, corrosive environment, and with big load change gradient, temperature change gradient of the shaft, which is easy to crack even to fracture phenomenon and major accidents. The reason is that poor mechanical stress state is emerged in the high cycle fatigue, creep, and stress corrosion cracking, especially in the conditions of high temperature and strong corrosion also complex motion of the rotor ultimately leads to crack of shaft. The crack is in the opened, closed, sometimes opened, and sometimes closed states. The vibration is intensified when the crack is open, the deflection of the shaft is larger than the deflection of the crack, also the phase changes. The crack has no effect on the vibration of the rotor when the crack is in closed state.

When the stress in the crack area is self-weight or other radial load, the crack is periodically opened or closed that cause the complex vibration. The non-symmetry of rigidity caused by crack is not only the function of the depth of the crack, but also the function of the position of the shaft and the running time which is nonlinear vibration with two times, three times and high frequency harmonics of the rotating shaft, at the time the phase change, the crack propagation, the stiffness decreases and the amplitude of the vibration components increases. The harmonic resonance emerge when the working trajectory through 1/2, 1/3, 1/5, at the critical speed of the rotor during the boot process, the phase mutate at critical time, the axis trajectory is double ellipse or irregular.

5. Loosening of mechanical components

Due to the low installation quality in the turbo pump system, the $C3$, $K3$, $C4$, $K4$ in Fig. 2.2 change because of engine vibration that temperature causes inconsistent contraction of different materials, stress relaxation of fixed bolts for long-term storage and seals which result the loose of shell connection, bearing block, inducer and centrifugal wheel, and intensified vibration until to damage. When loosening occurs, periodic jump of the shell bearing block is caused under the action of unbalance force which leads to the change of system's stiffness and with impact response, and K, C will change in the model (2.2) which cause the change in vibration mode and natural frequency, when the loose gap increases, the vibration intensifies and the impact effect increases, which is easy to cause the resonance of the natural frequency of each order, as known nonlinear vibration. Characteristics are: the vibration frequency is precise 1/2 and 1/3, ..., 1, 2, 3, ... times of speed frequency which is expressed as fractional harmonic resonance, the other main characteristics are directional especially in the loose direction, different critical speeds in horizontal and vertical directions, vibration form can jump and the vibration will suddenly increase or decrease when the speed increases or decreases which is not continuous.

6. Fluid seal excitation

In turbo pump rotor system, various forms of sealing device are adopted in order to meet the need of isolation between the working medium, but fluid is sealed in the eccentricity of rotor relative to the fixed part which will cause sub synchronous precession that will result the destruction of the turbo pump rotor.

In the labyrinth seal, the sealing backlash is not uniform due to the error of manufacture, installation and operation, when rotor lean in sealed cavity if the rotor is in the vortex state due to the initial disturbance, the periodic variation of the seal gap between the rotor and the seal and the pressure in the seal chamber will also periodically change which will stimulate the rotor to aggravate the eddy and make the rotor failure.

On the other hand, kinetic energy of the fluid which flows into the sealed cavity cannot be completely lost. One flow along the axial direction and another rotate along the circumferential direction of the rotor. The uneven circumferential gap between rotor and the seal cavity, or the change of circumferential gap as time due to rotor precession motion, will cause inhomogeneous pressure distribution in the sealing cavity, and that the resultant force of the distributed pressure creates transverse component which is perpendicular to the displacement of the rotor which lead to the precession motion and instability of rotor.

The characteristics of fluid seal excited vibration are that: the vortex frequency is less than 0.6–0.9 times of the rotational speed, an ellipse axis orbit, positive precession and vibration frequency is sub-harmonic that is less than 0.5 times of speed frequency, which is always with one time, two times and higher harmonics, the instability of the vibration and phase, the disorder of the axis orbit, and the divergence.

For high speed gas turbine, when the rotor bends or turbine eccentricity which causes the uneven gap between tip and turbine casing in the circumferential, there is the sum of the axial forces except couple which acts on the blade, but also transverse power that is perpendicular to the displacement of the rotor which leads to the precession motion and instability of rotor, the vibration characteristics of which are similar to the fluid seal vibration.

7. Pump cavitation

During the working process of the pump, when static pressure of somewhere is below the cut-off saturation vapor pressure at that temperature that bubbles emerges and volume expanse, bubble condenses into liquid with volume shrinkage in the high pressure zone which leads the increasing pressure to the huge hydraulic shock. It is known as the pump cavitation that cracks and erosion forms by the alternating pressure shocks on the impeller surface. When cavitation occurs, the formation, growth, and rupture process of bubble will reach tens of thousands of times per second, the local pressure can reach hundreds of MPa, which results in the decline of flow rate, outlet pressure, and efficiency, until the cutout and impeller damage. Vibration extends to the whole rotor system, self-excited oscillation emerges. Low frequency oscillation of pump outlet pressure, wide spectrum of mechanical vibration, and noise which is easy to cause the resonance of the rotor system, so pump cavitation occurs several times in the development and production process. In the using, it will also induce pump cavitation in such condition (low pump pressure). It should strictly control the cavitation in the practical operation.

Characteristics of cavitation are: vibration performance is shown as broad spectrum of mechanical vibration and with a sudden, mild noise and "baba" sound when starts, vibration intensified and accompanied with detonation when severe, low frequency oscillation of pump outlet pressure, capacity, efficiency, and lift decreased rapidly.

8. Bearing damage

The bearings used in the turbo pump are the ball bearing which bear the radial and axial load, work in the serious environment and usually appear as the mode of the fatigue spalling, damage, fracture, and so on. Stiffness, damping, and external exciting force of rotor system all changed when bearing fail which lead to the instability of rotor system, intensified vibration, and axial float, at last the friction between rotor and other parts are severe which lead propellant channeling cavity to explode. Therefore, it should detect the bearing damage before the rotor loses stability.

Characteristics of rolling bearing failure are: intensified vibration, wide vibration frequency which is from below the low frequency (1 kHz) to high frequency even to very high frequency (thousands of Hz), accompanied with noise and impact, vibration of blade passing frequency and its harmonics and fractional frequency emerges, and temperature rise.

The impact vibration will generate at the surface fatigue spalling of the bearing parts, one is called "passing vibration" that low frequency pulsation when rolling body sequentially rolls over the defects of working surface which is subjected to repeated impact, and the corresponding frequency is known as "passing frequency" which can be obtained through the bearing parameter, general below 1 kHz. It is difficult to diagnose because of the big random noise and fluid dynamic noise. The other is natural vibration, including the natural frequency of each part and the acceleration sensor, and with 1–60 kHz frequency band. When the transient impact occurs and the natural vibration is excite, so the bearing fault and fault element can be judged according to the impulse frequency.

9. Pulsation of fluid pressure

Pulsation of fluid pressure is the phenomenon that the fluid flows by some disturbance to cause the variation of pressure. The reason for which contains in the pump fed liquid rocket system is that limited blades of centrifugal pumps, combustion instability of gas generator, tank pulsation of pressure caused by longitudinal vibration, hydraulic impact of propellant transport network, combustion instability of combustion chamber, leak, etc. Now, the technology has been quite mature to suppress combustion instability of the middle and low frequency in liquid rocket system, but pressure pulsation of the pipeline also happens when fault occurs such as plug of leakage, throttle, and liquid container. The resonance happens which is disastrous when the pressure pulsation is the same as the natural frequency of the piping system and the component, and the pulsation of pressure also affects the combustion process of incendiary agent to lead the unstable working process.

The frequency of the pressure pulsation is lower generally under the 1 kHz. The excitation force is the pressure pulsation excitation in the formula (2.2).

The characteristic frequency of the rotor response to the pressure fluctuation is the pressure pulsation frequency. When the pressure pulsation frequency and the natural frequency of the rotor coincide, which will lead instability of rotor system and vibration intensification, and the vibration direction is along axial direction which emerges the vibration of the blade passing frequency.

10. Structure resonance

The turbine pump rotor system is coupled with the engine. The high and low frequency unstable combustion during the engine work cause the severe vibration which is transmitted to the turbine pump system through the connecting device, and the pipeline vibration of the propellant conveying system will also be transmitted to the turbine pump system by coupling, the K, C, and R in the type (2.2) will alter which cause the vibration mode of the turbine pump rotor system. When the frequency of external excitation force is the same as natural frequency of the rotor system, the structural resonance of the rotor system will lead catastrophic accidents.

11. Gear damage

The gears in the turbo pump generally work in high speed, high power, and corrosive medium, and are close to the high temperature turbine and two pumps with great temperature difference. Damage is mainly due to the high speed of gear pitch circle, the large centrifugal force, the large meshing surface contact stress, severe fatigue or spalling caused by meshing surface lubrication and poor cooling. The large change of load causes large load impact on the gear transmission during the start-up of the engine, which is prone to bending fatigue and broken teeth. To the turbine pump, the main failure of the gear box part is the "pitting" caused by the free peeling of the gears and the "broken teeth" caused by the bending fatigue and impact.

Gear box is with quality and elasticity, and the meshing gear pair is a vibration system. When the gear rotates, the transmission force periodically changes with impact which causes the vibration of gear. Under the condition of no fault, gear transmission also vibrates which is known as conventional meshing vibration with approximate sine meshing waveform of vibration, frequency based on meshing frequency and harmonic component.

When the gear failure, change of vibration amplitude caused by fluctuation of tooth surface load that emerges complex frequency modulation and amplitude modulation vibration signal. For "pitting" and "broken teeth faults", the feature of the spectrum characteristics is mainly the rotation frequency, which is shown in Fig. 2.4.

Vibration excitation force which is mainly based on fc will intensify the vibration of the turbo pump system; it will cause the rotor system instability especially when the natural frequency of the rotor system is coincident.

Fig. 2.4 Typical spectrum of
gear fault

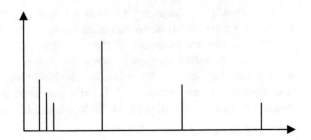

2.4.3 Seal Components

Seals are widely used in the engine as a key component of the pneumatic and
hydraulic system. The space and rocket accident which are caused by the failure of
the seal are disastrous. The main failure mode is leakage, wear, and physical damage.

(1) Sealing mechanism of seal

The sealing performance of static seals are relied on the assembly of
pre-deformation and pressure transferred to medium, realized by the contact pressure and adhesion that seal adhere to the metal surface, as shown in Fig. 2.5.

The sealing force P:

$$p = p_0 + kp_j + f(t, T) \tag{2.5}$$

Formula:

P_0 The contact pressure generated by the pre-deformation is caused by the
        fastening bolts

P_j The pressure of working medium

$f(t, T)$ The adhesion force between the seal and the metal surface is a function of
        time t and temperature T

Fig. 2.5 The sealing theory

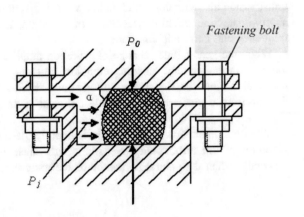

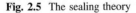

k The pressure transfer coefficient, which is related to the sealing material, $k = \mu/(1 - \mu)$, is the Poisson coefficient of the sealing material.

Reliable sealing must meet:

$$P \geq P_j \cos \alpha \qquad (2.6)$$

Formula:

α The angle between the tangent line of curvature of the contact surface and the sealing material

It just shows that:

1. When the pre-tightening force $P0$ reduce, the sealing force P reduce. The seal fail as the sealing force P reduce to a certain degree. The main cause of this failure is vibration and metal stress relaxation.
2. The vary of aging, corrosion, and temperature of the sealing material causes the change of k, μ, $f(t, T)$ and the decreases of sealing force, which lead to the failure of seal.
3. It may fail that the sealing force will not be guaranteed when the medium pressure P_j increases.

The seal material is mainly rubber, metal parts, and fluorine plastic part in the rocket engine, and fluorine plastic part is in the majority which is not easy to aging because of its special molecular structure. External shock to seal is small in the storage condition, only affected by assembly stress. Therefore, it cannot be considered long-term storage, the aging of fluoride and mechanical damage.

(2) Aging mechanism of rubber seal

The phenomenon is known as aging that high molecular compound is influenced by internal and external factors in using which causes the original mechanical properties and appearance shape change as time goes on. Aging is divided into two kinds, one is the main chain of the degradation that makes the molecules small, the performance of which becomes sticky, the other is to be oxidized, that peroxide is formed and rubber becomes brittle because of excessive cross-linking. The factors that affect the aging are: cross-linking deformation of the internal polymer and filler, environment (chemical and biological), and physical (light, heat, moisture, air and so on), which not only works alone but also in the form of coupling function. For rubber materials, the aging damage deformation is:

$$\varepsilon = Ktb \qquad (2.7)$$

Formula:

ε Accumulation of aging damage deformation of rubber materials;

b   Constant related to variety;

K   Aging rate, $K = Ae^{-\frac{B}{T}}$, *A*, *B* is constant, *T* is temperature (K).

Thus it can be seen from the formula, temperature and time are the biggest impact on rubber aging. In dynamic environment such as dynamic sealing and vibration, heat produced by friction leads the rising temperature that accelerates the aging and reduces the rubber life.

The aging test of a certain type of engine seal showed that: the aging speed of the early stage is fast, but the aging of the rubber seal is not consistent to some certain.

(3) Principle of metal stress relaxation

Experiments have proved that: the leakage of seal is not caused by aging of the seal but the loosening of the fastener. Therefore, Variety of the relaxation and metal fasteners force problems should be mainly considered during the analysis of seals' performance.

Stress relaxation is the process that the stress reduces as time. The relationship between stress relaxation and time is:

$$\Delta\sigma/\sigma = K1\mathrm{Ln}(1 + rt) \tag{2.8}$$

Formula:

$\Delta\sigma$ Variation of stress

σ Stress

$K1, r$ Relaxation constant

t Relaxation time.

It can be known: the fastener will automatically release in the storage and using, and the preload of the spring will automatically reduce. Therefore, measures should be taken to prevent some fasteners loose, that the fastener and the spring preload should be retighten before using. It should take measures to prevent vibration in the process of transport and using, because vibration will impose additional load on the force of the fastener and accelerate the stress relaxation.

It is similar with the failure mechanism and the analytic procedure of seal such as the leakage and function fault in other parts such as valve, regulator, filters, piping, pressure vessels, and power actuating device, except the defect structure.

2.5   Standard Failure Pattern of the LRE

The occurrence and development of fault are a process which supply time for the fault detection and diagnosis. In addition that faults are always shown in signs, there are early signs even to sudden failure which provide possibility and evidence for fault detection and diagnosis. Sign is represented in many aspects, sometimes

Table 2.1 Standard failure mode of thrust chamber

Failure mode		Fault characteristics and performance													
		Combustion chamber pressure		Injection drop		Combustion chamber vibration			Engine thrust		Cooling jacket outlet temperature		Venturi pressure		
		Drop	Fluctuation	Enlarge	Fluctuation	High frequency	Low frequency	Heavy intensity	Drop	Fluctuation	Drop	Deviation from normal values	Drop	Fluctuation	Rise
1	Injector plug	1		1					1			1		1	
2	Nozzle ablation	1	1		1					1					1
3	Throat ablation	1							1				1		
4	High frequency unstable combustion		1			1		1		1				1	
5	Abnormal cooling of cooling jacket											1			
6	Low frequency unstable combustion		1	1			1	1		1	1	1		1	

Note 1 indicates that the corresponding feature has occurred, 0 which means that the corresponding feature does not occur

Table 2.2 Standard mode of common failure of turbine pump system

Failure mode			Fault characteristics and performance																		
			0.00–0.39fr	0.40–0.49fr	0.50fr	0.51–0.99fr	1.0fr	2.0fr	3.0–5.0fr	Oddfr	>5.0fr	Meshing frequency	Roar	Component resonance frequency	Axial vibration	Gradual change	Mutation	Variation with load	Variation with speed	Main monitoring site	Pump outlet pressure
I	1	Component mass eccentricity					0.90	0.05	0.05				1	1		1			1	1	
	2	Blade fracture					0.90	0.05	0.05				1	1			1		1	1	1
	3	Shaft bending					0.90	0.05	0.05				1	1	1	1		1	1	1	1
II	4	Rotor radial rub	0.10	0.05	0.05	0.10	0.30	0.10	0.10	0.10	0.10		1	1	1	1		1	1	1	1
	5	Rotor axial rub	0.05	0.05	0.05	0.05	0.30	0.20	0.10	0.10	0.10		1	1	1	1		1	1	1	1
III	6	Shaft crack					0.40	0.20	0.20		0.20		1	1		1		1	1	1	
	7	Shaft rigidity is not equal						0.80	0.20				1	1		1		1			
	8	Bearing damage	0.10	0.10			0.40	0.20	0.20				1	1	1		1		1	1	
IV	9	Body connection loose	0.20	0.20			0.30	0.10	0.10	0.10			1	1			1		1		1
	10	Pressure fluctuation	0.20	0.20			0.10	0.10	0.30		0.10		1	1	1				1		1
	11	Structural resonance	0.20	0.20			0.50	0.10					1	1	1		1		1	1	1
V	12	Rotor component loosing	0.40	0.40			0.10		0.10				1	1			1	1	1	1	1
	13	Cavitation	0.50	0.30			0.05	0.05	0.05	0.05	0.05		1	1			1	1	1		1
	14	Loose bearing	0.70				0.20		0.05	0.05			1	1		1	1	1	1	1	
VI	15	Gear damage					0.20					0.80			1						

Note fr the working frequency of the rotor, the vibration spectrum of the space is 0; 1 represents that the corresponding feature is true, the space is 0, which means that the corresponding characteristic is false

Table 2.3 Standard failure mode of gas generator

Failure mode	Fault characteristics and performance													
	Chamber pressure		Injection drop		Combustion chamber vibration			Turbine speed		Outlet temperature of turbine		Engine thrust		
	Drop	Fluctuation	Enlarge	Fluctuation	High frequency	Low frequency	Intensity	Drop	Fluctuation	Drop	Deviation from normal values	Drop	Fluctuation	Rise
1 Injector plug	1		1					1		1		1		
2 Nozzle ablation	1	1		1					1		1	1		
3 Propellant mixing uneven		1							1		1		1	
4 Low frequency unstable combustion		1	1			1	1		1				1	

Note 1 indicates that the corresponding feature has occurred, 0 which means that the corresponding feature does not occur

Table 2.4 Regulator standard fault mode

Failure mode		Fault characteristics and performance						
		Upstream pressure	Downstream pressure		Switch status indicator		Characteristic frequency	
		Normal deviation	Drop	Normal deviation	Normal	Abnormal	Medium flow	Leak
I II III IV V VI 1	Internal leakage		1	1	1		1	
2	Stagnation (switch failure)	1		1		1		1
3	Regulation failure	1		1	1		1	
4	External leakage		1	1	1		1	1

Note 1 indicates that the corresponding feature has occurred, 0 which means that the corresponding feature does not occur

Table 2.5 Valve standard fault mode

Failure mode		Fault characteristics and performance						
		Upstream pressure	Downstream pressure		Switch status indicator		Characteristic frequency	
		Normal deviation	Drop	Normal deviation	Normal	Abnormal	Medium flow	Leak
I II III IV V VI 1	Internal leakage	1	1	1	1		1	
2	Stagnation (switch failure)	1		1		1		1
3	External leakage	1	1	1	1		1	1

Note 1 indicates that the corresponding feature has occurred, 0 which means that the corresponding feature does not occur

explicit and sometimes implicit. It is not possible and necessary to detect all these signs in the monitoring and diagnosis. Some are known as the feature of fault which can character the faults. Features can be the fault performance of the research

Table 2.6 Standard failure mode of pipe line

Failure mode		Fault characteristics and performance							
		Upstream flow		Downstream flow		Upstream pressure		Downstream pressure	
		Increase	Drop	Increase	Drop	Increase	Drop	Increase	Drop
Jam	I		1		1	1			1
	II								
	III								
	IV								
	V								
	VI								
Leak	1								
	2				1				1

Note 1 indicates that the corresponding feature has occurred, 0 which means that the corresponding feature does not occur

Table 2.7 Standard failure mode of pressure vessel

Failure mode		Fault characteristics and performance				
		Pressure		Maintain pressure		Characteristic frequency
		Deviation from normal values	Drop	Normal	Drop	
I II III IV V VI 1	Leak	1	1		1	1
2	Deformation					1
3	Safety valve failure	1		1		

Note 1 indicates that the corresponding feature has occurred, 0 which means that the corresponding feature does not occur

Table 2.8 Fault mode of filter standard

Failure mode		Fault characteristics and performance							
		Upstream flow		Downstream flow		Upstream pressure		Downstream pressure	
		Increase	Drop	Increase	Drop	Increase	Drop	Increase	Drop
I II III IV V VI 1	Jam		1		1	1			1
2	Leak				1				1
3	Filter element damage	1			1			1	

Note 1 indicates that the corresponding feature has occurred, 0 which means that the corresponding feature does not occur

object, but also can be the fault performance of research object in the whole system which should be more consideration in the establishment of the condition monitoring and diagnosis system. The features shown in failure modes are also many, and it is not only consuming machine time but also not necessarily good if these features are using directly for diagnosis. So it is necessary to make appropriate and necessary choice of the fault feature, but also make the dimension as less as possible. Therefore, it is necessary to establish the standard fault model of the engine which is not only beneficial to the online monitoring and diagnosis system, but also facilitate the development, production, and use. According to the above analysis, the standard failure modes of the engine components are established as shown in Tables 2.1, 2.2, 2.3, 2.4, 2.5, 2.6, 2.7 and 2.8.

The above table is the initial value of fault standard mode, which should be developed and improved in the practical application and should be amended accordingly for specific equipment and system.

Chapter 3
Analysis Methods of Failure Model for LRE

3.1 Introduction

LRE is a very complicated system working under severe environment and the requirement for the system is very high. Therefore, it is necessary to research the fault analysis, detection, and diagnosis. Fault analysis, especially the fault phenomenon, mode, and mechanism of the analysis, is the basis of fault detection and diagnosis. There are mainly three kinds of methods for the fault analysis. As the most basic method, the first one is to accumulate all kinds of fault phenomenon and patterns and analyze the reason based on the ground test. However, it is impossible to get a complete fault connotation due to the limited number of tests. The second one is carrying out fault simulation experiment on the test stand and accumulating fault data. In principle, this is the most effective and reliable way to reflect the fault process and analyze the fault mechanism. Unfortunately, it is limited by the conditions of equipment, funding, risks, and test times. Some faults are not physical reoccurrence or not allowed to be tested at all. The third one is the mathematical computer simulation of fault state. This is one of the most economical ways and most important steps in fault analysis, the principal means to analyze the fault mode. The key factor of this method is to get the appropriate mathematical model and the effective calculation method [47–56]. On the basis of previous test data and experience, this chapter establishes the steady state work model, dynamic model, and fault model of LRE according to the dynamic equations of liquid rocket propelling system and lays a foundation for the fault simulation, analysis, detection, and diagnosis of LRE.

© Springer-Verlag Berlin Heidelberg and National Defense Industry Press 2016
W. Zhang, *Failure Characteristics Analysis and Fault Diagnosis for Liquid Rocket Engines*, DOI 10.1007/978-3-662-49254-3_3

3.2   Working Process of LRE

The whole working process of LRE can be roughly divided into these steps: propellant the rocket with propellant, ignite the first-stage engine, ignite the second-stage engine and shutdown the first-stage engine, shutdown the second-stage engine and cope with emergency.

1. Filling propellant

   The first step is to fill propellant into the tank from the tank on the ground.

2. Ignite and start up the engine

   Each LRE has the similar starting up process. After issuing the ignition instruction, the starting valve of incendiary agent opens firstly and then the oxidant starting valve opens. Under the pressure of tank pressurization and liquid, the incendiary agent and oxidant fill the propellant transporting pipelines, thrust chamber, the cooling channels, and liquid collector of precombustion chamber. The powder starter begins to work after electricity and produces the gas to drive the turbine to work. The centrifugal pump works and pressurizes the propellant. The bipropellant goes into the thrust chamber and gas generator and combusts spontaneously. The gas from gas generator replaces powder gas generator to drive the turbine pump to work continuously under the rated condition. The oxidant evaporator and the gas cooler supply pressurized gas to oxidant tank and incendiary agent tank, respectively, to pressurize them. The thrust from thrust chamber increases rapidly to exceed the rocket weight and makes it take off.

3. The steady work of engine

   This is the principal period, in which the engine completes the flying under the rated condition and these parameters such as the engine thrust, flow, and turbine pump revolving speed reach the rated values.

4. Shutdown

   The engine shuts down after the task. Each engine has the same shutdown procedure. Firstly the secondary cut-off valve of oxidant is closed to make the turbine working medium supply system to stop working and thus the turbine pump stops working. Then the two main valves are closed to make the propellant supply system to stop working or stop supplying propellant to thrust chamber to terminate the thrust.

3.3   Model of Steady State Process for LRE

The LRE is always affected by various interference factors and works under the fault conditions rather than the rated conditions. In order to control the working process of LRE more accurately and ensure the reliable performance index, it is necessary to establish the mathematical model of LRE. Since the LRE consists of some basic components and subsystems such as liquid pipeline, gas pipeline, pressurization system, thrust chamber, and gas generator corresponding model in the steady state should be analyzed by the establishment of every basic component.

3.3.1   Analysis of Liquid Flow in the Pipeline

The flow of propellant in the engine is equal to that of viscous incompressible fluid in the circular tube. It is known that the viscous liquid flow exists in two kinds of states. The first one is the laminar flow and the other one is the turbulent flow. The flow characteristics, calculation methods of the route loss, and the local loss are different from the states of the flow [57–60].

It is known that the laminar flow and turbulent flow can be judged by Reynolds number, which is defined by

$$Re = \rho v d / \mu = v d / v$$

where

μ	dynamic viscosity of the fluid;
$v = \mu/\rho$	kinematic viscosity of the fluid;
ρ	fluid density;
v	fluid velocity;

When $Re \leq 2000$, the liquid flows in the laminar state. Comparatively, the liquid flows in the turbulent state when $Re \geq 2000$.

For the flying rocket, the flow state of the fluid in the pipeline is influenced by the flight height, tilt angle, and acceleration. Assume that: (1) the liquid is incompressible and the flow is uniform one-dimensional flow; (2) the pipe is rigid; (3) there is no heat exchange between the flow and the environment. As shown in Fig. 3.1, the flow in a uniform section circular tube can be described by

$$p_2 - p_1 = R\frac{\dot{m}^2}{\rho} - L\rho g \left(\frac{a + \sin\theta}{g} \right) \tag{3.1}$$

Fig. 3.1 Fluid flow in
circular tube

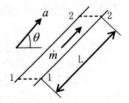

where

p_1, p_2 pressures of the sections 1-1 and 2-2;

L pipe length between the sections 1-1 and 2-2 along the axis;

$\dot{m}$ fluid mass flow;

θ tilt angle of ballistic trajectory;

a axial acceleration of the rocket or missile;

ρ density of fluid medium;

$\frac{a + \sin\theta}{g}$ axial overload coefficient, expressed in nx;

R flow resistance coefficient along L, including the route resistance coefficient and the local resistance coefficient;

g gravity acceleration. $g = g_0 R_0^2 / (R_0 + H)^2$. g_0, the ground gravity acceleration (9.80665 m/s^2). R_0, earth radius (6371 km). H, flight altitude.

In the ground test, $a = 0$, $g = g_0$ and the route resistance coefficient and the local resistance coefficient are generally determined by experiment tests.

3.3.2   Working Characteristic Equation of Engine Parts

Each engine component must complete a specific function in the system. The working characteristic equation is the mathematical description of its function [61].

1. Centrifugal pump

The working characteristic of centrifugal pump can be described by its static characteristic equation

$$\Delta p_o = \left[A_o \left(\frac{\dot{m}_o}{\rho_o} \right)^2 + B_o \left(\frac{\dot{m}_o}{\rho_o} \right) n + C_o n^2 \right] \rho_o \left(\frac{D_o}{\overline{D}_o} \right)^2 \tag{3.2}$$

$$\Delta p_f = \left[A_f \left(\frac{\dot{m}_f}{\rho_f} \right)^2 + B_f \left(\frac{\dot{m}_f}{\rho_f} \right) n + C_f n^2 \right] \rho_f \left(\frac{D_f}{\overline{D}_f} \right)^2 \tag{3.3}$$

where

A, B, C	Pump characteristic coefficient;
Subscript "o"	Oxidant;
Subscript "f"	Incendiary agent;
N	Speed of the pump;
D	Outer diameter of impeller.

The power equation of centrifugal pump is

$$N_o = \left[E_o \left(\frac{\dot{m}_o}{\rho_o} \right)^2 n + F_o \left(\frac{\dot{m}_o}{\rho_o} \right) n^2 + G_o n^3 \right] \frac{\rho_o}{\bar{\rho}_o} \left(\frac{D_o}{\bar{D}_o} \right)^5 \qquad (3.4)$$

$$N_f = \left[E_f \left(\frac{\dot{m}_f}{\rho_f} \right)^2 n + F_f \left(\frac{\dot{m}_f}{\rho_f} \right) n^2 + G_f n^3 \right] \frac{\rho_f}{\bar{\rho}_f} \left(\frac{D_f}{\bar{D}_f} \right)^5 \qquad (3.5)$$

where E_f, F_f, G_f are the pump's power characteristic coefficients.

2. Turbine

The turbine's working characteristic can be represented by its power and flow characteristic.

$$N_t = \Delta h_t \dot{m}_t \eta_t \qquad (3.6)$$

where

$\dot{m}_t$	Gas mass blowing turbine;
Δh_t	Gas' adiabatic expansion power in turbine nozzle;
η_t	Turbine efficiency.

$$\Delta h_t = \frac{k}{k-1} RT_t \left[1 - \left(\frac{p_{t,c}}{p_g} \right)^{\frac{k-1}{k}} \right] \qquad (3.7)$$

where

$p_{t,c}, p_g$	Turbine's inlet and outlet pressure;
RT_t	Fluid's kinematic viscosity;
k	Fluid's density;

$$\eta_t = a \frac{u}{c} - b \left(\frac{u}{c} \right)^2 \qquad (3.8)$$

$$\frac{u}{c} = \frac{\pi D n}{60 \sqrt{2 \Delta h_t}} \qquad (3.9)$$

where

a, b Obtained by the test;
D Turbine's average diameter;
n Revolving speed.

$$\dot{m}_t = \frac{A_{t,t}}{A_{t,t} + A_{j,t}} G_f \tag{3.10}$$

$$G_f = \begin{cases} \Gamma_g \dfrac{p_g A_{g,t}}{\sqrt{RT_g}} & \text{Overcritical flow} \\[3ex] \sqrt{2g \dfrac{k}{k+1} \dfrac{\left(p_g\right)^2 A_{g,t}}{RT_g} \left[\left(\dfrac{p_2}{p_g}\right)^{\frac{2}{k}} - \left(\dfrac{p_2}{p_g}\right)^{\frac{k+1}{k}}\right]} & \text{Subcritical flow} \end{cases} \tag{3.11}$$

where

$A_{t,t}, A_{g,t}$ Areas of turbine nozzle and sonic nozzle;
G_f Gas flow from gas generator;
P_2 Turbine back pressure;
Subscript g Generator.

3. Ordinary pipeline equation

The working characteristic of pipeline and liquid throttling components is

$$p_1 - p_2 = R\frac{\dot{m}^2}{\rho} \tag{3.12}$$

where

R Local and frictional resistance coefficient;
Subscript 1, 2   Upstream and downstream.

The working characteristic of gas throttling components (throttle ring and nozzle) is determined by its flow characteristic

$$\dot{m} = C_d A p_i^* z / \sqrt{RT_i^*} \tag{3.13}$$

When $p_e/p_i^* \le \left(\frac{2}{k+1}\right)^{\frac{k}{k-1}}$, the throat velocity reaches sound velocity and

$$z = \sqrt{k\left(\frac{2}{k+1}\right)^{\frac{k+1}{k-1}}} \tag{3.14}$$

When $p_e/p_i^* > \left(\frac{2}{k+1}\right)^{\frac{k}{k-1}}$, the gas flows with subsonic velocity and

$$z = \sqrt{\frac{2k}{k-1}\left[\left(\frac{p_e}{p_i^*}\right)^{\frac{2}{k}} - \left(\frac{p_e}{p_i^*}\right)^{\frac{k+1}{k}}\right]} \qquad (3.15)$$

where

A Throat areas of throttle ring or nozzle;
p_e, p_i^* outlet static pressure and inlet total pressure of throttle ring or nozzle;
C_d flow coefficient.

4. Cavitation tube

The working characteristic of cavitation tube is determined by its flow characteristic, namely

$$\dot{m}_v = C_{dv}\frac{\pi d^2}{4}\sqrt{2\rho(p_{v,i} - p_s)} \qquad (3.16)$$

where

C_{dv} Flow coefficient;
$p_{v,i}$ Inlet pressure;
p_s Propellant's saturated vapor pressure.

5. Thrust chamber

The thrust chamber is the component that sprays, mixes, combusts incendiary agent and oxidant, and produces gas of high temperature and high pressure which accelerates and converts to mechanical energy to generate thrust. It is mainly composed of combustion chamber and nozzle.

The characteristic of energy conversion in the combustion chamber is described by its characteristic velocity equation

$$\dot{m}_{c,o} + \dot{m}_{c,f} = \frac{p_c A_t}{c_{th}^* \eta_c} \qquad (3.17)$$

where $\dot{m}_{c,o}, \dot{m}_{c,f}$ are the flux of incendiary agent and oxidant into the combustion chamber. The theoretical characteristic velocity c_{th}^* is determined by the combustion chamber pressure, component ratio r_c, and specific enthalpy h_p of the propellant into the combustion chamber.

$$c_{th}^* = f(p_c, r_c, h_p)$$

$$r_c = \frac{\dot{m}_{c,o}}{\dot{m}_{c,f}}$$

$$h_p = \frac{r_c c_{p,o} T_o + c_{p,f} T_t}{r_c + 1}$$

(3.18)

where

$c_{p,o}$, $c_{p,f}$ Ratios of specific heat of incendiary agent and oxidant, respectively;

T_o, T_f Temperatures of incendiary agent and oxidant into the combustion chamber, respectively.

The specific impulse characteristic of thrust chamber is

$$I_{s,ch} = \eta_c \eta_f c_{th}^* C_{f,th}$$

(3.19)

Theoretical thrust $C_{f,th}$ is the function of nozzle area ratio ε and gas specific heat ratio k

$$C_{f,th} = f(\varepsilon, k)$$

(3.20)

The gas specific heat ratio is determined by combustion chamber pressure, component ratio, and nozzle area ratio

$$k = f(p_c, r_c, \varepsilon)$$

(3.21)

The thrust characteristic of thrust chamber is as follows:

$$F = I_{s,ch}(\dot{m}_{c,o} + \dot{m}_{c,f})$$

(3.22)

6. Gas generator

The working characteristic of gas generator is mainly determined by the characteristic of energy conversion.

$$c_p T_t = f(p_g, r_g, \eta_g)$$

(3.23)

$$r_g = \frac{\dot{m}_{g,o}}{\dot{m}_{g,f}}$$

(3.24)

7. Evaporator and cooler

The evaporator and cooler are all called as heat exchanger, whose working characteristic is determined by the heat transfer and flow characteristics, namely

$$Q = KA\Delta T_m \tag{3.25}$$

where

Q Actual exchanged heat;

A Heat exchange area;

K Overall heat transfer coefficient measured by test;

ΔT_m Logarithmic mean temperature difference, determined by

$$\Delta T_m = \frac{\left(T_{og} - T_{el}\right) - \left(T_{e.g.} - T_{ol}\right)}{\ln \frac{T_{og}-T_{el}}{T_{og}-T_{ol}}}$$

The effective heat exchange coefficient of heat exchanger is

$$\varepsilon = \frac{Q}{Q\text{max}}$$

where Q denotes the actual exchanged heat and Qmax denotes the maximal possible exchanged heat. The outlet temperature of the heat exchanger is

$$T_{el} = T_{ol} + \varepsilon\left(T_{og} - T_{ol}\right)$$

In the equations above, the subscripts g and l represent hot fluid and cool fluid, respectively, while subscripts o and e denote the inlet and outlet, respectively.

The flow characteristic is given by

$$p_o - p_e = \xi \frac{(\dot{m}/A)^2}{2\rho} \tag{3.26}$$

where

p_o, p_g Inlet and outlet pressure of serpentine tube;

$\dot{m}$ Flux;

A Flux area of heat exchanger;

ρ Average of inlet and outlet;

ξ Resistance coefficient measured by test.

8. Tank

The tank's working characteristic can be determined by its pressurization pressure performance, namely

$$p = \frac{R}{VC_p} \left[WTC_p \right] \qquad (3.27)$$

where
p Pressure;
W Weight;
R Constant of mixed gas;
T Temperature;
V Volume;
C_p Specific heat.

All these parameters are related to the pressurized gas in tank at any time.

In the steady work state, the pressure of the tank which is a system with complex gas composition can be expressed as

$$P = \frac{R}{VC_p} \left[W_0 T_0 C_{p0} + \int_0^t G_2 T_2 C_{p2} \mathrm{d}t + \int_0^t G_1 T_1 C_{p1} \mathrm{d}t - A \int_0^t p \mathrm{d}\overline{V} - \int_0^t (q_1 + q_2) \mathrm{d}t \right]$$

$$(3.28)$$

where
$W_0 T_0 C_{p0}$ Initial heat of the gas in the air pillow;
$\int_0^t G_1 T_1 C_{p1} \mathrm{d}t$ Heat of the self-pressurized gas;
$\int_0^t G_2 T_2 C_{p2} \mathrm{d}t$ Heat in the pressurized nitrogen;
$A \int_0^t p \mathrm{d}\overline{V}$ Consumed heat by the mixed gas expanding to work in the tank;
$\int_0^t (q_1 + q_2) \mathrm{d}t$ Heat loss from the pressurized gas to the wall in the tank.

3.3.3   Parameter Balance Model of Engine

Parameter balance model is the mathematical expressions which can describe the relationship between the work parameters of each engine component, including such basic formula as flow balance, pressure balance, and power balance.

1. Flow balance

Flow balance means that the engine flow is equals to the flow's sum of thrust chamber and gas generator.

The engine oxidant flow is

$$\dot{m}_o = \dot{m}_{c,o} + \dot{m}_{g,o} = \frac{r_c}{1+r_c}\dot{m}_c + \frac{r_g}{1+r_g}\dot{m}_g \qquad (3.29)$$

The engine incendiary agent flow is

$$\dot{m}_f = \dot{m}_{c,f} + \dot{m}_{g,f} = \frac{r_c}{1+r_c}\dot{m}_c + \frac{r_g}{1+r_g}\dot{m}_g \qquad (3.30)$$

where $\dot{m}_c$ is the total flow of combustion chamber, $\dot{m}_g$ is the total flow of gas generator, and

$$\dot{m}_c = \dot{m}_{c,o} + \dot{m}_{c,f} \qquad (3.31)$$

$$\dot{m}_g = \dot{m}_{g,o} + \dot{m}_{g,f} \qquad (3.32)$$

The engine's total flow balance is

$$\dot{m}_T = \dot{m}_o + \dot{m}_f = \dot{m}_c + \dot{m}_g \qquad (3.33)$$

2. Pressure balance

The pressure balance denotes that the pump's outlet pressure is summation of the combustion chamber pressure and the pressure drop between the pump's outlet and the sprayers' outlets in the thrust chamber.

$$p_{c,o} = p_c + \sum \Delta p_o \qquad (3.34)$$

$$p_{f,o} = p_c + \sum \Delta p_f \qquad (3.35)$$

where $p_{c,o}$ and $p_{c,f}$ denote the outlet pressure of oxidant pump and incendiary agent pump. For the secondary system with the cavitation tube, the pressure balance is described by

$$p_{b,o} = p_{s,o} + \sum \Delta p_{g,o} \qquad (3.36)$$

$$p_{b,f} = p_{s,f} + \sum \Delta p_{g,f} \qquad (3.37)$$

where p_b is the pressure at the branch of the main pipeline and the secondary system.

3. Power balance

Power balance denotes that the turbine power is equal to the power summation of the driving pumps, namely

$$c_{pg}T_t\left[1-\left(\frac{p_{et}}{p_g}\right)^{\frac{k-1}{k}}\right]\eta_t\dot{m}_t = \frac{\Delta p_{p,o}\dot{m}_o}{\rho_o\eta_{p,o}} + \frac{\Delta p_{p,f}\dot{m}_f}{\rho_f\eta_{p,f}} \tag{3.38}$$

In the normal condition, the model parameter of LRE is balanced. When some faults occur, the parameter balance may be destroyed or the balance point may be moved. Therefore, the parameter balance model is also an important basis the condition monitoring of the LRE.

3.3.4   Fault Characteristic Equation of Engine Components [55, 62–64]

When any fault occurs, the working characteristics of LRE will change accordingly. The fault mathematical model can be obtained by introducing the change of the working characteristics of system or components caused by fault into the engine mathematical model. The accuracy degree of mathematical model is greatly related to the description of the states and change rules of parameters in the actual fault state. Here the main engine fault characteristics are described mathematically as follows:

1. Description of the supply pipeline fault

The faults of propellant supply pipeline are generally shown as weld leakage, filter and hole blockage, propellant valve opening not in place and so on. The propellant leakage may cause the flow imbalance, while the pipe blockage may cause flow resistance variation.

(1) Propellant leakage

Because the LRE works in the environment of high temperature, high pressure, and strong vibration, the fatigue damage may occur in the pipeline and the propellant leakage may happen. Generally, the propellant leak quantity depends on the crack area and the pressure difference between the pipeline sides, namely

$$G_l = \mu A\sqrt{\rho(p_1 - p_2)} \tag{3.39}$$

where
G_l Propellant's leakage quantity;
μ Flow coefficient;
A Crack area;
ρ Propellant's density;
p_1 Inner pressure at leakage;
p_2 Outer pressure at leakage.

Obviously, if the propellant leaks to the atmosphere environment, $p_2 = p_a$, where pa is the atmospheric pressure. The product μA is determined by the crack shape and size. When the leaked medium is gas, the leakage quantity can be calculated by

$$G_l = \frac{A_n \cdot A}{\sqrt{R_1 T_1}} p_1 \tag{3.40}$$

where $R_1 T_1$ is the power capability of the leaked gas, A is the crack area, and A_n is determined by the flow state as follows

$$A_n = \begin{cases} \sqrt{k \left(\frac{2}{k+1} \right)^{\frac{k+1}{k-1}}} & \text{Overcritical} \\ \sqrt{\frac{2k}{k-1} \left[\left(\frac{p_2}{p_1} \right)^{\frac{2}{k}} - \left(\frac{p_2}{p_1} \right)^{\frac{k+1}{k}} \right]} & \text{Subcritical} \end{cases} \tag{3.41}$$

where k is the specific heat ratio of leaked gas.

For the propellant leakage, the fault effect can be described by the change rule of the crack area, namely

$$A = A(f) \tag{3.42}$$

where f denotes fault factor, the specific form of $A(f)$ is determined by the development rule of the described fault. For example, when the crack appears at the moment of t_0 and linearly expands to the area of A_m at the moment of t_m, the crack area can be described as

$$A(f) = \begin{cases} 0 & t \leq t_0 \\ \frac{t-t_0}{t_m-t_0} A_m & t_0 < t < t_m \\ A_m & t \geq t_m \end{cases} \tag{3.43}$$

The propellant leakage may cause the flow imbalance in supply pipeline. Therefore, when describing the propellant leakage, the leakage effect should be added in the corresponding flow balance equation, namely

$$G_2 = G_1 - G_l \tag{3.44}$$

where
G_2 Downstream flux at the leakage crack;
G_1 Upstream flux at the leakage crack.

For the case that the propellant leaks into other supply pipelines, the pipeline flux is given by

$$\widetilde{G}_2 = \widetilde{G}_1 + G_1 \tag{3.45}$$

where

$\widetilde{G}_2$ Downstream flux at the leakage in point;

$\widetilde{G}_1$ Upstream flux at the leakage in point.

(2) Pipeline blockage

The filter and holes of the propellant supply pipeline may be blocked by the potential impurities in the propellant and may increase the flow resistance. For a segment of supply pipeline, the relationship between flux and pressure drop is determined by the flow resistance, namely

$$p_1 - p_2 = \xi G_2 \tag{3.46}$$

where ξ is the flow resistance coefficient. If the pipeline's flux sectional area is A and the flow resistance is described by the concentrated loss, the flow resistance coefficient is given by

$$\xi = \frac{1}{\rho \cdot (\mu A)^2} \tag{3.47}$$

The pipeline blockage decreases the pipeline's flux sectional area. Therefore, the pipeline blockage can be described by the flux sectional area variation, namely

$$A = A(f) \tag{3.48}$$

where f represents the fault factor. The form of $A(f)$ is determined by the development rule of the fault. When the pipeline with the initial flux sectional area A_0 at the moment of t_0 and linearly blocks to the area of A_m at the moment of t_m, the flux sectional area can be described as

$$A(f) = \begin{cases} A_0 & t \le t_0 \\ A_0 - \frac{t-t_0}{t_m-t_0} A_m & t_0 < t < t_m \\ A_0 - A_m & t \ge t_m \end{cases} \tag{3.49}$$

Define $f = \frac{A_m}{A_0}$, then

$$A(f) = \begin{cases} A_0 & t \le t_0 \\ A_0\left(1 - \frac{t-t_0}{t_m-t_0}f\right) & t_0 < t < t_m \\ A_0(1-f) & t \ge t_m \end{cases} \tag{3.50}$$

Obviously, the defined fault factor is the percentage of blocked flux sectional area and the fault effect can be described by flow coefficient, namely

$$R(f) = \frac{1}{\rho \cdot [\mu A(f)]^2} \tag{3.51}$$

(3) Valve fault

The valve fault is mainly shown as dislocation and the fault effect is equivalent to the change of the valve flow resistance. Therefore, the valve fault can be described by referring to the pipeline blockage. If the valve normally opens with the flux sectional area of A_M and opens with the flux sectional area of A_m when fault occurs, the fault factor is defined as

$$f = \frac{A_M - A_m}{A_M} \tag{3.52}$$

Then

$$A(f) = \begin{cases} 0 & t \leq t_0 \\ A_M \left(1 - \frac{t-t_0}{t_m-t_0} f\right) & t_0 < t < t_0 + \delta \\ A_M(1 - f) & t \geq t_0 + \delta \end{cases} \tag{3.53}$$

where t is the valve's open time. The fault can be described by the flow coefficient as

$$R(f) = \frac{1}{\rho \cdot [\mu A(f)]^2} \tag{3.54}$$

Obviously, the fault factor f represents the loss of valve's flux sectional area.

(4) Heat exchanger (evaporator and cooler) fault

The heat exchanger fault is mainly shown as blockage and leakage which may cause the heat exchange efficiency decrease. The heat exchanger fault can be described according to the pipeline blockage and leakage fault and the decrease of heat exchange efficiency.

① Blockage

When blockage occurs, the flux area may decrease and the flow resistance may increase. If the heat exchanger is blocked with the initial flux area of A_0 at the moment of t_0 and linearly develops to A_m at the moment of t_m, the fault factor can be described as

$$f = \frac{A_0 - A_m}{A_0} \tag{3.55}$$

Then

$$A(f) = \begin{cases} 0 & t \le t_0 \\ A_0\left(1 - \frac{t-t_0}{t_m-t_0}f\right) & t_0 < t < t_0 + \delta \\ A_0(1-f) & t \ge t_0 + \delta \end{cases} \tag{3.56}$$

The fault can be described as

$$R(f) = \frac{1}{\rho \cdot [\mu A(f)]^2} \tag{3.57}$$

When the flow resistance increases, the outlet pressure, the outlet temperature, and the exchanged heat decrease. If the rated exchanged heat is Q_0 and deceases to Q_m at t_m, the fault factor can be described as

$$f = \frac{Q_0 - Q_m}{Q_0} \tag{3.58}$$

② Leakage

The heat exchanger leakage can be described referring to the pipeline leakage. It is shown as that the pressure and temperature at the medium outlet and the exchanged heat drops. Detailed description can be seen in the section of the combustion chamber and gas generator fault.

2. Fault description of turbo pump system

Although many reasons may induce turbo pump fault and causes similar changes of the pump's efficiency and work characteristics. Therefore, the fault effect of turbine pump system can be described by the variation of pump efficiency and work characteristics.

(1) Centrifugal pump fault effect

The centrifugal pump fault always causes the efficiency decrease and can be described as

$$\eta_p = \eta_p(f) \tag{3.59}$$

$$\eta_P(f) = \begin{cases} \eta_0 & t \le t_0 \\ \eta_0\left(1 - \frac{t-t_0}{t_m-t_0}f\right) & t_0 < t < t_m \\ \eta_0(1-f) & t \ge t_m \end{cases} \tag{3.60}$$

where $f = \frac{\Delta\eta}{\eta_0}$.

(2) Turbine fault effect

The turbine's fault effect is comprehensively described as

$$\eta_T = \eta_T(f) \tag{3.61}$$

Due to the efficiency decrease, there is

$$\eta_T(f) = \begin{cases} \eta_0 & t \le t_0 \\ \eta_0\left(1 - \frac{t-t_0}{t_m - t_0}f\right) & t_0 < t < t_m \\ \eta_0(1 - f) & t \ge t_m \end{cases} \tag{3.62}$$

where $f = \frac{\Delta\eta}{\eta_0}$.

3. Fault description of the combustion chamber and gas generator

The combustion chamber and gas generator fault mainly include the sprayer blocking and ablation, ablation of the wall and throat of combustion chamber, etc. The pipe blockage or leakage caused by these faults can be described by the equation of pipeline leakage and blockage. When the combustion chamber wall is ablated to broken, the leakage of propellant used for regenerative cooling must be considered. When the combustion chamber throat is ablated, the flux sectional area increases. Define the fault factor by

$$f = \frac{\Delta A_T}{A_{T,0}} \tag{3.63}$$

where ΔA_T and $A_{T,0}$ denote the increase of throat area caused by ablation and the initial area, respectively. The throat variation can be described as

$$A_T(f) = A_{T,0}[1 + \delta(f)] \tag{3.64}$$

and the fault effect is

$$\delta(f) = \begin{cases} 0 & t \le t_0 \\ \frac{t-t_0}{t_m - t_0}f & t_0 < t < t_m \\ f & t \ge t_m \end{cases} \tag{3.65}$$

4. Fault description of propellant tank

Lots of components and pipe joints are installed on the propellant tank. The main fault components are concentrated on overflow valve and pressurization pipeline and the main fault are the function fault of overflow valve, blockage, and leakage of pressurization pipeline. All these faults may cause the raise and fallen of tank pressure and corresponding raise and fallen ratio can be used for the tank fault description.

$$f = \frac{\Delta P}{P_0} = \frac{P - P_0}{P_0}$$

where ΔP, P, P_0 are the pressure difference, the actual pressure and the rated pressure of the tank.

$$P_t = P_t(f) \tag{3.66}$$

$$P_t(f) = \begin{cases} P_0 & t \leq t_0 \\ P_0\left(1 - \frac{t-t_0}{t_m-t_0}f\right) & t_0 < t < t_m \\ P_0(1-f) & t \geq t_m \end{cases} \tag{3.67}$$

where t_0 and t_m denote occurring time and terminating time of the fault, respectively.

3.3.5 Steady State Model of the First-Stage Engine

According to the analysis above, the steady state numerical model of the first-stage engine system can be built by the combination of the model of the components. The steady state model, actually a group of nonlinear equations, includes the parallel work of four engines, the propellant supply system, autogenous pressurization system, the influence of the internal and external interference factors, and the faults. By meeting the flow balance, pressure balance, and power balance of the whole system, the equation groups include 145 performance parameters and 85 internal and external interference factors. The performance parameters are expressed by x (j) and a_k and the interference factors are denoted by D_j. Detailed description of these parameters are listed in Appendix A.

3.3.6 Steady State Model of the Second-Stage Engine

The second-stage power system is composed of a host engine and a vernier engine system and corresponding mathematical model is the large nonlinear equation groups that can reflect the working performance of the main engine and vernier engine, including the propellant supply system, turbo medium supply system, autogenous pressurization system, and influence of the internal and external interference factors. By meeting the flow balance, pressure balance, and power balance of the whole system, the equation groups include 71 performance parameters and 51 internal and external interference factors. The performance parameters are expressed by x_j and a_k and the interference factors are denoted by D_j. Detailed description of these parameters are listed in Appendix B.

3.4   Dynamic Model of LRE [55, 62–64]

The dynamic process of LRE includes the processes of normal startup and shutdown, overload in test and the abnormal transition caused by faults. Because the fault appears and develops mainly in this period, the dynamic process must be analyzed deeply to reveal the causes of engine fault, fault mode, testing method, monitoring means, etc.

3.4.1   Thrust Chamber

In the thrust chamber the propellant is sprayed, nebulized, mixed, combusted to become gas with high temperature and high pressure. By accelerating and expanding the gas through the nozzle and jet out at high-speed jet, the chamber gets thrust.

1. Dynamics equation of thrust chamber

The mathematical model of combustion chamber is to establish the relationship between the combustion chamber pressure P_c and the flux of propellant components $\dot{m}_{oc}$ and $\dot{m}_{fc}$. Following assumptions are needed for the derivation.

(1) The propellant burns and becomes combustion gas at the time lag of T_c in the chamber;
(2) The combustion gas is ideal gas;
(3) The pressure wave instantly spreads in the combustion chamber, namely, the pressure of every point is equivalent at the same time. The temperature and gas constant of all points are constant.

In the combustion chamber, the conservation of mass should be met. At the moment of t the change rate of gas quality $dm_c(t)/dt$ is equal to the quality flux difference of generated gas and the exhausted gas from the nozzle, namely

$$\frac{dm_c(t)}{dt} = \dot{m}_{mc}(t) - \dot{m}_{mmte}(t)$$

$$= \dot{m}_{oc}(t) + \dot{m}_{fc}(t) - \Gamma_c A_{tc} \frac{P_c(t)}{\sqrt{RT_c(t)}}$$

where

$\dot{m}_{oc}(t), \dot{m}_{fc}(t)$ Flux of oxidant and flux into the combustion chamber, respectively;

           k is polytropic index of the gas;

$$\Gamma_c = \sqrt{k\left(\frac{2}{k+1}\right)^{\frac{k-1}{k+1}}}$$

A_{tc} Throat area of thrust chamber;

$P_c(t)$ Gas pressure in the combustion chamber.

According to the gas balance equation, $m_c(t) = \frac{V_c P_c(t)}{RT_c(t)}$, where V_c is the volume of combustion chamber.

Calculating the derivation of t $\frac{dm_c(t)}{dt} = \frac{V_c}{RT_c(t)} \cdot \frac{dP_c(t)}{dt} - \frac{V_c P_c(t)}{[RT_c(t)]^2} \cdot \frac{d[RT_c(t)]}{dt}$

RT_c can be regarded as a function of mixing ratio r_c

$$RT_c \approx a_0 + a_1 r_c + a_2 r_c^2 + a_3 r_c^3$$

In this instance,

$$\frac{dm_c(t)}{dt} = \frac{V_c}{RT_c} \cdot \frac{dP_c}{dt} - \frac{V_c P_c(t)}{(RT_c)^2} \cdot \frac{\partial(RT_c)}{\partial r_c} \cdot \frac{dr_c}{dt}$$

$$= \dot{m}_{oc}(t) + \dot{m}_{fc}(t) - \dot{m}_{moc}(t)$$

$$\frac{V_c}{RT_c} \cdot \frac{dP_c(t)}{dt} - \frac{V_c}{(RT_c)^2} \cdot \frac{\partial(RT_c)}{\partial rr_c} \cdot \frac{dr_c}{dt} \cdot P_c(t) + \Gamma_c A_{tc} \frac{P_c(t)}{\sqrt{RT_c}} = \dot{m}_{oc}(t) + \dot{m}_{fc}(t)$$

Combining the definition of mixing ratio

$$r_c = \frac{m_{oc}}{m_{fc}}$$

$$\frac{dr_c}{dt} = \frac{d}{dt}\left(\frac{m_{oc}}{m_{fc}}\right) = \frac{1}{m_{fc}} \frac{dm_{oc}}{dt} - \frac{m_{oc}}{m_{fc}^2} \frac{dm_{fc}}{dt} = \frac{1}{m_{fc}}\left[\dot{m}_{oc}(t) - \frac{m_{oc}}{m_{fc}} \cdot \dot{m}_{fc}(t)\right]$$

$$= \frac{1}{m_{fc}}\left[\dot{m}_{oc}(t) - r_c \dot{m}_{fc}(t)\right]$$

and

$$m_c(t) = \frac{V_c P_c(t)}{RT_c(t)} = m_{oc} + m_{fc} = (r_c + 1)m_{fc}$$

$$\frac{1}{m_{fc}} = (1 + r_c) \cdot \frac{RT_c(t)}{V_c P_c(t)}$$

The following equation exists

$$\frac{dr_c}{dt} = (1 + r_c) \cdot \frac{RT_c}{V_c P_c(t)} \left[\dot{m}_{oc}(t) - r_c \dot{m}_{fc}(t) \right] \tag{3.68}$$

and

$$\frac{dP_c(t)}{dt} = \frac{1}{V_c} \left[RT_c + \frac{\partial(RT_c)}{\partial rr_c}(1 + r_c) \right] \dot{m}_{oc}(t)$$

$$+ \frac{1}{V_c} \left[RT_c - \frac{\partial(RT_c)}{\partial r_c}(1 + r_c)r_c \right] \dot{m}_{fc} - \frac{1}{V_c} \sqrt{RT_c} \Gamma_c A_{tc} P_c(t) \tag{3.69}$$

$$\Gamma_c = \sqrt{k \left(\frac{2}{k+1} \right)^{\frac{k+1}{k-1}}} \tag{3.70}$$

$$RT_c \approx a_0 + a_1 r_c + a_2 r_c^2 + a_3 r_c^3 \tag{3.71}$$

where
k polytropic index of the gas;
a_0, a_1, a_2, a_3 Constant determined by the test.

Equations (3.68) and (3.69) constitute the dynamics equation of the thrust chamber.

2. Introduction of the fault factors

The Introduction method of thrust chamber's fault factors is as same as that of steady state's fault factors. For example, A_{tc} in Eq. (3.69) is $A_{tc}(f)$ when the throat is ablated.

3.4.2 Gas Generator

The dynamic equation of gas generator's fault process can be obtained by referring to that of thrust chamber's fault process. The detailed presentation is listed as the following:

$$\frac{dP_g(t)}{dt} = \frac{1}{V_g} \left[RT_g + \frac{\partial(RT_g)}{\partial rr_g}(1 + r_g) \right] \dot{m}_{og}(t)$$

$$+ \frac{1}{V_g} \left[RT_g - \frac{\partial(RT_g)}{\partial r_g}(1 + r_g)r_g \right] \dot{m}_{fg}(t) - \frac{1}{V_g} \sqrt{RT_g} \Gamma_g A_{tg}(f) P_g(t)$$

$$\tag{3.72}$$

$$\frac{dr_g}{dt} = (1 + r_g) \frac{RT_g}{V_g P_g(t)} \left[\dot{m}_{og}(t) - r_g \dot{m}_{fg}(t) \right] \tag{3.73}$$

where

V_g Combustion chamber volume in the gas generator;
P_g Pressure;
$\dot{m}_{of}, \dot{m}_{og}$ Flux of oxidant and fuel;
r_g Component ratio.

The calculations of Γ_g and RT_g can refer to Eqs. (3.70) and (3.71).

3.4.3   Turbine Pump System

1. Balance equation of turbine pump power

   According to the theorem of momentum moment

$$\left(\frac{\pi}{30}\right)^2 Jn \frac{dn}{dt} = N_t - N_O - N_f - N_\mu \tag{3.74}$$

where

J Rotaty inertia of turbine pump;
n Speed;
N_t, N_o, N_f, N_μ Powers of turbine, oxidant pump, fuel pump, and servo, respectively.

2. Power equation of the pump

   For the oxidant pump (P_0):

$$N_o = (P_{opo} - P_{ipo}) m_o / \eta_o \rho_o \tag{3.75}$$

   For the fuel pump (P_f)

$$N_f = (P_{opf} - P_{ipf}) \dot{m}_f / \eta_f \rho_f \tag{3.76}$$

3. Delivery head of pump

$$\Delta P_o = P_{opo} - P_{ipo} = A_o n^2 - B_o n \dot{m}_o - C_o \dot{m}_o^2 - D_o \frac{d\dot{m}_o}{dt} - E_o \frac{dn}{dt} \tag{3.77}$$

$$\Delta P_f = P_{opf} - P_{ipf} = A_f n^2 - B_f n\dot{m}_f - C_f \dot{m}_f^2 - D_f \frac{d\dot{m}_f}{dt} - E_f \frac{dn}{dt} \qquad (3.78)$$

where A is the influence of impeller on the delivery head of pump, B is the shape of flow passage, C is the pressure loss and inlet pipe, D is the influence of flux and E is the speed on the inertia component of delivery head. They are determined by the following equations:

$$A = \frac{2.75 \times 10^{-3}}{(1+a)} D_1^2 \left(\frac{D_2^2}{D_1} - 1 \right)$$

$$B = \frac{1.67 \times 10^{-2}}{(1+a)} \left(\frac{1}{\varphi_2 b_2 tg\beta_2} - \frac{1}{\varphi_1 b_1 tg\beta_1} \right)$$

$$C = \xi_H + \frac{1}{2\rho} \left(\frac{1}{F_H^2} - \frac{1}{F_{Bx}^2} \right)$$

$$D = \sum \frac{\Delta S_{iH}}{2\pi r_i \rho \phi b_{2k} \sin\beta_i} + \sum \frac{\Delta S_{iHx}}{gF_{iBx}} + \sum \frac{\Delta S_i}{gF_{iH}}$$

$$E = \frac{\pi\rho}{60g} \sum D_i \Delta S_1 \cos\beta_i$$

where

α	Decrease ratio of the delivery head due to finite impeller blades;
$\varphi = 1 - \frac{\delta_k i}{2\pi r r \sin\beta}$	Shrinkage coefficient;
δ_k	Blade thickness;
β	Angle between relative velocity and circumferential velocity;
F	Channel area between the impeller blades;
Δ_S	Micro length unit of fluid flow in the impeller;
D	Diameter of selected micro circular unit;
g	Gravity acceleration;
b_{2k}	Blade width at the outlet of impeller;
i	Index of blade.

4. Introduction of fault factors

(1) There are many reasons for turbine fault such as the moment decrease caused by the breakage of the blade and its tip, the blockage, and ablation of the turbo nozzle. The introduction method of the fault factors is as same as that of steady state's fault factors. For example η_t in Eq. (3.18) should be replaced by $\eta_t(f)$.

(2) Mainly including structure breakage, inner fault and the ablation, the pump faults can be expressed by their influence on the pump efficiency. Therefore, the introduction method of the pump fault factors is as same as that of steady

state's fault factors. For example, η_0 and η_f in Eqs. (3.75) and (3.76) should be substituted by $\eta_0(f)$ and $\eta_f(f)$, respectively.

3.4.4   Liquid Pipeline System

The main parameters in the liquid pipeline are the fluid's flux and pressure and corresponding dynamic equations is their relationships with time. There are two basic assumptions: (1) The fluid is one-dimensional and incompressible and there is no heat exchange with environment; (2) The pipeline is rigid. According to *Navier–Stokes* equations,

1. Pipe with uniform section

$$p_1 - p_2 = \zeta \frac{\dot{m}^2}{2\rho} + \lambda \frac{\mathrm{d}\dot{m}}{\mathrm{d}t} - p_e \tag{3.79}$$

where

p_1, p_2	Inlet and outlet pressure of the selected pipe, respectively;
$\dot{m}$	Fluid mass flow;
ρ	Density of fluid medium;
ζ, λ	Flow resistance coefficient and route loss coefficient($\lambda = L/gA$);
A	Sectional area of the pipeline;
p_e	Pressure change caused by external forces and $p_e = L\rho g((a+\sin\theta)/g)$;
g	Gravity acceleration. $g = g_0 R_0^2/(R_0 + H)^2$. g_0 denotes the ground gravity acceleration (9.80665 m/s^2). R_0 denotes the earth radius (6371 km). H denotes the flight altitude.
L	Distance (pipe length) between the selected sections along the axis;
θ	Tilt angle of ballistic trajectory;
a	Axial acceleration of the rocket;
$(a + \sin\theta)/g$	Axial overload coefficient, expressed in nx;

In ground experiment and testing, $a = 0$, $g = g_0$ and Eq. (3.79) is reduced to be the general flow equation.

2. Pipe with nonuniform section

$$p_1 - p_2 = \left(\zeta + \frac{1}{A_2^2} - \frac{1}{A_1^2}\right)\frac{\dot{m}^2}{2\rho} + \lambda\frac{d\dot{m}}{dt} - p_e \tag{3.80}$$

where A_1 and A_2 are the inner sectional areas of the two corresponding sections.

3. Cavitation venturi tube

$$p_i - p_s = K_s\dot{m}^2 + \lambda\frac{d\dot{m}}{dt}$$

$$K_s = \frac{1}{2\rho A_t^2 C_d^2}, \quad C_d = \frac{1}{\sqrt{1 + \zeta A_t^2 - \left(\frac{A_t}{A_i}\right)^2}}, \tag{3.81}$$

where

K_s Cavitation coefficient;
C_d Flux coefficient;
p_i, p_s Inlet pressure and saturation vapor pressure;
A_t, A_i Area of throat and inlet.

Let $\alpha = (p_i - p_{out})/(p_i - p_s)$. When $\alpha \geq \delta$, where δ is the cavitation redundancy and determined by test, the cavitation has been happened in the tube.

4. Oxidant pipeline

From the Oxidant tank to the straight pipe outlet, the pressure is determined by

$$p_{To} - p_{wt} = \xi_{o0}\frac{\dot{m}_{o0}^2}{2\rho_0} + \lambda_{o0}\frac{d\dot{m}_{o0}}{dt} - p_{o0}^H \tag{3.82}$$

where p_{o0}^H is the pressure change caused by external forces, $p_{o0}^H = L_{o0}\rho_o g n_x$.

The outlet pressure of each branch pipe is equivalent and the following equations are the same for each engine.

From the straight pipe outlet to the pump inlet:

$$p_{wt} - p_{ipo} = \xi_{o1}\frac{\dot{m}_o^2}{2\rho_o} + \lambda_{o1}\frac{d\dot{m}_o}{dt} - p_{o1}^H \tag{3.83}$$

From the pump to the branch of subsystem:

$$p_{pop} - p_o^l = \xi_{o2}\frac{\dot{m}_o^2}{2\rho_o} + \lambda_{o2}\frac{d\dot{m}_o}{dt} - p_{o2}^H \tag{3.84}$$

From the branch to the pipeline before the oxidant sprayer:

$$p_o^l - p_{oc} = \xi_{o3} \frac{\dot{m}_o^2}{2\rho_o} + \lambda_{o3} \frac{\mathrm{d}\dot{m}_{oc}}{\mathrm{d}t} - p_{o3}^H \tag{3.85}$$

At the segment between the head chamber and the sprayer:

$$p_{oc} - p_c = \xi_{o4} \frac{\dot{m}_{oc}^2}{2\rho_o} + \lambda_{o4} \frac{\mathrm{d}\dot{m}_{oc}}{\mathrm{d}t} - p_{o4}^H \tag{3.86}$$

For the subsystem pipeline when the cavitation happens:

$$p_o^l - p_{bo} = \xi_{o5} \frac{\dot{m}_{o2}^2}{2\rho_o} + \lambda_{o5} \frac{\mathrm{d}\dot{m}_{o2}}{\mathrm{d}t} \tag{3.87}$$

Pressurization pipeline:

$$p_{oc} - p_{bo} = \xi_{o6} \frac{\dot{m}_{o3}^2}{2\rho o} + \lambda_{o6} \frac{\mathrm{d}\dot{m}_{o3}}{\mathrm{d}t} \tag{3.88}$$

5. Fuel pipeline

The pipeline structure from the fuel tank to each engine is the same and their performance parameters are consistent. The dynamics models of different structures are presented as the following:

From the tank to the pump inlet.

$$p_{Tf} - p_{ipf} = \xi_{f1} \frac{\dot{m}_f^2}{2\rho_f} + \lambda_{f1} \frac{\mathrm{d}\dot{m}_f}{\mathrm{d}t} - p_{f1}^H \tag{3.89}$$

From the pump to the branch of subsystem:

$$p_{opf} - p_f^l = \xi_{f2} \frac{\dot{m}_f^2}{2\rho_f} + \lambda_{f2} \frac{\mathrm{d}\dot{m}_f}{\mathrm{d}t} - p_{f2}^H \tag{3.90}$$

From the branch to the sprayer of main pipeline:

$$p_f^l - p_{fc} = \xi_{f3} \frac{\dot{m}_{f1}^2}{2\rho_f} + \lambda_{f3} \frac{\mathrm{d}\dot{m}_{f1}}{\mathrm{d}t} - p_{f3}^H \tag{3.91}$$

For the sprayer:

$$p_{fc} - p_c = \xi_{f4}\frac{\dot{m}_{fc}^2}{2\rho_f} + \lambda_{f4}\frac{d\dot{m}_{fc}}{dt} - p_{f4}^H \tag{3.92}$$

From the partition pipeline to the sprayer:

$$p_f^l - p_{fc} = \xi_{f5}\frac{\dot{m}_{f3}^2}{2\rho_f} + \lambda_{f5}\frac{d\dot{m}_{f3}}{dt} - p_{f5}^H \tag{3.93}$$

For the subsystem pipeline when the cavitation happens:

$$p_f^l - p_{bf} = \xi_{f6}\frac{\dot{m}_{f2}^2}{2\rho_f} + \lambda_{f6} \cdot \frac{d\dot{m}_{f2}}{dt} \tag{3.94}$$

6. Flux balance

$$\dot{m}_{o0} = \dot{m}_o^I + \dot{m}_o^{II} + \dot{m}_o^{III} + \dot{m}_o^{IV} \tag{3.95}$$

$$\dot{m}_o = \dot{m}_{o1} + \dot{m}_{o2} \tag{3.96}$$

$$\dot{m}_{o1} = \dot{m}_{oc} + \dot{m}_{o3} \tag{3.97}$$

$$\dot{m}_f^I = \dot{m}_f^{II} = \dot{m}_f^{III} = \dot{m}_f^{IV} = \dot{m}_{f1} + \dot{m}_{f2} + \dot{m}_{f3} \tag{3.98}$$

$$\dot{m}_{fc} = \dot{m}_{f1} + \dot{m}_{f3} \tag{3.99}$$

where superscript I, II, III, and IV denote the index of engine. The extension engine is not mentioned above and corresponding model can be obtained by referring to Eq. (3.95).

7. Introduction method of fault

The expression of pipeline equation and introduction of fault factors are similar to the steady state process model. Only one or two fault factors need to be considered in simulation.

The main faults of liquid pipeline are blockage and leakage. The blockage decreases the flux area A and in the dynamic equation $A(f)$ should be used to replace A. For leakage, Fig. 3.2 illustrates the mechanism and the model is shown as follows:

By referring Fig. 3.2, the fault equation of pipeline can be given by

Fig. 3.2 Leakage model of liquid pipeline

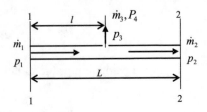

$$p_1 - p_3 = \zeta_{1-3}\frac{\dot{m}_1^2}{2\rho} + \frac{4L_{1-3}}{\pi d_1^2 g}\frac{\mathrm{d}\dot{m}_1}{\mathrm{d}t} - p_{1-3}^H$$

$$p_1 - p_3 = \zeta_{1-3}\frac{\dot{m}_1^2}{2\rho} + \frac{4L_{1-3}}{\pi d_1^2 g}\frac{\mathrm{d}\dot{m}_1}{\mathrm{d}t} - p_{1-3}^H \qquad (3.100)$$

$$p_3 - p_4 = \zeta(A(f))\frac{\dot{m}_2^2}{2\rho}$$

$$\dot{m}_1 = \dot{m}_2 + \dot{m}_3. \qquad (3.101)$$

where p_4 is the environment pressure at leakage and $\zeta(A(f))$ is the resistance coefficient function. In this instance,

$$p_1 - p_2 = \frac{l}{L}\zeta\frac{\dot{m}_1^2}{2\rho} + \frac{L-l}{L}\zeta\frac{\dot{m}_3^2}{2\rho} + \frac{4L}{\pi d_1^2 g}\frac{\mathrm{d}\dot{m}_1}{\mathrm{d}t} - p_L^H \qquad (3.102)$$

3.4.5 Autogenous Pressurization System

Autogenous pressurization is a very complicated process, in which the exchanges of heat and mass happen. The heat exchange includes the pneumatic heating of high-speed airflow and the heat exchange of the pressurized gas with the pipe wall, tank wall, and flow level. The propellant shake can also increase the heat exchange between the gas and tank wall. For the oxidant tank, the dissociated oxidant will condense when it encounters the cold gas, wall, and fluid level in the tank. Besides, the propellant may evaporate when the hot pressurized gas contacts its surface. At some time mass exchange reduces the pressurized gas. In addition, about 10 % of carbon is contained in the gas and has no pressurization result. In summary, pressurization process is extremely complex and is affected by many factors. To simplify the calculation, following assumptions are given:

(1) The carbon content in the pressurized gas is removed;
(2) The temperature in the tank is uniform;
(3) The gas composition in the tank is uniform;
(4) The pressurized gas is insoluble in propellant;

(5) The mass exchange, caused by the condensation of pressurized gas and the propellant evaporation, and corresponding exhausted and absorbed heat are not considered;

(6) The heat exchange of the pressurized gas with the pressurization pipe is considered as forced convective heat transfer. The heat exchange of the pressurized gas with the tank wall and the liquid level is considered as natural convective heat transfer. The heat exchange of the atmosphere with the tank is considered as high-speed air flow heat transfer. The radiation is not considered. The heat exchange calculation in the pipe and tank only considers the physical characteristics of oxidant and fuel.

1. Equation of evaporator and cooler

The evaporator and cooler are components involving the countercurrent heat transfer and are mainly composed of shell and serpentine pipe. A kind of fluid flows in the serpentine pipe. The other flows in the shell and out of the serpentine pipe and exchanges heat with the liquid in the pipe. The heat parameters of the hot and cold medium flowing on the two sides of the wall change when heat exchange happens between these two kinds on medium. The heat storage in the wall also changes. To simplify the calculation, following assumptions are given:

(1) The heat transfer coefficient of two kinds of fluid with the wall is constant and the total heat transfer coefficient is invariable;

(2) The outer wall has good heat preservation character and the heat loss is zero;

(3) The heat capacity of middle wall is very small and the heat storage can be ignored;

(4) In the same section the temperature and velocity distribution are uniform and the medium in the pipe flow only along the axis. No internal circulation exists;

(5) The medium and the metal wall exchange heat in radial direction and the axial heat transferring is not considered.

The mathematical model of heat transfer process can be derived based on the above assumptions. A micro unit dl of the fluid in the serpentine pipe is considered for the heat equilibrium and subscripts I and II denote the fluid in the pipe and in the serpentine pipe, respectively. Apparently, the heat storage variation of fluid in serpentine pipe equals the summation of the heat transferring from pipe to serpentine pipe, heat of fluid flowing into serpentine pipe, and the heat of fluid flowing out of serpentine pipe, namely

$$K \cdot \frac{A}{L} dl \cdot (T_2 - T_1) + \dot{m}_1 \cdot C_1 \cdot T_1 - \dot{m}_1 C_1 \cdot \left(T_1 + \frac{dT_1}{dl} \cdot dl \right) = \frac{M_1}{L} \cdot C_1 \cdot dl \frac{dT_1}{dt}$$

It can be simplified as

$$\frac{\partial T_1}{\partial t} + \frac{L\dot{m}_1}{M_1} \cdot \frac{\partial T_1}{\partial l} = -K \cdot \frac{A}{\rho_1 M_1 C_1}(T_2 - T_1)$$

For the fluid in the serpentine pipe:

$$\frac{\partial T_2}{\partial t} - \frac{L\dot{m}_2}{M_2} \cdot \frac{\partial T_2}{\partial l} = -K \cdot \frac{A}{\rho_2 M_2 C_2}(T_2 - T_1)$$

where
T Fluid temperature (K);
L Effective length of heat exchange pipe (m);
A Heat transfer area of the middle wall (m^2);
G Flux (m^3/s);
C Fluid's specific heat (kJ/(m^{3}K));
M Fluid's volume in the pipe (m^3);
K Total heat transfer coefficient of heat exchange pipe (W/(m^{2}K));
t Time(s).

Because only the outlet temperature of the heat exchanger is considered and the temperature distribution along the length of heat exchange pipe can be ignored, the above equation can be simplified as

$$\frac{\partial T_{1c}}{\partial t} + \frac{L\dot{m}_1}{M_1} \cdot (T_{1c} - T_{1r}) = -K \cdot \frac{A}{\rho_1 M_1 C_1}(T_{2c} - T_{1c})$$

$$\frac{\partial T_{2c}}{\partial t} + \frac{L\dot{m}_2}{M_2} \cdot (T_{2c} - T_{2r}) = -K \cdot \frac{A}{\rho_2 M_2 C_2}(T_{2c} - T_{1c})$$

where
T_{2c}, T_{2r} Temperature of fluid II at the inlet and outlet of heat exchanger;
T_{1c}, T_{1r} Temperature of fluid I at the inlet and outlet of heat exchanger.

The evaporator's equation:

$$\frac{\partial T_{olcz}}{\partial t} + \frac{L_{oz}\dot{m}_{o3}}{M_{olz}} \cdot (T_{olcz} - T_{olrz}) = -K_{oz} \cdot \frac{A_{oz}}{\rho_{olz} M_{olz} C_{olz}}(T_{o2cz} - T_{olcz}) \quad (3.103)$$

$$\frac{\partial T_{o2cz}}{\partial t} + \frac{L_{oz}\dot{m}_{o2z}}{M_{o2z}} \cdot (T_{o2cz} - T_{o2rz}) = -K_{oz} \cdot \frac{A_{oz}}{\rho_{o2z} M_{o2z} C_{o2z}}(T_{o2cz} - T_{olcz}) \quad (3.104)$$

The cooler's equation:

$$\frac{\partial T_{f1cz}}{\partial t} + \frac{L_{fz}\dot{m}_{f1z}}{M_{f1z}} \cdot \left(T_{f1cz} - T_{f1rz}\right) = -K_{fz} \cdot \frac{A_{fz}f_z}{\rho_{f1z}M_{f1z}C_{f1z}}\left(T_{f2cz} - T_{f1cz}\right) \quad (3.105)$$

$$\frac{\partial T_{f2cz}}{\partial t} + \frac{L_{fz}\dot{m}_{f2z}}{M_{f2z}} \cdot \left(T_{f2cz} - T_{f2rz}\right) = -K_{fz} \cdot \frac{A_{fz}}{\rho_{f2z}M_{f2z}C_{f2z}}\left(T_{f2cz} - T_{f1cz}\right) \quad (3.106)$$

2. Equation of pressurization pipeline

There is a collector before each pressurization pipeline. To simplify the calculation, here the outlet temperature of the collector is considered as the average of the four inlet temperature.

The pressurization pipeline is considered as a heat exchanger here. The outside air is considered as fluid II, while the inside one is the fluid I. The heat transfers through the pressurization pipeline. Due to the good heat preservation performance of the pressurization pipeline, the heat transferred through the pressurization pipeline is small and has few influences on the outside gas outside. Therefore, the outside gas temperature is basically constant and the equation of pressurization pipeline can be obtained as

$$\frac{\partial T_{1c}}{\partial t} + \frac{L\dot{m}_1}{M_1} \cdot \left(T_{1c} - T_{1r}\right) = -K \cdot \frac{A}{\rho_1 M_1 C_1}\left(T_2 - T_{1c}\right) \quad (3.107)$$

where

T_{1c},   Temperature of pressurized gas at the inlet and outlet of pressurization
T_{1r} pipe;
T_2 Gas temperature outside the pipe.

The pressurization pipeline of oxidant tank:

$$\frac{\partial T_{olcg}}{\partial t} + \frac{L_{og}\dot{m}_{olg}}{M_{olg}} \cdot \left(T_{olcg} - T_{olrg}\right) = -K_{og} \cdot \frac{A_{og}}{\rho_{olg}M_{olg}C_{olg}}\left(T_{o2g} - T_{olcg}\right)$$

$$(3.108)$$

The pressurization pipeline of fuel tank:

$$\frac{\partial T_{f1cg}}{\partial t} + \frac{L_{fg}\dot{m}_{f1g}}{M_{f1g}} \cdot \left(T_{f1cg} - T_{f1rg}\right) = -K_{fg} \cdot \frac{A_{fg}}{\rho_{f1g}M_{f1g}C_{f1g}}\left(T_{f2g} - T_{f1cg}\right) \quad (3.109)$$

Flux equation:

$$\dot{m}_{olg} = \dot{m}_{o3}^{I} + \dot{m}_{o3}^{II} + \dot{m}_{o3}^{III} + \dot{m}_{o3}^{IV} \quad (3.110)$$

$$\dot{m}_{f1g} = \dot{m}_{f1z}^{I} + \dot{m}_{f1z}^{II} + \dot{m}_{f1z}^{III} + \dot{m}_{f1z}^{IV} \qquad (3.111)$$

3. Tank equation

For the state equation of ideal gas $p\overline{V} = WRT$, Multiply C_p at both sides $p\overline{V}C_p = WRTC_p$, then

$$p\overline{V}C_p = R[WTC_p]$$

$$= R\left[W_0T_0C_{p0} + \int_0^t G_1T_1C_{p1}\mathrm{d}t + \int_0^{t2} G_2T_2C_{p2}\mathrm{d}t - A\int_0^t p\mathrm{d}\overline{V} - \int_0^t (q_1 + q_2)\mathrm{d}t\right]$$

where

p	Gas pressure in the tank;
R	Pressurized gas constant in the tank;
$\overline{V}$	Pressurized gas volume in the tank;
C_p	Constant volume specific heat of pressurized gas in the tank;
W_0	Weight of the initial pressurized gas;
T_0	Temperature of the initial pressurized gas;
C_{p0}	Constant volume specific heat of initial pressurized gas;
W_1	Weight of the autogenous pressurized gas;
T_1	Temperature of the autogenous pressurized gas;
C_{p1}	Constant volume specific heat of autogenous pressurized gas;
W_2	Weight of gas for supplementing pressure in the gas bottle;
T_2	Temperature of gas for supplementing pressure in the gas bottle;
C_{p2}	Constant volume specific heat of gas for pressure supplementing in the gas bottle;
A	Thermal equivalent;
$q_1 + q_2$	Heat loss of autogenous pressurized gas exchanging with the propellant and wall in the tank.

Because dynamic model is devoted on the description of the transition process during the steady state period and caused by the fault, several items in the above formula, such as $W_0T_0C_p$, are constant when reaching a steady state. In addition, the interval from a steady state to another steady state under the faults is very short, generally about 0.5 s. Therefore, the following assumptions are given:

(1) In the transition process, the nature of the pressurized gas in the tank does not vary, namely the constants R, C_p, and C_{p1} do not change over time;

(2) The occupied volume of the pressurized gas does not change over time.

In this instance, the previous equation can be simplified after derivation

$$C_p(pd\overline{V} + \overline{V}dp) = R\left[G_1T_1C_{p1} + (q_1 + q_2) - Apd\overline{V}\right]$$

Import the volume variation and flux relationship, it can be simplified as

$$\frac{C_p\overline{V}}{R}dp = G_1T_1C_{p1} + (q_1 + q_2) - \left(Ap + \frac{C_p\overline{V}}{R}\right) \cdot \dot{m}$$

Equation of oxidant tank is given by

$$\frac{C_{op}\overline{V}_o}{R_o}dp_o = \dot{m}_{o1g}T_{o1}C_{op1} + (q_{o1} + q_{o2}) - \left(A_op_o + \frac{C_{op}\overline{V}_o}{R_o}\right)\dot{m}_{o0} \qquad (3.112)$$

$$\dot{m}_{f0} = \dot{m}_f^{I} + \dot{m}_f^{II} + \dot{m}_f^{III} + \dot{m}_f^{IV} \qquad (3.113)$$

Equation of fuel tank is given by

$$\frac{C_{fp}\overline{V}_f}{R_f}dp_f = \dot{m}_{f1g}T_{f1}C_{fp1} + (q_{f1} + q_{f2}) - \left(A_f \cdot p_f + \frac{C_{fp}\overline{V}_f}{R_f}\right) \cdot \dot{m}_{f0} \qquad (3.114)$$

$$\dot{m}_{f0} = \dot{m}_f^{I} + \dot{m}_f^{II} + \dot{m}_f^{III} + \dot{m}_f^{IV} \qquad (3.115)$$

3.4.6 Dynamic Model of the First-Stage Engine

The dynamic characteristics of a pump-pressurized LRE can be accurately descri-bed by the above equations. Since each equation has a very clear physical meaning, it is convenient to consider the influence of a variety of fault factors on the engine's dynamic characteristics. There are only 56 independent physical variables in the basic equations. The dynamic model of the first-stage engine can be obtained by sorting the equations in Appendix A and the result is detailedly shown in Appendix C.

3.4.7 Dynamic Model of the Second-Stage Engine

The dynamic model of the second-stage engine can be set up by referring that of the first-stage engine and can be seen in Appendix D.

Chapter 4
Fault Characteristic Analysis of LRE

It is known that the faults will occur in the LRE with the increase of the service time. The high reliability of LRE represents that the system may fulfill the task successfully with very low probability to go wrong. However, the structure defect and parameter mismatch may accelerate the development speed of the LRE to the design limitation and may make the system appear failures ahead of the design limitation. It is shown that the reason of the faults may be deduced by the performance of the fault, since any fault has its own expression.

Generally, the expression of fault is called fault mode. Corresponding study on the modes and characters is an important way to recognize and understand the fault mechanism. The deep understanding of the faults may provide great help for monitoring and diagnosis of LRE with suitable method. In this instance, based on the analysis on the phenomenon, mode and mechanism of the fault, the mechanism of the fault is deeply investigated and the standard fault mode of LRE is established to seize the difference of the faults.

4.1 Characteristic Analysis of Failure Patterns in Steady State

In order to study the feature of fault, the mathematical models of all kinds of components and widgets of LRE can be established according to the principle of modularization. The model of the first-level LRE can be obtained by the combination of the component models. Generally, the state of the components will be changed when the adjustable parameter of the model is adjusted. In this instance, the steady state model of LRE with faults can be built by changing the parameter of the normal model. As listed in Table 4.1, nine parameters which can be directly collected are selected to describe the work state of LRE.

© Springer-Verlag Berlin Heidelberg and National Defense Industry Press 2016
W. Zhang, *Failure Characteristics Analysis and Fault Diagnosis
for Liquid Rocket Engines*, DOI 10.1007/978-3-662-49254-3_4

Table 4.1 Parameters of the steady state

Serial number	Measured parameters Θ_i	Deviation rate ε_i	Notes
1	Engine oxidant flow	Deviation rate of engine oxidant flow	A
2	Engine fuel flow	Deviation rate of engine fuel flow	B
3	Oxidant pressure before the insufflation	Deviation rate of pre-oxidant pressure	C
4	Pressure at the branch fuel main pipeline	Deviation rate of pressure of combustion agent road branch	D
5	Turbine pump speed	Deviation rate of turbine pump speed	E
6	Pump outlet pressure	Deviation rate of pump outlet pressure	F
7	Fuel pump outlet pressure	Deviation rate of fuel pump outlet pressure	G
8	Oxidant tank pressure	Deviation rate of oxidant tank pressure	H
9	Fuel tank pressure	Deviation rate of fuel tank pressure	I

For the sake of calculation, define

$$\varepsilon_i = \frac{\Delta_i}{\Theta_i^N} = \frac{\left|\Theta_i^F - \Theta_i^N\right|}{\Theta_i^N}, \quad i = 1, 2, \ldots, 9 \tag{4.1}$$

where

Θ_i^N, Θ_i^F normal value and actual value of parameter i;

Δ_i corresponding deviation value;

ε_i corresponding deviation rate.

In the following simulation analysis picture, the deviation rates are orderly denoted by A, B, C, D, E, F, G, H, and I.

4.1.1   Numerical Solution of the Steady State Model

It could simulate one fault and obtain the change of the characteristic parameters under action of the fault by changing the corresponding value of the failure factor and solving the fault steady state numerical model.

The LRE's steady state is shown by the following nonlinear equations:

$$F_l(\theta) = 0, \quad l = 1, 2, \ldots, 143$$

where $\theta = [x_j]T$ is the parameter rector.

When there is only at most one fault at any time, the numbers of the independent parameters are less than that of the nonlinear equations. In this case, solution of the equations can be taken as solving the following constrained optimization or unconstrained optimization problem:

$$\min_{\theta}\left\{J \equiv \sum_l \eta_l F_l^2(\theta)\right\} \tag{4.2}$$

where $1 \geq \eta_l \geq 0$ is the weight coefficient determined by the physical significance, degree of importance, and the possibility of the faults in a complex manner, satisfying $\sum_l \eta_l = 1$. If the lth equation does not go wrong, the equations would be set up. In this case, $\eta_l = 0$ may be chosen and equation l is the constrain condition in the optimization problem.

Many methods can be applied to solve the optimization problem [69–71], such as coordinate transforming method, interior point interpolation method, Newton method, variable metric method, ant colony algorithm, evolutionary algorithm, etc. Aiming to the constrained optimization problem, the Lagrangian relaxed method can also be used to transform the constrained problem into unconstrained problem. For example, the nonlinear least squares optimization function lsqnonlin in the Matlab optimization toolbox can be used to solve the obtained equation.

4.1.2 Simulation Analysis of Steady State Feature

All the factors containing the interference factors and failure factors are comprehensively considered in the above equations. In case of ignoring the failure factors, the simulation of steady state of the LRE can be accomplished by solving the equations.

The parameters of LRE such as entrance pressure, propellant density, stir change of turbopump of the pump, and other factors have effects on the LRE and are analyzed by the nonlinear least squares optimization method under the normal working condition of LRE. The results of calculation are consistent with that of the experiment.

The following presents the comparison of the small deviation equation and the nonlinear equation under the interference factors.

The small deviation equation is a simplified algorithm in steady state calculation. It utilizes the linear difference equation to calculate approximately when the interference factors change slightly. It is applied on many fields due to its simpleness. Here is an example. When the entrance pressure of the oxidant pump of an LRE changes from 0.4903 to 0.9022 MPa and the entrance pressure of the fuel pump changes from 0.2942 to 0.39227 MPa, the small deviation equation and the linear difference equation are used for calculation. The comparative results in this work and that in the literature [55] are shown in Tables 4.2 and 4.3 and prove the validity of the nonlinear equation models built in this work.

Table 4.2 Simulation results of the main parameters

Name	Unit	The calculation results in this paper	Small deviation calculation results	Calculation results in literature [55]
The engine thrust	kN	13.1696	15.30	13.1983
Ratio of propellant element		9.5788×10^{-2}	8.45×10^{-2}	9.6254×10^{-2}
Turbine pump speed	r/min	9.9839	29.03	10.3467
Engine oxidant flow	kg/s	5.7334	6.02	5.7491
Engine fuel flow	kg/s	−1.0962	−0.61	−1.1018
Generator oxidant flow	kg/s	1.6936×10^{-2}	1.84×10^{-2}	1.7010×10^{-2}
Generator fuel flow	kg/s	1.2120×10^{-2}	3.02×10^{-2}	1.2295×10^{-2}
Cooling jacket flow	kg/s	−0.9746	−0.56	−0.9640
The evaporator flow	kg/s	6.1168×10^{-3}	6.84×10^{-2}	6.1566×10^{-3}
Oxidant pump pressure increment	MPa	-8.9668×10^{-2}	-6.28×10^{-2}	-8.9040×10^{-2}
Fuel pump pressure increment	MPa	4.4274×10^{-2}	8.14×10^{-2}	4.4971×10^{-2}
Pre-oxidant pressure	MPa	2.0690×10^{-1}	3.30×10^{-2}	3.3702×10^{-1}
Pressure of combustion agent road branch	MPa	5.1600×10^{-2}	8.53×10^{-1}	5.2319×10^{-2}

The inlet pressure of the oxidant pump changes in the range 0.4903 MPa to 0.9022 MPa

Table 4.3 Simulation results of the main parameters

Name	Unit	The calculation results in this paper	Small deviation calculation results	Calculation results in literature [55]
The engine thrust	kN	0.9063	0.70	0.8991
Ratio of propellant element		-2.4463×10^{-2}	-3.03×10^{-2}	-2.4560×10^{-2}
Turbine pump speed	r/min	−6.5553	−6.40	−6.6543
Engine oxidant flow	kg/s	−0.4506	−0.39	−0.4540
Engine fuel flow	kg/s	0.7697	0.64	0.7760
Generator oxidant flow	kg/s	-3.6315×10^{-3}	-3.64×10^{-4}	-3.9709×10^{-3}
Generator fuel flow	kg/s	1.3890×10^{-2}	1.59×10^{-2}	1.3881×10^{-2}
Cooling jacket flow	kg/s	0.6545	0.56	0.6551
The evaporator flow	kg/s	6.1815×10^{-5}	3.64×10^{-5}	6.0499×10^{-5}
Oxidant pump pressure increment	MPa	-6.8542×10^{-3}	-6.87×10^{-3}	-7.0229×10^{-3}
Fuel pump pressure increment	MPa	-3.3730×10^{-2}	-3.65×10^{-2}	-3.3810×10^{-2}
Pre-oxidant pressure	MPa	2.0782×10^{-3}	8.88×10^{-4}	1.9684×10^{-3}
Pressure of combustion agent road branch	MPa	5.9140×10^{-2}	6.77×10^{-2}	5.9050×10^{-2}

Fuel pump inlet pressure changes from 0.2942 to 0.39227 MPa

4.1.3 The Numerical Method Based on the Hopfield Neural Nets

The Hopfield net is typical feedback neural net. It consists of a series of neurons, as shown in Fig. 4.1. Every neuron could be simulated by the nonlinear amplifier as shown in Fig. 4.2. In Fig. 4.2, u_i and V_i are the input and output of the neurons, respectively. The typical relationship of the input and output is the nonlinear function relationship such as Sigmold function.

In the feedback neural network, the input data decides the original state of the feedback system. The network evolves according to the dynamic equation and the system gradually converges to the equilibrium after a series of state transition. This equilibrium state is the output result calculated by the feedback neural network and also is the low-lying minima of the network. So the network could be used to solve constrained optimization problems: the energy function of the network corresponds to the objective function of the optimization problem and one equilibrium point of the network system corresponds to one local minima of the problem.

Fig. 4.1 Hopfield neural network structure diagram

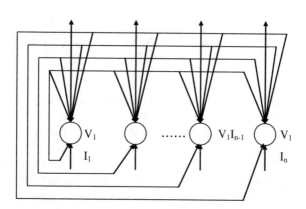

Fig. 4.2 Hopfield network simulation of amplifier

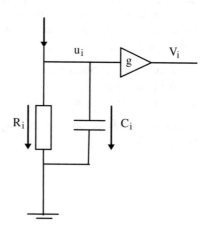

The Hopfield neural nets are divided into continuous type and discrete type. The continuous Hopfield could be described by the following nonlinear differential equations:

$$\frac{dx_j(t)}{dt} = -a_jx_j + \sum_{i=1}^{n} \omega_{ji}\varphi(x_i) + \theta_j, \quad j = 1, 2, \ldots, n \tag{4.3}$$

$$y_i = \varphi(x_i)$$

where

$x_j(t)$	the state of the neuron j at t moment
a_j	real constant
ω_{ji}	the connection weight of the neurons in the network
$y_i = \varphi(x_i)$	excitation function
$\theta = (\theta_1, \theta_2, \ldots, \theta_n)$	threshold vector (the external input of the system)

Equation (4.3) is the dynamic equation of the Hopfield nets.

Lyapunov energy function of Hopfield is defined as follows:

$$E = -\frac{1}{2}\sum_{j=1}^{n}\sum_{i=1}^{n} w_{ji}x_ix_j + \sum_{j=1}^{n} a_j \int_0^{x_j} \varphi_j^{-1}(x)dx - \sum_{j=1}^{n} \theta_jx_j \tag{4.4}$$

Theorem *for the continuous Hopfield net, if ω_{ji} is equal to ω_{ij} and the excitation function $\varphi(x_i)$ is monotone increasing function, then the network solution path in state space is always towards the direction of reducing energy E and stable equilibrium point of network is the minimum point of energy E.*

The theorem guarantees that, given a set of weights of ω_{ji}, the continuous Hopfield network can converge to the stable point from any initial state. Each stable point of the system can be considered as the minimum point of energy function or the attrahens point of dynamic system.

The substance of the optimization problem using Hopfield network is to solve the initial value problem of nonlinear constant differential coefficient equation. The solving steps are as follows:

1) Lyapunov function energy E of network is constructed according to the practical problems;
2) Write network dynamic equations.

Because the Hopfield network is a gradient system, its dynamic equation is given by Eq. (4.5)

$$\frac{dx_j}{dt} = -\frac{\partial E}{\partial y_j}$$

$$y_j = \varphi(x_j) \tag{4.5}$$

3) Choose the right initial value $x_j(0)$ ($j = 1, 2, \ldots, n$) to make the network to evolve until convergence according to the dynamic equation (the classical algorithm of solving differential equations is available for obtaining convergence value in numerical simulation).

The steady state mathematical model of missile engine can be converted into the following form:

$$f_j(x_1, x_2, \ldots, x_n) = 0 \quad j = 1, 2, \ldots, n \tag{4.6}$$

where $f_j (\cdot)$ is quadratic continuously differentiable function, remembering

$$h_j(x) = |f_j(x)| \quad j = 1, 2, \ldots, n \tag{4.7}$$

Except x that can make $f_j(x) = 0$, $h_j(x)$ is also quadratic continuously differentiable, remembering

$$H(x) = [h_1(x), h_2(x), \ldots, h_n(x)]T; \quad F(x) = [f_1(x), f_2(x), \ldots, f_n(x)] \tag{4.8}$$

Considering the unconstrained extreme value problem

$$\min: J(x) = \frac{1}{2} H^{\mathrm{T}}(x) H(x) = \frac{1}{2} \sum_{j=1}^{n} h_j^2(x) \tag{4.9}$$

Because of $J \geq 0$, if x^* is the solution of nonlinear equation Eq. (4.6), x^* can make Eq. (4.9) reach the minimum value of 0. Conversely, if x^* make the $J(x)$ reach the minimum value of 0, then $h_j(x) = 0$, $j = 1, 2, \ldots, n$. Therefore, x^* is also the solution of Eq. (4.6). It suggests that the solution set that makes Eq. (4.9) to reach the minimum of 0 is equal to that of Eq. (4.6).

The literature [72] utilizes the Chua-Lin [73] circuit to construct the Hopfield net for solving unconstrained extreme value problem of Eq. (4.6). The network is an improvement on the Hopfield net. The improved network not only can improve the convergence speed, but also can easily implement with hardware. The dynamic equation is

$$\frac{dx}{dt} = -\frac{1}{C} I \tag{4.10}$$

where $x = (x_1, x_2, \ldots, x_n)T$, C is a small real constant, $I = (I_1, I_2, \ldots, I_n)$ (I_i is the current of the port i) can be determined by the following expressions:

$$I_j = \sum_{i=1}^{n} I_i \mathrm{sgn}(f_i(x)) \frac{\partial f_i(x)}{x_j}$$

$$I_i = \frac{1}{R} h_i(x)$$

(4.11)

where the R is a little real constant, usually taking $R = C$.

The literature [72] also proves that, for any given initial value $x(0)$, the network will converge to a stable point; namely, if there are multiple solutions, we can choose more a few initial values to obtain all solutions of the nonlinear equations. Smaller the value of the parameter R, convergence is relatively faster. Meanwhile, the steady state value of the improved Hopfield, as well as the other with planning method to solve nonlinear equations, is probably the solution of det $F(x) = 0$. Namely, it can bring the added roots that can be removed by the method of trying the root.

Matlab 6.1 is used to calculate via the above algorithm and the calculation steps are as follows:

Solve the expression of Jacobi matrix $\mathrm{DF} = \frac{\partial F}{\partial x}$

$$\frac{\partial F}{\partial x} = \begin{bmatrix} \frac{\partial f_1(x)}{\partial x_1} & \frac{\partial f_2(x)}{\partial x_1} & \cdots & \frac{\partial f_n(x)}{\partial x_1} \\ \frac{\partial f_1(x)}{\partial x_2} & \frac{\partial f_2(x)}{\partial x_2} & \cdots & \frac{\partial f_n(x)}{\partial x_2} \\ \cdots & \cdots & \cdots & \cdots \\ \frac{\partial f_1(x)}{\partial x_n} & \frac{\partial f_2(x)}{\partial x_n} & \cdots & \frac{\partial f_n(x)}{\partial x_n} \end{bmatrix}$$

(4.12)

Take the engine working parameter value at the rated conditions as the initial value $x(0)$ and set R for a little real number. The expression of differential equation Eq. (4.10) can be obtained and solved via Eqs. (4.6)–(4.12).

Validate the solution obtained from the above step by Eq. (4.6).

Setting $R = 1e-13$, the Jacobi matrix DF can be solved by the Jacobi function in Matlab 6.1 and the differential equations can be solved using ode23 function. The calculation results for static characteristics of an engine are shown in Table 4.4. The values in the table are the corresponding parameters deviation at working condition. As the calculation results can be seen, the steady state fault simulation algorithm based on improved Hopfield network is effective, and lays a foundation for failure mode and effect analysis.

4.1.4 Analysis of Engine Steady State Fault

For the steady state numerical model of engine, a particular fault can be simulated to obtain the changeable characteristic parameters under the action of the fault factor by changing the value of the fault factor. The solving method can not only utilize the above method, but also the other methods such as intelligent optimization

Table 4.4 Engine parameters and their calculation results

Number	Name	Parameter	Symbol	Units	Calculation results
1	Engine oxidant flow	$x(1)$	$\dot{m}_o$	Kg/s	0.98028755345313
2	Engine fuel flow	$x(2)$	$\dot{m}_f$	Kg/s	0.96423774258471
3	Combustion chamber oxidant flow	$x(3)$	$\dot{m}_{oc}$	Kg/s	0.98586561825816
4	Gas generator oxidant flow	$x(4)$	$\dot{m}_{of}$	Kg/s	0.98822682872277
5	Gas generator fuel flow	$x(5)$	$\dot{m}_{ff}$	Kg/s	0.98428228916301
6	Cooling jacket flow	$x(6)$	$\dot{m}_{fj}$	Kg/s	0.96302646559345
7	The evaporator flow	$x(7)$	$\dot{m}_{ou}$	Kg/s	0.99161882292946
8	Combustion chamber pressure	$x(8)$	p_c	MPa	0.97962272050371
9	Pump boost pressure	$x(9)$	Δp_{po}	MPa	0.97560096541433
10	Fuel pump supercharging pressure	$x(10)$	Δp_{pf}	MPa	0.96699048075003
11	Pressure before combustion chamber oxidant injection	$x(11)$	p_{oc}	MPa	0.97778794517963
12	Pressure of combustion agent road branch	$x(12)$	p_f^l	MPa	0.96881697825804
13	Turbine pump speed	$x(13)$	n	转/分	0.98196555313179
14	Chamber fuel flow	$x(14)$	$\dot{m}_{fc}$	Kg/s	0.96293139008050
15	Baffle flow	$x(15)$	$\dot{m}_{fg}$	Kg/s	0.950787462633652
16	Total propellant flow of engine	$x(16)$	$\dot{m}$	Kg/s	0.978975658559007
17	The turbine gas flow	$x(17)$	$\dot{m}_t$	Kg/s	0.984842886665194
18	Subsystem coefficient of oxygen	$x(18)$	a_c		1.004141597794560
19	Mixing ratio	$x(19)$	r_m		1.022423476408484
20	Oxidant pump power	$x(20)$	P_o	KW	0.961900964192472
21	Fuel pump power	$x(21)$	P_f	KW	0.932467115026254
22	Turbine power	$x(22)$	P_t	KW	0.947248451521356
23	Turbine gas calorific value	$x(23)$	RT	KJ/Kg	1.000291964972655
24	Thrust	$x(24)$	F	KN	0.976089329639182
25	Sonic nozzle flow	$x(25)$	$\dot{m}_j$	Kg/s	1.000500921901213

method. Analysis of the main steady state fault using nonlinear least squares optimization is as follows [72]:

(1) Tank pressurization system fault

The tank pressurization system fault will cause tank pressure deviation from normal. The percentage of the pressure rising and pressure dropping can be used to simulate pressure conduit leakage, pressure catheter obstruction, and an overflow valve falsely opening and closing.

(2) The main valve fault

The possible fault of main valve is not fully open, leading to increase in local pressure loss. The value of the pressure loss is directly related to the closed angle and flow rate. Generally, pressure loss can be expressed as the function of closed angle and flow rate.

For the oxidizer: $h_j \times 9 = f$ (closed angle, $x_2(9)$), for the fuel, $h_j \times 22 = f$ (closed angle, $x_2(22)$).

The values of a_9 and a_{10} of the formula can be adjusted in the calculation to make the corresponding main valve open to different degrees. Suppose the main valve fully opens at zero degree, a_k is as follows at the fault condition:

$$a_k' = a_k \left(\frac{1}{\cos^4 \theta} \right), \quad k = 9, 10$$

where a_k is normal value, θ is the opening angle of the valve, the other a_k is the value at normal working condition.

(3) The turbine pump fault

Pump efficiency loss can be simulated by changing the interference factors D_1 and D_2. The value of efficiency loss is $\Delta D_k/D_k$ ($k = 1, 2$), D_k is normal value. The turbine gas leakage fault can be simulated by adjusting the parameter value of a_{20} in the equation.

(4) Thrust chamber fault

The fuel and oxidant injection device blocking fault can be simulated by adjusting the values of a_{11} and a_{13} in the formula. The fuel injection device blocking fault will mainly cause the combustion chamber mixture ratio to increase, the engine and the deputy system to be at rising working condition, but the engine thrust to reduce.

The combustion chamber throat ablation can be reflected by adjusting the interference factors D_{15}. The fault may mainly cause the engine propellant flow rate to increase significantly, propellant mass mixing ratio to increase slightly, and the pressure in thrust chamber to reduce.

(5) Gas generator pipeline fault

The fault caused by oxidizer and fuel pipeline blocking can be simulated by adjusting the parameters a_{17} and a_{18} in the equation. The leaks of gas generator at fuel entrance can be reflected by adjusting the values of parameters a_1 and a_{19} in the equation.

(6) Evaporator fault

The fault caused by the decrease of thermal efficiency of evaporator can be simulated by adjusting the parameters a_{105} and a_{106} in the equation. The leakage of pressurized pipeline after evaporator can be simulated by adjusting the value of the parameter a_{121}. The above fault mainly causes fuel tank pressurization pressure to drop.

(7)   The cooling device fault

The fault caused by the decrease of thermal efficiency of cooling device can be simulated by adjusting the parameters a_{125} and a_{126} in the equation. The leakage of pressurized pipeline after cooling device can be simulated by adjusting the value of the parameter a_{141}. The above fault mainly causes oxidizer tank pressurization pressure to drop.

(8)   Restrictor fault

The fault caused by the restrictor jam of oxidizer and fuel systems can be simulated by adjusting the parameters a_9 and a_{12} in the equation.

The parameter variation of partial fault mode is calculated through the simulation model as shown in Fig. 4.3.

4.1.5   Secondary Engine Steady State Fault Analysis Based on Ant Colony Algorithm

1. Ant colony optimization algorithm

Since the middle of 1990s, many scholars are inspired from the mechanism of biological evolution and put forward many new methods to solve complex optimization problems, such as genetic algorithm, simulated annealing algorithm, tabu search algorithm, ant colony algorithm, immune algorithm, etc. These methods provide a new way to solve complex problems. Especially, the new heuristic algorithm—ant colony algorithm has attracted more and more attention of scholars both at home and abroad due to the characteristics of distributed concurrency, positive feedback, strong robustness, fast convergence speed and easy to obtain the global optimal solution and become a hot spot of heuristic algorithm research [74–76].

Ant colony algorithm, derived from the nature of the biological world and proposed by Italian scholars Dorigo and Maniezzo first in the early 1990s, is a new type of bionics algorithm and has a great achievement on the complex system optimization problems showing its superiority in solving complex optimization problems, especially in the discrete optimization problem. Ant colony algorithm is particularly suitable for the uncertain search in the solution space of the discrete optimization problems and has been effectively applied to the TSP problem, quadratic assignment problem, dispatching problem, graph coloring problem, and many other classic combinatorial optimization problems. Thus it becomes a potential evolutionary algorithm to solve the combinatorial optimization of NP-hard problem.

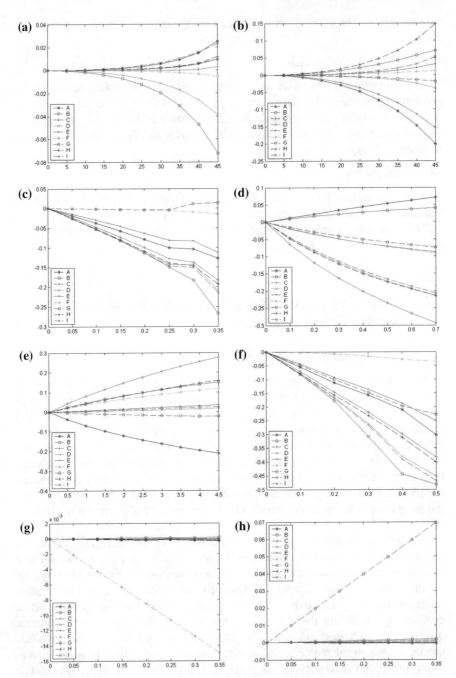

Fig. 4.3 The changes of parameters of each fault mode (Abscissa is fault degree and ordinate is the influence degree of corresponding parameters). **a** Fault of fuel main valve. **b** Simultaneous fault of the two main valves. **c** Decline of oxidant pump efficiency. **d** Throat ablation of thrust chamber. **e** Oxidizer nozzle blocking of combustion chamber. **f** Fuel pipeline leakage of gas generator. **g** Decline of thermal efficiency of evaporator. **h** Decline of thermal efficiency of cooler

The main features of ant colony algorithm are positive feedback and distributed calculation. Combined with some heuristic algorithm, the positive feedback process can make this method quickly find a good solution and easily implement parallel calculation.

Artificial ants swarm algorithm is proposed for imitating real ant colony behavior. Biomimetics find that ants transfer information by a substance called pheromone after meticulous observation. In the process of movement, ants leave the pheromone that they can feel in the path and guide the direction of movement. Therefore, the collective behavior of ant colony composed of a large number of ants shows a positive feedback phenomenon: the more ants go through a path, the greater the probability of others to choose the path will be. The ants exchange the information between individuals to achieve the purpose of searching for food.

Dorigo presents the ant colony algorithm as follows: as Fig. 4.4 shows, A is the nest, E is the food source, H and C are the obstacles. Due to the obstacles, if the ant wanted to reach from A to E, or return to A from E, it can only go around obstacles from H or C. Figure 4.4a shows the distance between each points. Suppose 30 ants leave from A to B, 30 ants from E to D per unit time, the hormone content quality (called pheromones) left by the ants of 1 and the stay time of the material of 1 for convenience. At the initial time, due to no information existing on the path of BH, BC, DH, and DC, the ants located in B and E can randomly choose the path. They choose BH, BC, DH, or DC in the same probability from the statistics view as shown in Fig. 4.4b. After a time unit, the amount of information on the path BCD is twice of that on the path BHD. At the moment of $t = 1$, there will be 20 ants from B and D to C, and 10 ants from B and D to H. With the time going by, the ants will choose the path BCD in the increasing probability. Eventually, the ants will choose path BCD completely, as shown in Fig. 4.4c, and find the shortest path from the nest to food source. Thus, the exchange of information between individual ants is a positive feedback process.

Fig. 4.4 The principle diagram of the colony algorithm. a distances between each points; b pheromones of each path when t = 0; c pheromones of each path when t = 1

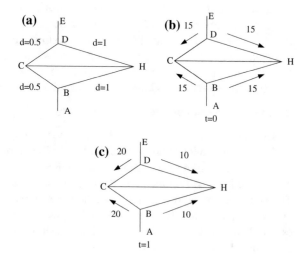

In the multidimensional function optimization problems, solving nonlinear equations can be transformed as follows:

$$\max Z = f(x), \quad x \in [x_0, x_f]$$

where $f(x): R_s \rightarrow R$ is the known multidimensional function, $[x_0, x_f]$ is the known s dimension solution space, and take a position vector x_0 for each of these elements those are less than the corresponding element of the position vector x_f.

To achieve the group search process of the ant colony algorithm, the following principles of the transfer probability are constructed. Suppose the artificial ants of m group and the number of ants in each group of s (i.e., equal to the solution space dimension of the objective function). The ants of each group randomly are located somewhere in the uniform area $(n_1 n_2 \ldots n_i \ldots n_s)$ in the solution space $[x_0, x_f]$, the ant groups between regions realize the state transition according to the probability that is defined as follows:

$$P_{ij} = \begin{cases} (\tau_j)^\alpha (\eta_{ij})^\beta, & if \quad \eta_{ij} > 0 \\ 0, & if \quad \eta_{ij} < 0 \end{cases} \quad i,j \in \{1,2,\ldots,(n_1 n_2 \ldots n_i \ldots n_s)\} \text{ and } i \neq j$$

where τ_j is the attraction intensity of the jth uniform area in the solution space, η_{ij} is the expected value and defined as $\eta_{ij} = f_j \max - f_i \max$, namely ant colony is the maximum difference of the objective function in the spatial location that is searched in the jth and ith uniform areas in the solution space. Suppose parameters $\alpha, \beta > 0$ for the heuristic factor. They are the effects on the state transition probability of the attraction intensity τ_j and the expected value η_{ij} of all areas in the solution space during state transition process of ant group, respectively.

Visibly, each group of ants either locally make random search from the current area in the solution space to other area according to the above rules, or locally make random search in the current area. Then, the optimization problem of multidimensional function $f(x)$ can recur to continual moving of the m group of ants in the $n_1 n_2 \ldots n_i \ldots n_s$ uniform area of the solution space $x \in [x_0, x_f]$ and local random search in some area. Once the number m of ant group is large enough, the way of optimization is equivalent to the finite random search at the guidance of priori knowledge of a group of ants for the function $f(x)$ in the domain $[x_0, x_f]$, and eventually converge to approximate global optimal solution of the problem [76–78].

2. The choice of characteristic parameters

The purpose of the steady state simulation is to acquire the particular fault mode of the different faults, analyze the effects of engine performance of various internal and external interference factors. Due to the inherent characteristics of liquid missile engine, it is impossible to directly measure the values of all the parameters. The possible way is to take the parameters measured in the original system as the evaluating parameters for fault state and utilize the indirect parameters when necessary. The 14 parameters in the equations, as shown in Table 4.5, are chosen in the following steady state fault simulation.

Table 4.5 The second-level engine characteristic parameters

Number	The measured parameters	Parameters (deviation) rate	Expression
1	Engine oxidant flow	Deviation rate of engine oxidant flow	A
2	Engine fuel flow	Deviation rate of engine fuel flow	B
3	Main engine pre-oxidant pressure	Deviation rate of main engine pre-oxidant pressure	C
4	Pressure of main engine combustion agent road branch	Deviation rate of pressure of main engine combustion agent road branch	D
5	The host turbine pump speed	Deviation rate of the host turbine pump speed	E
6	The host oxidant pump outlet pressure	Deviation rate of the host oxidant pump outlet pressure	F
7	Host fuel pump outlet pressure	Deviation rate of host fuel pump outlet pressure	G
8	Antioxidant multipass pressure	Deviation rate of antioxidant multipass pressure	H
9	Fuel multipass pressure	Deviation rate of fuel multipass pressure	I
10	Turbine pump speed	Deviation rate of turbine pump speed	J
11	Oxidant pump outlet pressure	Deviation rate of oxidant pump outlet pressure	K
12	The fuel pump outlet pressure	Deviation rate of the fuel pump outlet pressure	L
13	The pressurization pressure of oxidant gas tank	Deviation rate of the pressurization pressure of oxidant gas tank	M
14	The pressurization pressure of fuel gas tank	Deviation rate of the pressurization pressure of fuel gas tank	N

3. **Simulation and analysis of secondary engine steady state fault**

For a particular fault, changing the related parameter a and interference factor D of the secondary engine in the nonlinear equations and solving the equations, the corresponding characteristic parameters can be obtained. We can get the simulation results of all fault effects through ant colony optimization algorithm. Figures 4.5, 4.6, 4.7, and 4.8 show the decrease of main oxidant turbine pump efficiency, the blocking of the main engine thrust chamber oxidizer nozzle, the decrease of the thermal efficiency of the evaporator, and the decrease of thermal efficiency of the cooler, respectively.

4.1.6 Analysis of Engine Fault Feature Based on Evolutionary Calculation

As known from the above research, the change of the parameters has strong non-linear characteristics in the engine working process, especially stronger in the process of starting, transit, and shutdown, and in the case of failure and abnormity.

Fig. 4.5 Host pump
efficiency decline

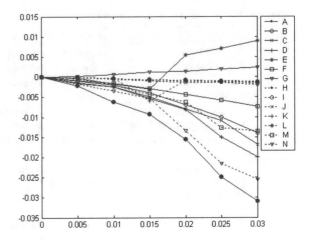

Fig. 4.6 The host thrust
chamber oxidizer nozzle
blockage

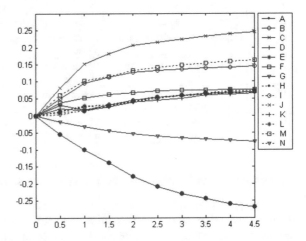

Fig. 4.7 Thermal efficiency
of evaporator decline

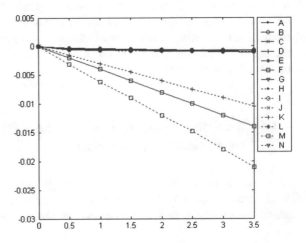

Fig. 4.8 Thermal efficiency
of cooler decline

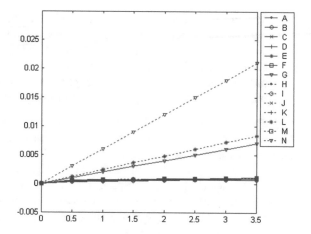

The work model of the system is all nonlinear equations under the condition of normal working, fault state, and fault interference. In the fault monitoring and diagnosis system, the method of nonlinear analysis is mainly adopted. The common optimization methods can only get local optimal solution. The random heuristic optimization method such as the simulated annealing technique and the Monte Carlo method have been used to solve the problem of global optimization. Using these optimization technologies theoretically can find the global optimal solution, but actually has some disadvantages of low search efficiency, limited ability to jump out of local optimal point, and sometimes powerless to solve the actual ultrahigh-dimensional optimization problem. However, bionics evolution calculation technology can be used to solve the ultrahigh-dimensional optimization problem [76]. Evolutionary calculation is applied to such fields as solution space search and machine learning due to its characteristics of solving global optimization, implicit parallel calculation, randomization, adaption, and robustness. Evolutionary calculation has been widely used in computer science, engineering, management science, social science, and other fields, especially suitable for the optimization problem of large-scale, nonlinear, multipolar value, even no objective function. The following uses the evolutionary calculation method to analyze the steady state fault feature.

1. **The basic principle of evolutionary calculation [79–80]**

Evolutionary calculation method is a kind of adaptive global optimization probability search algorithm to simulate biological evolution process in the natural environment. This method produces a variety of optimization calculation model based on the simulation of biological evolution mechanism. The most typical models are the genetic algorithm, evolution strategy, evolutionary programming, and genetic programming.

(1) Genetic algorithm: It is first proposed by J. Holland professor in the University of Michigan, and stems from the study of natural and artificial adaptive

systems in the 1960s. De Jong has conducted amounts of numerical function optimization experiments based on the ideas of genetic algorithm in 1970s. On the basis of a series of research, Goldberg has summarized and formed the basic framework of genetic algorithm in the 1980s. With the in-depth research and successful application of the genetic algorithm, it has been attracted widespread attentions and introduced in many fields.

Genetic algorithm is a kind of global search method for complicated problem with its characteristics of simple, general, robust, and efficient parallel. It can simulate the evolution process of the nature from low level to high level, automatically acquire and accumulate the knowledge in the process of search, and control the search process adaptively to obtain the optimal solution. Genetic algorithm starts from an arbitrary initial group and makes this group evolve into the better area of the search space until to reach the optimal solution by random selection, crossover, mutation, etc.

The main characteristics of genetic algorithm are as follows:

Encode feasible solution through the appropriate method;
Select the probability operation based on the fitness of the individual;
Use the cross operator for the restructuring technology of the individual;
Use random mutation technology for mutation;
Be suitable for searching the discrete space.

(2) Evolution strategy: evolution strategy is a kind of optimization algorithm and proposed by I. Rechenberg and H.P. Schwefel in the 1960s. The purpose to develop the algorithm is to solve multimodal nonlinear function optimization problem. Then, they have carried on this algorithm and ultimately formed a branch of evolutionary computation.

The differences between evolutionary strategy and genetic algorithm are that using the genetic algorithm is actually to map the solution space of original problem to bits string space, and then carry the genetic operation. It emphasizes the effects of changes of individual genetic structure on its own fitness. But using the evolutionary strategy is directly to carry on the solution space and emphasize the adaptability and diversity of behavior from the parent to the offspring and the adaptive step length adjustment in the search direction during the evolution process. The evolutionary strategy is mainly used for solving the numerical optimization problem. In recent years, genetic algorithm and evolutionary strategy have no obvious boundaries due to their mutual penetration.

(3) Evolutionary planning: Evolutionary planning, proposed by L.J. Fogel in the 1960s, is a finite state machine evolution model to solve the prediction problem. In this model, the machine state varies according to the uniform random distribution. Later, D.B. Fogel expands evolutionary planning ideas into the real space, making it to solve optimization problems in real space, and introduces the technology of normal distribution in its mutation operation.

Compared with genetic algorithm and evolution strategy, evolutionary planning mainly has the following features:

The simulation of biological evolution process using evolutionary planning focuses on the evolution of species, without using individual operators;

The choice of arithmetic in evolutionary planning focuses on the competition between the individuals in the group;

Evolutionary planning directly takes feasible solution as the form of the individual, not encoding the individual and considering the influence of random disturbance factors on the individual;

Evolutionary planning takes the optimization problem in real number space as the main object.

(4) Genetic programming: the thought of genetic programming is proposed by J.R. Koza in the early 1990s, and mainly uses the basic idea of the genetic algorithm to make the computer automatically programming. In the process of evolutionary calculation, the solution space is expressed using hierarchical structure.

1) The basic concept of evolutionary calculation

① Chromosome: it is the main carrier of genetic materials, composed of DNA (deoxyribonucleic acid) and protein. The chromosome corresponds to the data or array in evolutionary calculation.

② Gene: it is the DNA fragments with genetic effects, stores the genetic information, can accurately replicate, mutate, and control the biological traits by controlling the protein synthesis.

③ Locus: it is the location of genes in the chromosome.

④ Alleles: it is the values of genes in the chromosomes. Locus and gene determine the characteristics of chromosome and also determine the character of the individual traits.

⑤ Phenotype: it refers to the individual traits that are present.

⑥ Genetype: it consists of genetic that is closely related to the phenotype. The individual with the same kind of genotypes can have different phenotypes under different environmental conditions, so the phenotype is the result of the interaction between genotype and environment.

⑦ Individuals: it is the chromosome object processed by the genetic algorithm, also called genotype individuals.

⑧ Population: it is composed of a certain number of individuals.

⑨ Population Size: it is the number of individuals in the group.

⑩ Fitness: it is the adaptation to the environment for each individual.

⑪ Coding: it is the transformation from the phenotype to genotype, namely, transform parameters or solution in search space into chromosomes or individual in genetic space, also called the representation of problem.

⑫ Decoding: it is the transformation from the genotype to phenotype, namely, transform chromosome or individual to the parameters or solution in search space.

2) The basic theory of evolutionary calculation

   ① Mode theorem

Definition 4.1 Schemata: some similar modules. It describes a subset of individual coding string with similar structural characteristics in some locations. For binary encoding, the individual is a coding string that consists of the elements in the binary character set $V = \{0, 1\}$, while the schemata is composed of coding string that consists of the elements in the three-value character set $V+ = \{0, 1, *\}$, where "*" is the wildcard.

Definition 4.2 Schema Order: the confirmative number in the schemata H, remembering $o(H)$.

Definition 4.3 Schema definition length: the distance between the first confirmative location and the last one in the schemata H, remembering $\delta(H)$.

    It is proved that the number of the offspring sample of schemata H is as shown in Eq. (4.13) under the action of selection, crossover, and mutation of genetic operators.

$$m(H, t+1) \geq m(H, t) \cdot (f(H)/\bar{f}) \cdot [1 - P_c \cdot \delta(H)/(l-1) - o(H) \cdot P_m] \quad (4.13)$$

where l is the length of the string, $m(H, t)$ is the sample number that matches the schemata H in t, $f(H)$ is the average fitness of all samples in schemata H, $\bar{f}$ is the group average fitness, P_c is the crossover probability, P_m is the mutation probability. On this basis, the schema theorem can be concluded as follows:

Theorem 4.1 *Schema theorem: under the action of selection, crossover and mutation of genetic operators, the schema, with low order, short defining length and average fitness higher than the group average fitness, increases exponentially in the offspring.*

    Schema theorem ensures that the samples of better schema increase exponentially and lay a theoretical foundation for the genetic algorithm, which is the realizable optimization process for searching feasible solution.

   ② Building block hypothesis

Definition 4.4 Building block: the schema with low order, short defining length, and high fitness.

Hypothesis 4.1 Building block hypothesis

    Under the action of genetic operators, the models (building block) with low order, short defining length, and high average fitness combine mutually to generate the models with higher order, long defining length, and high average fitness and eventually produce the global optimal solution.

    Schema theorem ensures that the number of samples increase exponentially in the optimal solution model, which meet the needs of searching for the optimal solution, namely there is the possibility for the algorithm to find global optimal solution. And building block hypothesis points out that the algorithm has the ability to find the global optimal solution, namely, under the action of genetic operators, building block can generate the model with high order, long defining length, and high average fitness, so it can ultimately generate the global optimal solution.

③   Implicit parallelism of algorithm

In the operation process of the evolution, each generation processes M individuals. Because many different models are implied in an individual code string, the algorithm essentially processes more patterns. For binary coding, the number of effective mode processed by genetic algorithm is in direct proportion to the cube of group size M. That is to say, in the course of evolution, only M individuals are processed in each generation. Actually, the number of model parallel processed is in direct proportion to the cube of group size M. This implicit parallelism can make the algorithm to search out fast some great models.

④   Convergence of the algorithm

In the evolution of the genetic algorithm, the individual aggregate changes from one generation to another. Taking each generation group as a kind of state, the entire evolution is a Markov process. (Ideas can be generalized to other evolutionary algorithms).

Theorem 4.2 *The probability for the basic genetic algorithm converging to the optimal solution is less than 1.*

Therefore, using the basic genetic algorithm cannot guarantee the optimal solution. It is necessary to improve the basic genetic algorithm and a lot of improved genetic algorithms appear.

Theorem 4.3 *The probability of using genetic algorithm for reserving the best individual strategy converging to the optimal solution is 1.*

3)   The general process of evolutionary calculation

①   The basic components of evolutionary calculation

A. Coding method of chromosome

Evolutionary calculation is mainly to manipulate individual coding string. The common coding methods are binary coding method, the floating-point encoding method, and symbol encoding method. For the multiple parameters binary code, the substring is first obtained by binary coding, and then linked to generate a complete chromosome. Supposing the length of the substring of k, the range of the corresponding unsigned integer for the substring after decoding is $[0, 2k - 1]$. If a parameter is for $[U_{min}, U_{max}]$, the parameter encoded is u, the value of code is x, the process of encoding and decoding is

$$x = \frac{(u - U_{min})(2^k - 1)}{U_{max} - U_{min}} \quad \text{(encoding)} \tag{4.14}$$

$$u = \frac{(U_{max} - U_{min}) \cdot x}{2^k - 1} + U_{min} \quad \text{(decoding)} \tag{4.15}$$

Coding accuracy:

$$\Pi = \frac{(U_{max} - U_{min})}{2^k - 1}$$

B. Fitness function

Objective function $f(x)$ can be mapped to fitness function $F(x)$. For solving maximum value problem:

$$F(x) = \begin{cases} f(x) + C_{min}, & if \quad f(x) + C_{min} > 0 \\ 0, & if \quad f(x) + C_{min} \leq 0 \end{cases} \tag{4.16}$$

where C_{min} is a relatively small constant.

For solving minimum value problem:

$$F(x) = \begin{cases} C_{max} - f(x), & if \quad f(x) + C_{max} \\ 0, & if \quad f(x) + C_{max} \end{cases} \tag{4.17}$$

where C_{max} is a relatively large constant.

C. The genetic operation

In the genetic process and biological evolution, the "genetic," "hybrid," and "variation" process make the species with higher survival environment fitness have more opportunity to transmit to the next generation. For simulating this process, "selection operator," "cross operator," and "mutation operator" are used to weed out for individuals of a group to achieve the course of evolution.

Selection operator: selection operation is used to determine how to select which individual genetic group from the parent generation groups transmits to the next generation according to some way. Here are common selection operators such as proportion selection method, determine sampling selection, no playback random selection, ranking selection, random selection, etc.

Crossover operator: in the biological evolution, two homologous chromosomes form a new individual by mating and restructuring and produce new species. The crossover operation in evolutionary calculation is the two individuals to exchange some of its genes in some manner and to form two new individuals. Crossover operation is the main method to generate new individual and plays an important role in the evolutionary calculation.

Mutation operator: in the genetic and biological evolution, some genes mutate, due to some accidental factors, to produce new chromosomes and show new biological traits. Crossover operation is the main method to generate new individual and determines the global search ability of the algorithm. Mutation operation determines the local search ability of algorithm. The combination of crossover operator and mutation operator can jointly complete the global search and the local search, making the algorithm complete the optimization process.

4) The basic process of evolution algorithm in optimization calculation

Evolutionary calculation provides a universal model for solving complex system optimization problems. Its basic principle is the simulation for the process of biological evolution. The basic process is as follows:

① the initialization, randomly generate initial population $P(t)$.
② evaluate the fitness of group $P(t)$.
③ the operation of individual restructure: $P'(t) \leftarrow Recombination[P(t)]$.
④ the operation of individual varies: $P''(t) \leftarrow Mutation[P'(t)]$.
⑤ evaluate the fitness of $P''(t)$.
⑥ select and copy the operation individual: $P(t + 1) \leftarrow Reproduction[P(t) \cup P''(t)]$.
⑦ distinguish termination conditions. If the termination conditions do not meet, $t \leftarrow t + 1$, then switch to step ④ and continue the evolutionary operation; or, output the current optimal individuals and end algorithm.

Basic evolutionary algorithm is derived from the simulation of biological evolution process. Many restrictions are added in designing evolutionary algorithms, thus large amount of calculations are needed when solving the problems, and sometimes the global optimal solution are not obtained. Optimization algorithm is expected to find out all optimal solutions, and the basic evolutionary algorithm could not do it. The author uses basic evolutionary algorithm to solve the problem of the fault feature of the missile power system, but does not get satisfactory results due to the complexity of the problem and the large calculation scale. Therefore, many scholars propose a lot of improved algorithms, such as niche genetic algorithm, multiple-population evolutionary algorithm, a hybrid evolutionary algorithm, heuristic evolutionary algorithm (Differential Evolution, DE), etc., and acquire the good effect. Especially, DE evolutionary algorithm can find global optimal solution with fast convergence speed, less control parameters, easy to use, stable, reliable, wide applicable scope, etc. Therefore, DE evolutionary algorithm is used to solve the problem of fault feature of missile power system in the following.

5) DE evolutionary algorithm

The basic strategy of DE evolutionary algorithm is to take the difference of two randomly selected parameters vector as the source of random variation of the third parameter vector. In DE evolutionary algorithm, the number of D dimension parameter vector is NP

$$x_i, G, \quad i = 0, 1, 2, \ldots, NP - 1 \tag{4.18}$$

where G is the number of population, which is unchangeable in evolution; NP is the number of parameter vector, which is also unchangeable in evolution.

In the evolutionary calculation, the choice of the initial group shall be the random selection, and cover the whole parameter space. Generally, assume random decisions to be the uniform probability distribution. DE generates new parameter vector by adding the difference of two population vectors to the third population vector. If the new vector reduces the selected objective function of the previous vector, in the next generation the new vector will replace the previous vector, otherwise, the previous vector will be retained. In the actual DE modified algorithm, the basic principles are expanded. For example, destabilization parameter of an existing vector can be the summation of the difference of multiple vectors. In most cases, the vector newly obtained with destabilization parameter and the old vector should be mixed before comparing the value of objective function. The basic model of DE is as follows:

For each vector x_i, G, $i = 0, 1, 2, \ldots, NP - 1$, the vector after disturbance is

$$v_{i,G+1} = x_{r1,G} + F \cdot (x_{r2,G} - x_{r3,G}) \tag{4.19}$$

where r_1, r_2, $r_3 \in [0, NP - 1]$ are mutual unequal integers. F, as a real constant, is weight control factor with the value of $[0, 2]$, determining the magnification of differential variables $(x_{r2,G} - x_{r3,G})$.

$x_{r1,G}$ is a random selection vector of individual species to produce the vector $v_{i,G+1}$ by disturbance and not related to $x_{i,G}$. In this model, the disturbed vector is randomly selected and the disturbed part only contains a weighted difference vector. The basic two-dimensional optimization evolutionary process of DE algorithm is as shown in Fig. 4.9.

In order to increase the potential diversity of the perturbation parameter vector (species diversity), crossover operator is introduced. When the crossover operation ends, the vector is as follows:

Fig. 4.9 Basic two-dimensional evolution of DE algorithm

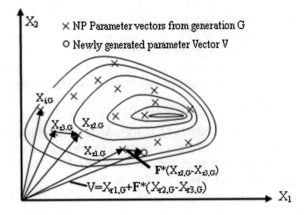

Fig. 4.10 Basic cross DE algorithm process

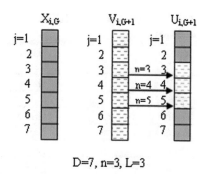

$$D=7, n=3, L=3$$

$$U_{i,G+1} = \left(u_{0i,G+1}, u_{1i,G+1}, \ldots u_{(D-1)i,G+1} \right) \tag{4.20}$$

while

$$u_{ji,G+1} = \begin{cases} v_{ji,G+1} & \text{for } j = \langle n \rangle_D, \langle n+1 \rangle_D, \ldots, \langle n+L-1 \rangle_D \\ x_{ji,G} & \text{for all other } j \in [0, D-1] \end{cases} \tag{4.21}$$

where $\langle \rangle_D$ is the operation of absolute value. N is a random integer in $[0, D-1]$. The integer L, drawn from $[1, D]$ according to the probability $\Pr(\Pr(L >= v) = (CR)v$, $v > 0)$, is the number of the parameter vector which needs to change. The range of CR which is a control parameter of design process is $[0, 1]$. The random variation of n and L makes the new vector $U_{i,G+1}$ be updated.

$U_{i,G+1}$ shall be compared with $x_{i,G}$. If the vector $U_{i,G+1}$ had a smaller objective function value, $x_{i,G+1}$ update for the $U_{i,G+1}$, otherwise, $x_{i,G+1}$ for $x_{i,G}$. Cross process is shown in Fig. 4.10.

2. **Analysis of fault feature on missile dynamic system based on evolutionary calculation**

(1) General process of fault feature analysis based on evolutionary calculation

The optimization problem, as shown in Eq. (4.2), can be solved using evolutionary calculation method. As mentioned previously, using evolutionary computation method to solve the optimization problem has the incomparable advantages over the other methods.

General process of fault feature analysis on missile dynamic system based on evolutionary calculation is as follows:

1) Analyze the characteristics of the subject and build the model;
2) Transform the model equation into optimization model, and choose the maximize or minimize according to the evolutionary calculation method;
3) Design the objective function and fitness function according to the evolutionary calculation method;
4) Design algorithm, write the program, and select operation parameter values;
5) Run the program;

Table 4.6 Simulation results of engine static characteristic main parameters using DE evolutionary algorithm

Name	Units	Calculation results in this work	Small deviation calculation results
The engine thrust	kN	13.1696	15.30
Ratio of propellant element		9.5788×10^{-2}	8.45×10^{-2}
Turbine pump speed	r/min	9.9839	29.03
Engine oxidant flow	kg/s	5.7334	6.02
The engine fuel flow	kg/s	−1.0962	−0.61
The generator oxidant flow	kg/s	1.6936×10^{-2}	1.84×10^{-2}
The generator fuel flow	kg/s	1.2120×10^{-2}	3.02×10^{-2}
Cooling jacket flow	kg/s	−0.9746	−0.56
The evaporator flow	kg/s	6.1168×10^{-3}	6.84×10^{-3}
Oxidant pump pressure increment	MPa	-8.9668×10^{-2}	-6.28×10^{-2}
Fuel pump pressure increment	MPa	4.4274×10^{-2}	8.14×10^{-2}
Pressure before oxidant injection	MPa	2.0690×10^{-1}	3.30×10^{-2}
Pressure of combustion agent road branch	MPa	5.1600×10^{-2}	8.53×10^{-1}

6) Analyze the results of calculation.

(2) Examples of fault feature analysis based on DE evolutionary algorithm

1) The static characteristic simulation

The simulation results of single engine static properties based on the DE evolutionary algorithm are shown in Tables 4.6 and 4.7.

In calculation, genetic algorithm parameters are as follows. The numbers of independent variable are $l = 25$, group size is $NP = 250$, crossover probability is $CR = 0.5$, weight control parameters is $F = 0.8$, maximum limit evolution algebra is $IG = 5000$.

2) The simulation of steady state fault features based on DE algorithm

The choice of parameters is the same as in Sect. 4.1.

The simulation of a particular fault can be carried on by changing the value of a certain fault factor in the steady state numerical model and solving steady state fault numerical model. In order to compare with the analysis of Sect. 4.1, the simulation is carried on for the following fault, as shown in Fig. 4.11.

Table 4.7 Simulation results of engine static characteristic main parameters using DE evolutionary algorithm

Name	Units	Calculation results in this work	Small deviation calculation results
The engine thrust	kN	0.9063	0.70
Ratio of propellant element		-2.4463×10^{-2}	-3.03×10^{-2}
Turbine pump speed	r/min	-6.5553	-6.40
Engine oxidant flow	kg/s	-0.4506	-0.39
The engine fuel flow	kg/s	0.7697	0.64
The generator oxidant flow	kg/s	-3.6315×10^{-3}	-3.64×10^{-4}
The generator fuel flow	kg/s	1.3890×10^{-2}	1.59×10^{-2}
Cooling jacket flow	kg/s	0.6545	0.56
The evaporator flow	kg/s	6.1815×10^{-5}	3.64×10^{-5}
Oxidant pump pressure increment	MPa	-6.8542×10^{-3}	-6.87×10^{-3}
Fuel pump pressure increment	MPa	-3.3730×10^{-2}	-3.65×10^{-2}
Pressure before oxidant injection	MPa	2.0782×10^{-3}	8.88×10^{-4}
Pressure of combustion agent road branch	MPa	5.9140×10^{-2}	6.77×10^{-2}

In calculation, genetic algorithm parameters are as follows. The numbers of independent variable are $l = 151$, group size is $NP = 1500$, crossover probability is $CR = 0.9$, weight control parameter is $F = 0.7$, maximum limit evolution algebra is $IG = 8000$.

From the above analysis, the method to analyze fault feature on missile dynamic system using the genetic algorithm is effective, which makes fault analysis and solution not affected by initial value and easy to get the global optimal solution. Due to the missile engine failure characteristics of nonlinearity, multiparameter, and strong coupling, the change of parameters in the calculation is discontinuous, large workload, and slow convergence. But the final calculation results are satisfied as shown in Fig. 4.11.

Evolutionary calculation is suitable for the optimization problem of large-scale, nonlinear, more extreme, even no objective function expression due to its characteristics of parallel, randomized, adaptive search features, robustness, and global optimization solution. Therefore, evolutionary computation method presented in this section is used to analyze the steady state fault features on missile dynamic system. The calculation results show that the method to analyze fault feature on missile dynamic system using the genetic algorithm is effective, which makes fault analysis and solution not affected by initial value and easy to get the global optimal solution, but the calculation speed is slow. This method is suitable for afterwards-fault analysis.

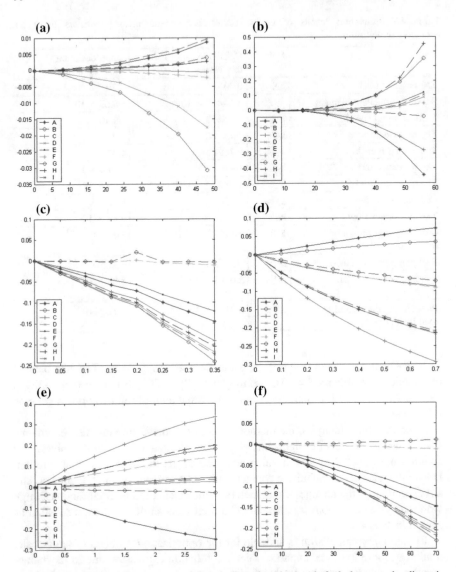

Fig. 4.11 The changes of parameters of each fault mode (Abscissa is fault degree and ordinate is influence degree of the corresponding parameters). **a** Fuel main valve fault effect. **b** The synchronous fault effect of two main valves. **c** Pump efficiency decline. **d** Thrust chamber throat ablation. **e** Combustion chamber oxidizer nozzle blocking. **f** Fuel subsystem blocking. **g** Pipe leakage after the evaporator. **h** Thermal efficiency of evaporator decline. **i** Oxidant main system restrictor jam. **j** Oxidant tank pressure decline

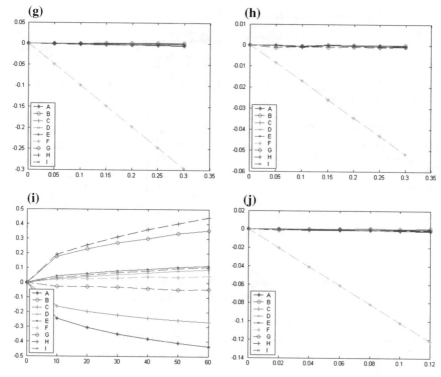

Fig. 4.11   (continued)

4.1.7   Acquisition of Engine Steady State Failure Mode

(1)   The system steady state fault mode

As is known from the above analysis, fault simulation can accurately describe the change of parameters on missile or rocket dynamic systems, and consequently comprehensively understand the behavior of dynamic system at fault condition. Fault mode set of dynamic system is formed on this basis of the fault simulation for the whole dynamic system or parts to facilitate the analysis, identification, and diagnosis of system faults. The model set is the implicit expression of dynamic system fault. On the basis of continuous improvement and correction for the system and component model, the model set is accurately taken as the standard fault knowledge, which has important significance for the product manufacture, production, finalization, detection, diagnosis, development of control system and the arrow fault diagnosis and monitoring system.

(2)   The component fault mode

Engine system steady state fault model includes the dynamics equations of all groups and components. The qualitative model of a group or component can be

Table 4.8 Fault mode by representation of the parametric

Fault mode	Position in Fig. 4.3	Test parametrics								
		A	B	C	D	E	F	G	H	I
Fuel main valve failure	(a)	1	−1	1	−1	1	−1	1	1	1
Two valves not fully open	(b)	−1	1	−1	−1	1	1	−1	1	1
Combustion chamber oxidizer nozzle blocking	(c)	−1	−1	−1	−1	−1	0	0	−1	−1
Thrust chamber throat ablation	(d)	1	1	−1	−1	−1	−1	−1	−1	−1
Pump efficiency decline	(e)	−1	1	1	0	1	0	0	1	0
Gas generator burners pipeline leakage	(f)	−1	−1	−1	−1	−1	0	−1	−1	−1
Thermal efficiency of cooler decline	(g)	0	0	0	0	0	−1	0	0	0
Thermal efficiency of evaporator decline	(h)	0	0	0	0	0	0	1	0	0

obtained using the dynamic steady state fault model and then according to the following steps, the fault mode of component can be obtained.

1) Select a component as object.
2) Determine the input and output parameters related to the component.
3) Form combination eigenvectors and get all the possible eigenvectors.

The qualitative method of parameters is as follows:

$$[x] = \begin{cases} 0 & x \leq x_L \\ \frac{x-x_L}{x_H-x_L} & x_L < x \leq x_H \\ 1 & x \geq x_H \end{cases}$$

x is the deviation rate of each parameter where the value of x_L is 0 and x_H is 5 %.

4) For each model, form the fault eigenvectors and fault mode using the steady state fault numerical model for calculation and analysis. Failure mode set of the first engine system is shown in Table 4.8.

4.2 Analysis of Dynamic Failure Mode

As is known from the research of steady state fault model, blocking pipeline, pipeline leak, thrust chamber throat ablation can be completely separated by the steady state fault analysis. Actually, some of the different faults can produce so similar parameter values for the working conditions of the system that it is failure to acquire and identify for the fault mode by steady state fault analysis. Therefore, the dynamic research with fault transient characteristics and fault transition effect analysis are conducted to obtain the engine fault mode in the transition [79–86].

4.2.1 The Numerical Method for Solving the Dynamic Model of Engine

Due to the fluid motion, mechanical movement and the chemical thermodynamic process in the engine, the combustion chamber, gas generator, and heat exchanger in the model having the characteristics of highly concentrated energy, dynamic equations of exchange components, high changeable frequency, sensitive to the integral step, and high calculation accuracy for the dynamic characteristics of fault condition, the nonlinear differential equations of engine at the starting, shutdown process and the fault transient process commonly are regarded as the stiff differential equations or usually referred to as morbid equations. In this case, the simulation calculation for the fault is actually to numerically solve the initial value problems of nonlinear differential equations. The methods to solve the equations include Runge–Kutta, Gill, Treanor and Bader et al. These methods have some disadvantages of small step size, slow convergence speed, low calculation accuracy, and poor stability. Runge–Kutta–Fehlberg (R-K-F) method can overcome the above disadvantages, with the characteristics of easy programming, simple calculation, and less computational effort. Initial value is determined based on the steady state model.

To use R-K-F method to solve the mathematical model, the mathematical model is first transformed into a standard solution:

$$\begin{cases} y_0^n = f_0(t, y_0, y_1, \ldots, y_{n-1}) \\ y_1^n = f_1(t, y_0, y_1, \ldots, y_{n-1}) \\ \vdots \\ y_{n-1}^n = f_{n-1}(t, y_0, y_1, \ldots, y_{n-1}) \end{cases}$$

The R-K-F method to solve the differential equations is as follows:

Suppose that the value of unknown function y_{ij} ($i = 0, 1, \ldots, n - 1; j = 0, 1, 2, \ldots, N$) at t_j points is known, the formula to calculate the value of unknown function $y_{i,j}$ $_{+1}$ at $t_j + 1 = t_j + n$ point is

$$y_{i,j+1} = y_{ij} + \left(\frac{25}{216} K_1 + \frac{1408}{2565} K_3 + \frac{2197}{4104} K_4 - \frac{1}{5} K_5 \right)$$

$$\bar{y}_{i,j+1} = y_{ij} + \left(\frac{16}{135} K_1 + \frac{6656}{12{,}825} K_3 + \frac{28{,}561}{56{,}430} K_4 - \frac{9}{50} K_5 + \frac{2}{55} K_6 \right)$$

where

$$K_1 = hf\left(t_j, y_{ij}\right)$$

$$K_2 = hf\left(t_j + \frac{h}{4}, y_{ij} + \frac{1}{4}K_1\right)$$

$$K_3 = hf\left(t_j + \frac{3}{8}h, y_{ij} + \frac{3}{32}K_1 + \frac{9}{32}K_2\right)$$

$$K_4 = hf\left(t_j + \frac{12}{13}h, y_{ij} + \frac{1932}{2197}K_1 - \frac{7200}{2197}K_2 + \frac{7296}{2197}K_3\right)$$

$$K_5 = hf\left(t_j + h, y_{ij} + \frac{439}{216}K_1 - 8K_2 + \frac{3680}{513}K_3 - \frac{845}{4104}K_4\right)$$

$$K_6 = hf\left(t_j + \frac{1}{2}h, y_{ij} - \frac{8}{27}K_1 + 2K_2 - \frac{3544}{2565}K_3 + \frac{1859}{4104}K_4 - \frac{11}{40}K_5\right)$$

and local truncation error is estimated to be

$$E(t_j, h) = \bar{y}_{i,j+1} - y_{i,j+1}$$

$$= \frac{1}{360}K_1 - \frac{128}{4275}K_3 - \frac{2197}{7540}K_4 + \frac{1}{50}K_5 + \frac{2}{55}K_6$$

(1) Select a step size h as general for interval/N;
(2) Calculate K_1–K_6, $E_r = E(t_j, h)$;
(3) If $E_r < \varepsilon_h$ or $h < 2h$ min, calculate $y_{i,j+1}$;
(4) If $j + 1 \geq N$, stop;
(5) $s = [\varepsilon_h/E_r/2]/4$.
(6) If $s < 0.75$ and $h > 2h$ min, then $h = h/2$ switch (2);
(7) If $s > 1.5$ and $2h < 2h$ max, then $2h = h$ switch (2).

where s is the step controlling parameter.

R-K-F method has some advantages like explicitness, easy program designing, higher calculation accuracy, and wide stability domain than Runge–Kutta, and is suitable for the simulation calculation of a stiff system of differential equations of similar engine fault state transition process. The equations also can be solved using the ode23t function in Matlab.

4.2.2 Analysis of Engine Dynamic Fault

During the dynamic change of engine state, the changes of parameters caused by some faults are great, but some are small. The size of changeable parameters has great influence on the detector, so the discussions of the dynamic fault mainly are classified according to the size of the changeable parameter.

Table 4.9 Dynamic failure monitoring parameters

Number	Parameters	Expression	Number	Parameters	Expression
1	Engine oxidant flow	m_o	6	Gas generator combustion chamber pressure	P_b
2	Engine fuel flow	m_f	7	Oxidiser tank pressure	P_o
3	Subsystem propellant mixing ratio	r_b	8	Fuel tank pressure	P_r
4	Engine propellant mixing ratio	r_c	9	The pump speed	n
5	The turbine outlet pressure	P_c			

Select the following nine variables, shown in Table 4.9, as monitor parameters in equations.

The common dynamic fault modes of the first-level engine system are calculated using R-K-F algorithm and the dynamic mathematical model aforementioned and the results are shown in Figs. 4.12, 4.13, 4.14, 4.15, 4.16, 4.17, 4.18, 4.19, 4.20, 4.21, 4.22, 4.23, 4.24, 4.25, 4.26, 4.27, 4.28, 4.29, 4.30, 4.31, 4.32, 4.33, 4.34, 4.35, 4.36, 4.37, 4.38, 4.39, 4.40, 4.41. Where Abscissa diagram is time $t(s)$, the vertical is the changeable percentages of the transient values x for various parameters and the rating value x^*, namely $\bar{x} = (x - x^*)/x^*$.

Figures 4.12, 4.13, and 4.14 show the transition characteristics of tank fault. Tank leakage and overflow valve not being closed make the corresponding tank pressure drop, supercharging pressure decrease, which make the pump inlet pressure less than the cavitation pressure and has impact on pump working condition. Moreover, tank pressurization pressure is too low to make the tank easy

Fig. 4.12 Fuel main valve fault effect

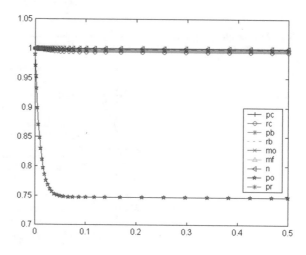

Fig. 4.13 Fuel tank overflow
valve failure

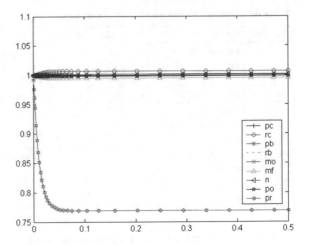

Fig. 4.14 Oxidiser tank
leakage

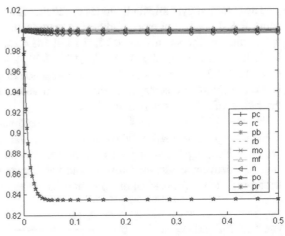

Fig. 4.15 Antioxidant start
valve not fully open

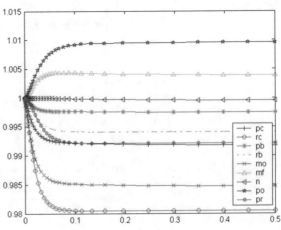

Fig. 4.16 Fuel start valve not fully open

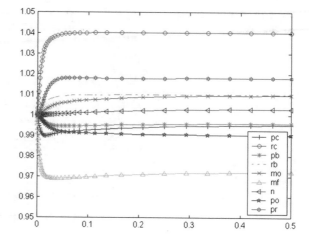

Fig. 4.17 Two start valve not fully open

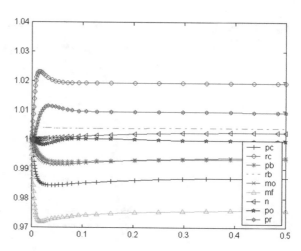

Fig. 4.18 The main valve not fully open

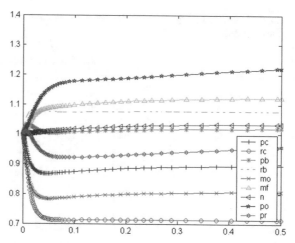

Fig. 4.19 Fuel main valve
not fully open

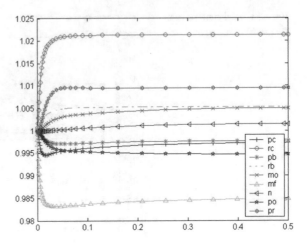

Fig. 4.20 Two main valve
not fully open

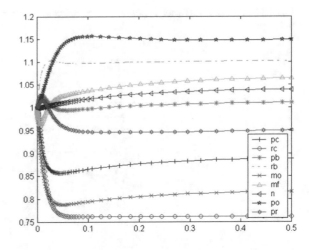

Fig. 4.21 Oxidant pump
efficiency decline

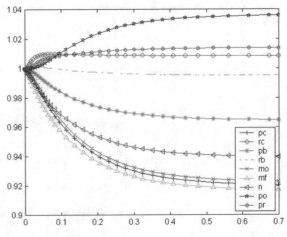

Fig. 4.22 Fuel pump
efficiency decline

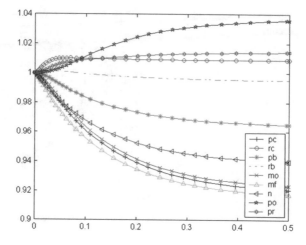

Fig. 4.23 Two pumps
efficiency decline

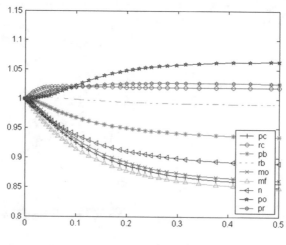

Fig. 4.24 Oxidant pump
cavitation

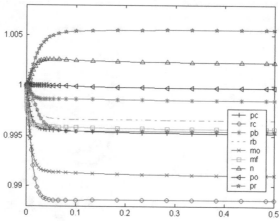

Fig. 4.25 Turbine efficiency decline

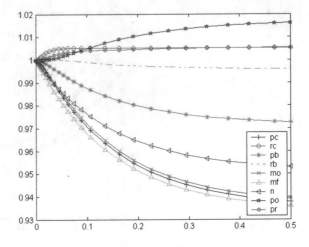

Fig. 4.26 Turbine inlet gas leakage

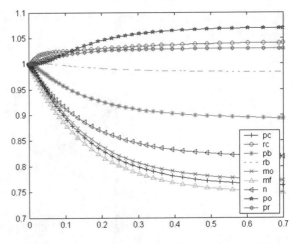

Fig. 4.27 Combustion chamber oxidant injection blocking

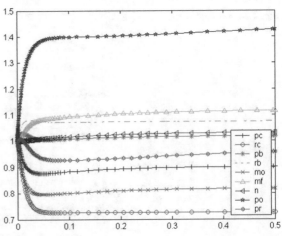

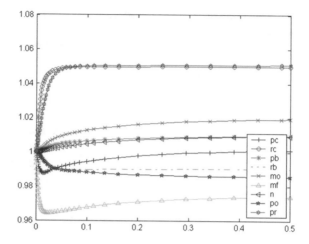

Fig. 4.28 Combustion chamber fuel injection blocking

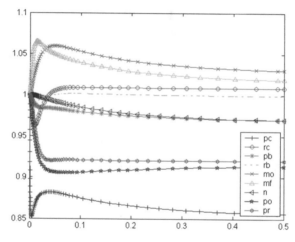

Fig. 4.29 Thrust chamber throat ablation

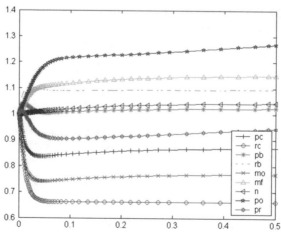

Fig. 4.30 Oxidant main restrictor jam

Fig. 4.31 Fuel main
restrictor jam

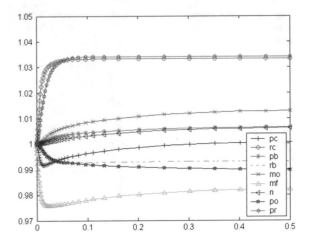

Fig. 4.32 Subsystem oxidant
cavitation tube obstruction

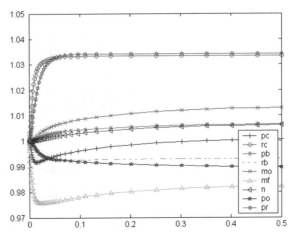

Fig. 4.33 Subsystem fuel
cavitation tube obstruction

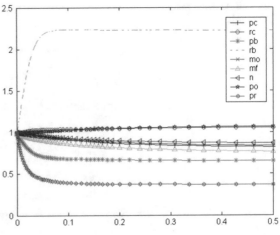

Fig. 4.34 Subsystem oxidant leakage

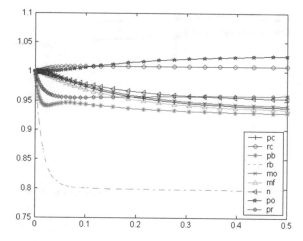

Fig. 4.35 Gas generator fuel entrance leakage

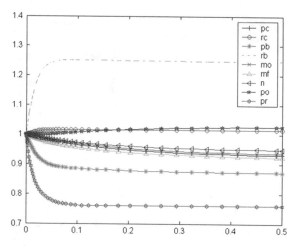

Fig. 4.36 Increase of heat transfer coefficient of the cooler

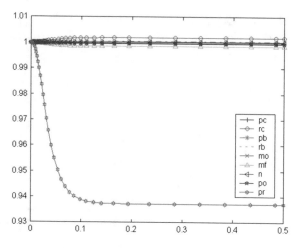

Fig. 4.37 Increase of heat transfer coefficient of the evaporator

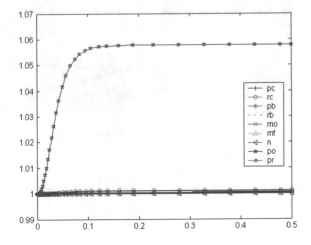

Fig. 4.38 Oxidant pressurization cavitation pipe plug

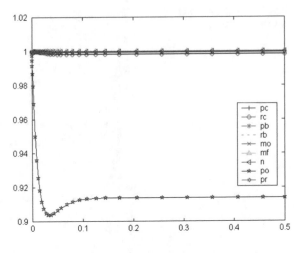

Fig. 4.39 Gas generator throat ablation

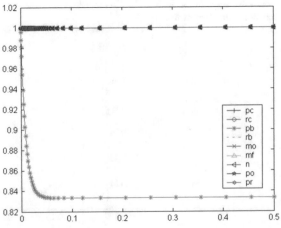

Fig. 4.40 Oxidant supercharging steam set leakage

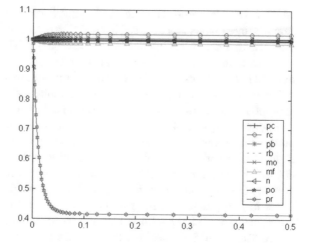

Fig. 4.41 Gas pressure pipe leakage

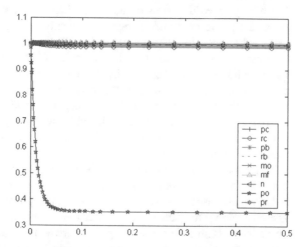

deformation even destruction by external forces. Tank leakage also makes the tank pressure decrease.

Figures 4.15, 4.16, and 4.17 show the transition characteristics of starting valve at half-open. Oxidant starting valve fault reduces the oxidant flow into the whole engine system, and lead to the fuel to increase relatively. The mixing ratio of gas generator and thrust chamber decreases, meanwhile, pump working conditions decrease, oxidant tank pressure increases, and the fuel tank pressure drops, which is contrary to the fuel starting valve fault. No matter what kind of fault, thrust chamber pressure always decreases. The parameters $p_c, \dot{m}_o, \dot{m}_f$ are sensitive to this fault, which can be used to monitor and separate this fault mode.

Figures 4.18, 4.19, and 4.20 show the transition characteristics of main valve. This fault makes the system output lower than that before fault condition, increases working condition of the turbine and component mixing ratio of deputy system, which is very dangerous and may cause some components to burn out. The interval from the three faults to a new stable condition is less than 0.4 s and the interval for dramatic changes of parameters is less than 0.04 s. The parameters $p_c, \dot{m}_o, \dot{m}_f$ are sensitive to this fault. The oxidant main valve fault has a greater influence on the output characteristics of main system and the fuel main valve fault has a greater influence on the output characteristics of subsystem. Therefore, choosing the appropriate monitoring parameter can separate the three faults modes qualitatively.

Figures 4.21, 4.22, 4.23, and 4.24 show the transitional characteristics of turbine pump fault. Generator pressures all decrease monotonously, oxidant flow in the main system and the fuel flow in combustion chamber are almost synchronous down. The reason why the effect of two pump faults are similar is that the efficiency of pump drops, the power system required increases, rotate speed decreases, and the heads of two pumps reduce. The interval for reaching a new stable condition is more than 0.5 s. Figure 4.24 is the fault transient process of oxidant pump cavitation. The figure indicates that the propellant mixing ratio of engine r_c, oxidant flow of engine, oxidant flow in subsystem, fuel flow of the engine, etc., are sensitive to oxidant pump fault and change dramatically among 0 and 0.2 s, which can be taken as the basis for separation of this fault.

Figures 4.25 and 4.26 show the transition characteristics of turbine fault. The figures indicate that the reason why the transition process of turbine fault is almost similar is that we built and solve the model by the power equation of the pump. To separate this fault, a more detailed and accurate mathematic model must be built and dynamic characteristics are analyzed by increasing the monitoring parameters of the turbine pump system. Then, this fault can be separated by analyzing the characteristic curve of pump.

Figures 4.27, 4.28, and 4.29 show the transition characteristics of thrust chamber fault. Figure 4.27 shows that oxidant flow of the engine decreases dramatically, but fuel flow increases relatively, which causes the propellant mixture ratio of combustion chamber to decrease, the power oxidant pump required to reduce, mixture ratio of subsystem to increase, the power capability of gas that enters into turbine to improve, and rotation speed of turbine pump to increase. After reaching the new stable working condition, the output of main system reduces, but the outputs of the subsystem and the turbine pump increase. Figure 4.28 shows that the engine fuel flow decreases dramatically, but the oxidant flow increases relatively, which causes the propellant mixture ratio of combustion chamber to increase, the power fuel pump required to decrease, mixture ratio of subsystem to reduce, the power capability of gas that enters into turbine to fall, and speed of turbine pump to reduce. After reaching the new stable working condition, the outputs of the main system, the subsystem, and turbine pump all reduce. Figure 4.29 shows that this fault increases engine propellant flow and the propellant mixture ratio of main system and subsystem, but causes the whole system to enter into the new working condition with low stability.

Figures 4.30 and 4.31 show the transition characteristics of main system restrictor jam fault. Figure 4.30 shows that the oxidant flow of the engine decreases dramatically, fuel flow increases relatively, which causes the propellant mixture ratio of combustion chamber to decrease, the power oxidant pump required to reduce, mixture ratio of subsystem to decrease, the power capability of gas that enters into turbine to fall, and rotation speed of turbine pump to reduce. After reaching the new stable working condition, the outputs of the main system, the subsystem, and turbine pump all reduce. Figure 4.31 shows that the engine fuel flow decreases dramatically, but the oxidant flow increases relatively, which causes the propellant mixture ratio of combustion chamber to increase, the power fuel pump required to decrease, the power capability of gas that enters into turbine to improve, and rotation speed of turbine pump to increase. After reaching the new stable working condition, the main system output reduces, but the outputs of the subsystem and the turbine pump increase. Therefore, the fault modes, as shown in Figs. 4.27, 4.28, 4.29, 4.30, and 4.31, can be separated absolutely by selecting appropriate monitoring parameters.

Figures 4.32, 4.33, 4.34, and 4.35 show the transient characteristics of subsystem fault. The situation changes fundamentally. The fuel leakage fault that occurs when the fuel enters into the gas generator from subsystem and oxidant filter blocking fault of subsystem shown in Figs. 4.32 and 4.21 respectively, can be separated according to the transient characteristics of the gas generator combustion chamber pressure $\bar{P}_b$, subsystem oxidant flow $\bar{m}_g$, pump speed $\bar{n}$, propellant mixture ratio $\bar{r}_b$ of subsystem, and the engine oxidant flow $\bar{m}_o$. Figure 4.32 indicates that the parameters $\bar{P}_b, \bar{r}_b$ and $\bar{m}_{fg}$ are sensitive to the fault. Their transitional characteristics are obviously different from those of others. Figure 4.23 shows that the parameters $\bar{m}_{og}, \bar{r}_b$ and $\bar{P}_b$ are sensitive to oxidant filter blocking of the subsystem, which can be taken as the monitoring parameters.

Figures 4.36, 4.37, 4.38, 4.39, 4.40, and 4.41 show the transient characteristics of the pressurization system fault. The heat transfer coefficient of the evaporator, the outlet temperature of the evaporator, and the tank pressure increase. Heat transfer coefficient of cooler increases, but the outlet temperature of the cooler and tank pressure reduces. The pressure of the gas generator is very sensitive to the throat ablation and sharply drops. Cavitation tube obstruction of oxidant pressurized pipeline reduces pressurizing gas flow and the tank pressure. Similarly, pressurized pipeline leakage also makes the corresponding tank pressure reduce. The above fault only reduces the pressure, but does not obviously change other parameters. The reason is that we build the dynamic equation of pump without considering the influence of pump cavitation, which meanwhile shows that the tank pressure has little influence on the whole system if only it meets the requirement of the minimum pressure of pump entrance.

Analysis of the second-level engine dynamic failure is similar to the first-level engine and not mentioned here.

4.3   Integrated Fault Analysis

(1) For steady state failure, the parameter variations caused by different faults are different, thus the fault mode can be determined according to the change of parameters.
(2) The speed of parameter variation caused by different faults is different. The faults that cause the parameter to vary fast can be easily detected in the early stages when the fault occurs, and the other fault such as fuel valve fault that causes the parameter to vary slow is hardly to detect in time unless it becomes serious.
(3) The change of the dynamic fault is similar to that of the steady fault. But the change of dynamic fault is faster and tends to another balance. Some fault parameters response obviously. But other fault parameters are little changeable.

To sum up, diagnosing the fault from the change of parameters should focus on the parameters those change most obviously and search for some auxiliary parameters (variation trend, parameter number, changeable degree, etc.), which can distinguish this fault from the others. From the perspective of guaranteeing equipment reliability and safety, diagnosing the fault should focus on the fault with the strong damage and fast development.

It is important to choose the sensor in parameter testing according to the need of diagnosis. In terms of sensitivity, the parameters change obviously require the sensor with low sensitivity, but the sensor with too low sensitivity cannot detect parameters those change little, while that with high sensitivity has influence on test results due to the high noise. Therefore, the different detection and diagnosis methods should be chosen according to different fault characteristics, but we do not know what and when the fault will happen in the actual detection and diagnosis, thus the best solution should be the combination of a variety of detection methods and diagnostic methods.

To choose appropriate detection and diagnosis methods, we should evaluate the fault. For example, it is concluded that the best combination ways of detection and diagnosis method can be obtained according to the destructive and development speed, and realize resource optimization as well as find the fault.

Usually, a sudden fault, with strong destructive, fast development speed, no obvious precursor or a detected precursor but not control can be predicted by the change of parameters at prophase. On this basis, we can shut down before fault occurs.

Gradual fault, due to there being a long process from the failure to spread, the change of the parameters can be detected and fault diagnosis can be achieved to control in time.

No matter sudden failure or gradual failure, there is a development process, but the duration is different. Therefore, it can sum up the general development of the fault, as shown in Fig. 4.42.

Fault development relies on the following assumptions:

(1) In engine working process, the working state of the components has been developing toward fault. From the reliability point of view, the life of the parts is certain and the working process such as vibration, wear and tear, etc., is toward the end of the life.
(2) The fault is mainly produced by such physical reasons as stress concentration exceeding maximum stress of the structure.

Therefore, different means of detection and diagnosis should be used for predicting and detecting the fault at the early stage. In the process of actual monitoring and diagnosis, measures should be taken as follows:

Analyze the condition data obtained from the engine and judge whether there is a fault. If there is a fault, measures should be adopted to control the fault according to the result of diagnosis. Otherwise, it should predict whether and when the fault occurs according to the current state, if possible, what the fault type is, and determine the implementation plan according to the results of the prediction. Diagnosis system is as shown in Fig. 4.43.

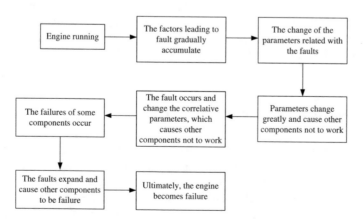

Fig. 4.42 The general development of fault

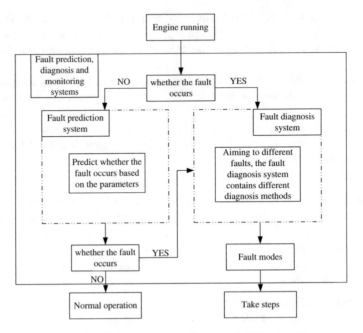

Fig. 4.43 Structure of engine fault detection, prediction, and diagnosis system

4.4   Separability and Detectability of Fault

The faults of engine can be reflected in the different measurement parameters and their expressions. Therefore, different fault detection and diagnosis methods can be adopted according to the different measurement parameters and expression forms. For different parameters and different diagnosis methods, it is difficult to give a uniform fault that can be detected and separated. The state space and the steady state mathematical model are used to describe the engine fault and the corresponding fault that can be detected and separated.

4.4.1   Separability of Fault

In state space, two different engine faults XF_1 and XF_2 correspond to two different states Y_1 and Y_2, with the corresponding measurement parameter vector of the X_1 and X_2. ε is the threshold which is determined by the statistical features of engine fault mode and the measured parameters. When $Y_1 \neq Y_2$, there is $d(Y_1, Y_2) > \varepsilon \geq 0$, fault XF_1 and XF_2 are separable in the state. When $X_1 \neq X_2$, there is $d(X_1, X_2) > \varepsilon \geq 0$, fault XF_1 and XF_2 are separable in measuring parameters.

4.4.2   The Detectability and Diagnostic Ability of Fault

In the state space, if the normal state of engine is separable, the engine system fault XF is detectable. If the XF can be diagnosed through a certain method, the fault XF is diagnosed.

Chapter 5
Fault Diagnosis of LRE Based on ANN

Due to the complex structure, unclear fault mechanisms and unknown disturbance in the operation process, etc., the LRE fault is nonstationary and uncertain and corresponding fault diagnosis is very difficult. Using the characteristics of dynamic signals to identify and classify the operation state, the diagnosis technology is gradually developing from the manual analysis identification to the intelligent automatic identification, and constructing the automatic intelligent diagnosis system is inevitable. However, the traditional mathematical simulation method involving a large number of nonlinear partial differential equations, is not conducive to real-time condition monitoring and fault diagnosis. However, the neural network making diagnosis rules implied in weight matrix has the characters of strong learning ability, nonlinear approximation capability, and so on is very suitable for the condition identification and fault diagnosis of LRE [83].

In the application of ANN on the fault diagnosis of LRE, the fault data used for the model training is very difficult to be acquired. Generally, the ground test and flight test are important ways of data acquisition. However, the fault information included in the test report cannot embody all kinds of fault modes since only several tests can be performed. Comparatively, diversified kinds of fault modes, in both test and flight process, can be simulated by the model of LRE. In this instance, the fault data obtained by the model of LRE established in the previous chapter is used in training the ANN classifier and several diagnosis methods based on the ANN developed.

5.1 Theory of ANN

Holding the capability of learning, memory, calculation, etc., ANN has permeated into all fields and corresponding application in the complex system, for example, in the fault diagnosis and condition monitoring, has been developed. As the basis of the ANN, corresponding theory devotes to deal with the objective things of the real

© Springer-Verlag Berlin Heidelberg and National Defense Industry Press 2016
W. Zhang, *Failure Characteristics Analysis and Fault Diagnosis*
for Liquid Rocket Engines, DOI 10.1007/978-3-662-49254-3_5

world in the same way of the natural neural system. Therefore, in the development of ANN, many kinds of ANN algorithm have been proposed.

5.1.1   Basics of ANN

(1) Neuron model

As Fig. 5.1 illustrated that the neuron model is a unit with multiple input and single output. According to the input, output, and the relationship between the internal state of neurons, neurons mathematical models are generally divided into discrete, continuous, difference equation model, and probability model. The continuous one is the most representative and is given by

$$y = f\left(\Sigma\left(\omega_{ij}x_i + \theta\right)\right) \tag{5.1}$$

where x_i denotes the input signal, ω_{ij} denotes the connection weights and θ denotes the threshold value, y denotes the output of the neuron and f denotes the excitation function. The step function, piecewise linear function and S shape function can be used for the excitation function. For example, the S shape function is given by

$$y = 1/(1 + \exp(-x)) \tag{5.2}$$

(2) Structure of ANN

A neural network is constructed by the connection of multiple neurons. If the network contains only one input layer and output layer, the network is called a single-layer network. Comparatively, if some middle hidden layers are involved between the input and output layers, the network is called a multilayer network. If the output of the node after the middle of network layer or in the layer is the input of the node in the layer, the network is called a feedback network. The network without feedback is called forward network and corresponding structure is shown in Fig. 5.2.

Fig. 5.1 Neuron model

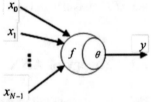

Fig. 5.2 Forward ANN

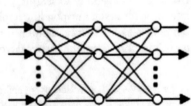

(3) Learning algorithm of ANN

In artificial ANN, the weight is the key to reflecting information storage, and is generally occupied through the network learning. At present, the learning algorithm of ANN can be classified as mentor learning and no mentor learning. In mentor learning, due to the given marked training mode samples, pattern recognizer can learn the mapping relationship from the n-dimensional feature space to explanation space and classification space. If the mechanism of producing mode has divided them into several categories in some sense, any way able to identify the location and distribution of these categories can be used to solve the problem. Accordingly, no mentor learning method tries to identify some specimen as clustering center.

At present, the multilayer forward ANN has been successfully used in many ways [83], the training of which is based on the steepest descent method of BP (back propagation) algorithm. In the application of condition monitoring and fault diagnosis, it is one of those which are the most successfully and widely used.

5.1.2 BP ANN and Improved Algorithm

(1) Basic BP algorithm

The typical three layer BP network structure is showed in Fig. 5.2. The learning process of network includes the forward calculation inside network and back propagation calculation of error. In the process of the forward calculation, the input information is handled step by step by the hidden layer from the input layer to the output layer, and the state of each layer's neurons only affects the state of next layer's neurons. If the desired output cannot be got in the output layer, the transferred error propagates backward, and modifies the state weights of each layer's neurons, so that makes the error signal minimal.

Take the incentive function of the forward network neurons as S shape function, the output of any node as the o_j, the input as net_j, then: $net_j = \sum_i W_{ji} \cdot o_i + \theta_j$, W_{ji} as the connection weights between node j and node i in the up layer, and also as the threshold of node (convenience to remember, θ_j is classified in W_{ji}).

Use mean square error function as the objective function:

$$E = \frac{1}{P}\sum_{p=0}^{P-1} E_p \quad E_p = \frac{1}{2}\sum_{l=0}^{L-1}(t_{pl} - y_{pl})^2 \tag{5.3}$$

where E denotes the average error of the system, E_p denotes the error of single sample p, p denotes the number of samples, t_{pl}, y_{pl} denotes the target output and actual output of output unit of p samples. According to the gradient steepest descent algorithm, the variation of weights is given by:

$$\Delta w_{ji} = -\alpha \frac{\partial E}{\partial w_{ji}} \tag{5.4}$$

$$o_j = f(\text{net}_j) \tag{5.5}$$

Moreover:

$$\text{net}_j = \sum w_{ji} \cdot o_i \tag{5.6}$$

Then:

$$\frac{\partial E}{\partial w_{ji}} = \frac{\partial E}{\partial \text{net}_j} \frac{\partial \text{net}_j}{\partial w_{ji}} = \frac{\partial E}{\partial \text{net}_j} \frac{\partial}{\partial w_{ji}} \sum w_{ji} \cdot o_i = \frac{\partial E}{\partial \text{net}_j} \cdot o_i$$

Define:

$$\delta_j = -\frac{\partial E}{\partial \text{net}_j} \tag{5.7}$$

Then:

$$\Delta w_{ji} = \alpha \cdot \delta_j \cdot o_i \tag{5.8}$$

1) When j denotes the output node, $o_j = y_j$

$$\delta_j = -\frac{\partial E}{\partial \text{net}_j} = -\frac{\partial E}{\partial y_j} \cdot \frac{\partial y_j}{\partial \text{net}_j} = (t_j - y_j) \cdot f'(\text{net}_j) \tag{5.9}$$

2) When j does not denote the output node,

$$\delta_j = -\frac{\partial E}{\partial \text{net}_j} = -\frac{\partial E}{\partial o_j} \cdot \frac{\partial o_j}{\partial \text{net}_j} = -\frac{\partial E}{\partial o_j} \cdot f'(\text{net}_j) \tag{5.10}$$

$$\frac{\partial E}{\partial o_j} = \sum_i \frac{\partial E}{\partial \text{net}_i} \cdot \frac{\partial \text{net}_i}{\partial o_j} = \sum_i \frac{\partial E}{\partial \text{net}_i} \cdot \frac{\partial}{\partial o_j} \sum_i w_{ji} \cdot o_i$$

$$= \sum_i \frac{\partial E}{\partial \text{net}_i} \cdot w_{ji} = \sum_i \delta_i \cdot w_{ji}$$

Namely:

$$\delta_j = -f'(\text{net}_j) \cdot \sum_i \delta_i \cdot w_{ji} \qquad (5.11)$$

When the excitation function is the S shape function, for the output layer:

$$\delta_j = f'(\text{net}_j) \cdot (t_j - o_j) = (t_j - o_j) \cdot o_j(1 - o_j) \qquad (5.12)$$

For the hidden layer:

$$\delta_j = o_j(1 - o_j) \sum_i w_{ji} \cdot \delta_j \qquad (5.13)$$

In order to improve the network convergence performance, add one impulse items to Eq. (5.8), namely:

$$\Delta w_{ji}(k+1) = \alpha \cdot \delta_j \cdot o_i + \eta \cdot \Delta w_{ji}(k) \qquad (5.14)$$

where k denotes the number of iterations, α denotes the vector, η denotes the constant inertia factor, and: $0 < \alpha < 1$; $0 < \eta < 1$.

(2) Improved algorithm of BP ANN [39, 84]

There are some problems in the practical application of BP network: First, the learning speed is slow, especially when the vector and the initialization weights are not selected inappropriately, the learning spread is more slow, and even oscillation and divergence appear at the time; Second, it is easily trapped in local minimum points, and cannot be up to the global optimal; Third, the new sample affects the learned sample, and the whole network needs to learn again. Thus, adopt the following improvement measures.

1) Self-adaptive vector

There is no theoretical basis about the suitable size of vector α so far in the BP algorithm. There is only the condition $0 < \alpha < 1$ in many documents, in which the bigger α is, the faster the learning speed is, but oscillation is prone to appear; the smaller α is, the slower the learning speed is.

According to the modification rule of the weight, in the j layer:

$$\omega_{ji}(k+1) = \omega_{ji}(k) + \alpha \cdot \delta_j \cdot o_i(k) + \eta \cdot [\omega_{ji}(k) - \omega_{ji}(k-1)] \qquad (5.15)$$

Entering the learning mode again, there should be:

$$o_i(k+1) = o_i(k) + \Delta o_i \qquad (5.16)$$

$$
\begin{aligned}
\mathrm{net}_j(k+1) &= \sum_j \omega_{ji}(k+1) \cdot o_i(k+1) \\
&= \sum_j [\omega_{ji}(k) + \alpha \cdot \delta_j \cdot o_i(k) + \eta \cdot \Delta\omega_{ji}(k)] \cdot (o_i(k) + \Delta o_i) \\
&= \mathrm{net}_j(k) + \alpha \cdot \delta_j \cdot \sum_j (o_i(k))^2 + \sum_j \omega_{ji}(k) \cdot \Delta o_i \\
&\quad + \alpha \cdot \delta_j \cdot \sum_j o_i(k) \cdot \Delta o_i + \sum_j \eta \cdot \Delta\omega_{ji}(k) \cdot (o_i(k) + \Delta o_i) \\
\Delta\mathrm{net}_j &= \sum_j \omega_{ji}(k) \cdot \Delta o_i \\
&\quad + \alpha \cdot \delta_j \cdot \sum_j (o_i(k))^2 + \alpha \cdot \delta_j \cdot \sum_j o_i(k) \cdot \Delta o_i + \eta \cdot \sum_j \Delta\omega_{ji}(k) \cdot (o_i(k) + \Delta o_i)
\end{aligned}
$$

$$(5.17)$$

where $\Delta o_i, \Delta w_{ji}(k)$ denotes the high order dimensionless of $o_i(k), w_{ji}(k)$, so that the equation can be further simplified as:

$$
\begin{aligned}
\Delta\mathrm{net}_j &= \sum_j \omega_{ji}(k) \cdot \Delta o_i \\
&\quad + \alpha \cdot \delta_j \cdot \sum_j (o_i(k))^2 + \alpha \cdot \delta_j \cdot \sum_j o_i(k) \cdot \Delta o_i + \eta \cdot \sum_j \Delta\omega_{ji}(k) \cdot o_i(k)
\end{aligned}
$$

$$(5.18)$$

Make that:

$$
u = \alpha \cdot \left(\sum_j (o_i(k))^2 + \sum_j o_i(k) \cdot \Delta o_i \right); \quad \lambda = \eta \cdot o_i(k)
$$

It is not hard to see, μ is related to the network scale and the output of the former layer. When α is fixed, the phenomenon $\mu \gg 1$ or $\mu \ll 1$ appears easily, where the former makes the algorithm weights easily oscillate due to too much change, and the latter makes convergence slow owing to too small change of the weights. To avoid the happening of this kind of situation, take μ as a constant value, then Eq. (5.18) becomes:

$$
\Delta\mathrm{net}_j = \sum_j \omega_{ji}(k) \cdot \Delta o_i + u \cdot \delta_j + \eta \cdot \sum_j \Delta\omega_{ji}(k) \cdot o_i(k) \tag{5.19}
$$

At this time:

$$
\alpha = \mu / \left(\sum_j (o_i(k))^2 + \sum_j o_i(k) \cdot \Delta o_i \right) \tag{5.20}
$$

At this time, vector becomes the function of the network scale and state. The vector α is determined by automatically calculating, which is called self-adaptive vector. μ is called control factor constant, making the learning process smoothly. In the actual calculation, Δo_i at the current moment can be substituted approximately by the moment before.

2) Absolute error equidistance approximation learning method

The main reason of the low vector of BP network and "too much learning" (training process accompanied by the objective function is further decreased, and the network may have learned some bizarre training mode, which leads to the deterioration of network generalization performance on the test set, called the phenomenon of too much learning) is that the objective function considers only the overall error effect from all the output node to all the training patterns determined by Eq. (5.2), and ignores the effect of individual error. Only when the sum of error effect is very small, the training success can be ensured. This will certainly need a lot of training cycles, and it is likely to make the network learning reduce the generalization performance of network because it is too close to some strange training mode, and network training cannot effectively curb the trend that some develop in the reversed direction of the output. Therefore, once reversed model output arises in the individual model, training cannot get rid of the state of the local minimum. In fault detection and diagnosis application, it is not necessary to demand strictly that the output state of the network is very close to the ideal output, but the requirements are to identify appropriate model. Therefore, it is also needed that the output error of the network reaches a set of range, set a minimum error as E_{min}. If the absolute error that a certain sample generates is less than E_{min}, error back propagation calculation is not conducted, and the error adopts the absolute error function, namely:

$$E = \left| t_{pi} - y_{pi} \right| \tag{5.21}$$

This is known as absolute error equidistance method. Error back propagation uses mean square error and function. Only the neuron nodes that the output error is greater than E_{min}, are involved in the correction of related rights and threshold.

5.2 Diagnostic Mechanism of ANN

In the process of monitoring and diagnosis of the equipment running condition, monitoring and diagnosis supply each other. Monitoring is aimed at finding fault as soon as possible, but diagnosis is with the purpose of finding out the cause of the problem. From this perspective, there is no essential difference between classification and monitoring/diagnosis. With the result of classification, on the one hand whether the running condition of the equipment is changed can be determined by the current running condition (monitoring), on the other hand the cause of fault can be found (diagnosis). Therefore, the importance of correct classification is obvious.

So far, in the process of monitoring and diagnosis, the proper methods of fault classification are being researched. A large amount of researches using traditional methods on pattern recognition, from K mean value method and clustering analysis method to fuzzy classification and expert system classification method, have been carried on, but these classification techniques are carried on according to the project designed before, and it is hard to do dynamic modification. These methods of the classification system are of great vulnerability. However, in the process of fault diagnosis, to meet the complexity of the field operation, the designed classification system should be easy for dynamic modification, have larger range, add new content and modify inappropriate characteristics. Therefore, using the neural network for fault classification receives more and more attention.

The essence of fault diagnosis is to realize the mapping from symptom space to fault space. Set x_{jn}, $(j = 1, 2, \ldots, k)$ corresponding to k characteristic parameters of the first n observation samples reflecting the equipment running status, y_{in}, $(i = 1, 2, \ldots, l)$ corresponding to l kind of fault modes of the first n samples. There are a total of N samples, $x_{jn} \in R_N$, $y_{in} \in R_N$, $(n = 1, 2, \ldots, N)$. The intrinsic relationship between the fault mode vector $Y = \{y_{in}, i = 1, 2, \ldots, l\}$ and characteristic parameter vector $X = \{x_{jn}, j = 1, 2, \ldots, k\}$ is showed with a function F, namely: $X = F(Y)$. When $N \to \infty$, the inverse function S of the function F exists, namely: $Y = S(X)$. So, the essence of the diagnosis problem is to determine the equivalence relation of the function $S(X)$ based on finite sample set, so that for any $\varepsilon > 0$, satisfying:

$$||S(X) - SS(X)|| = ||Y - YY|| < \varepsilon$$

where $YY = SS(X)$ denotes the model output, $Y = S(X)$ denotes the standard output, $||\cdot||$ denotes the norm defined in the sample space R, ε v denotes the mapping accuracy of the function $SS(X)$. The theorem is as following [85]:

Theorem 5.1 *For the problem of fault diagnosis, by certain mathematical transformation, $X \in [0,1]k$, $Y \in [0,1]l$ can be got. If mapping S is defined as the real continuous function from $[0,1]k$ to $[0,1]l$ and $[0,1]k$ is the closely subset of RN, mapping S can characterize the problem of fault diagnosis, namely: $Y = S(X)$.*

Theorem 5.2 *For fault diagnosis $Y = S(X)$, where $X \in [0,1]k$, $Y \in [0,1]l$, if there exists the mapping SS which makes arbitrary $X_0 \in [0,1]k$ infinite sample set $(X, Y)N$:*

$$\lim_{X \to X_0} SS(X) = S(X_0) = Y_0$$

The map SS established the mathematical model of diagnostic problems.

Inference If mapping $SS(X)$ uniformly approximates the real continuous function defined in the compact set on $[0,1]k$, the map SS established the mathematical model of diagnostic problems.

According to Theorems 5.1, 5.2 and inference, for a specific diagnosis problem, it can approximate the mapping relation of the diagnosis problem itself by training network using BP ANN.

5.3   Fuzzy Preprocessing of the Input Data for the ANN

Fault diagnosis of equipment can not only be discriminated to be of fault or free-fault, but should also be able to accurately portray "the extent of the belonging", which is the requirement of modern fault diagnosis. In equipment fault, there are many uncertain phenomenons, which are characterized by two aspects, of which one is the randomness, the other is the ambiguity. Randomness is caused by the uncertainty of fault causality, which can be studied by the statistical methods. Ambiguity refers to the rendered property in the middle transient process of the fault difference, which is a kind objective attributes of equipment malfunction. "Large vibration", "severe corrosion," and "serious ablation" are fuzzy concepts. How much of vibration quantity is big? What kind of ablation is serious? In different conditions, the evaluation standard to different equipment is different, and different people will draw different conclusions. Therefore, there are a lot of fuzzy problems in fault diagnosis. The fault fuzziness in large, complex systems equipments such as liquid missile engine is shown as follows:

(1)   The forms of the same fault are diverse, showed as different characteristics;
(2)   A variety of faults occur simultaneously and couple mutually;
(3)   The classification of the fault is fuzzy, that is to say, different faults are with similar and close characteristics.

Fault diagnosis determines the status of the equipment through the relationship between fault and symptom. Because, the fuzzy relation exists between the fault and symptom, fuzzy logic should be used to study the relationship, and fuzzy diagnosis should also be used for fault diagnosis.

By the former, the neural network can implement the nonlinear mapping from the fault symptoms set to the fault set, and realize the fuzzy diagnosis. Therefore, a conversion tool is necessary, which can both characterize fuzziness of fault diagnosis reasonably, and convert characteristic to neural network's input. This tool is the fuzzy mathematics. The most commonly used methods for fuzzy relations are the subordinate function method. The subordinate functions are defined as follows:

Fuzzy logic divides the space [0,1] into many subspaces, which, respectively, mean the different degrees that x_i belongs to the event A, known as the subordinate function $\mu A(x)$, namely

$$\mu A : X \Rightarrow [0, 1] \tag{5.22}$$

And set $A = \{x: \mu A(x) | x \in X\}$ is the fuzzy set in X. Generally speaking, characteristic curve determines the subordinate function. When x belongs to A, $\mu A(x) = 1$; When x does not belong to A, $\mu A(x) = 0$; When x cannot be considered exactly to belong to A, $\mu A(x)$ is a true value located in [0,1].

Subordinate function makes fuzziness to quantification, so that the uncertainty of things can be expressed and operated in the form of classical mathematical methods.

In practice, a set of fault information contains a variety of faults and the severity of various faults are fuzzy concepts. The reasonable expression to a fault information is to reflect the inherent nature of the information or malfunction in maximum. When a subordinate function is chosen as a prerequisite processing for the neural network input, the characteristics of the fault information should be fully reflected, and the emergence of "non" fault information should be inhibited, so that neural network can effectively classify the fault. Up half trapezoidal function can meet the above requirements, and therefore the following preprocessing function of neural network can be built.

$$\mu_A(x) = \begin{pmatrix} 0, & x \leq x_L \\ \frac{x-x_L}{x_H-x_L}, & x_L < x \leq x_H \\ 1, & x > x_H \end{pmatrix} \tag{5.23}$$

where $x_L \in R+$ denotes the tolerance values of the characteristics, that is to say, the contribution of the eigenvalues to the fault is not big when x is below, x_L. $x_H \in R+$ denotes the alarm value that equipment is corresponding to the characteristic.

5.4   Fault Diagnosis Method Based on BP ANN

Fault diagnosis method based on BP ANN includes the training and diagnosis of the network. The training and learning process of the network can be availably offline and online. For the complex systems, because there are many fault modes and characteristic parameters, the neural network structure is complex, and training time is long, online applications is not available, so the neural network training are conducted offline. The training and learning process of BP network include the forward calculation of internal network and back propagation of error, the purpose of which is to adjust the internal network connection weights to minimize network error. In fact, the learning process of the network is the acquisition process of diagnosis knowledge. At the end of the network training, the network weight means the special knowledge of the diagnostic object. Network parameters should be according to diagnosis object. The input can be corresponding to the fault feature vector, and the output can be corresponding to fault standard models. Diagnosis process of network is a forward calculation of the network. After network training is completed, the weight of the network storage the knowledge of the diagnostic object is contained in the weights of the network. Input the quarantining fault samples to network, through the forward calculation process of network, the information is transferred to the output layer, so that the output result is got, which is the diagnosis conclusion. If there is only one sample, this process will be very fast, which can be used for online monitoring and diagnosis.

The learning and training samples of the rocket motor ANN fault diagnosis are got by the transformation of simulation results of the engine fault feature analysis using up half trapezoidal function, and the target output is: when the corresponding fault sequence number is the same as the output unit serial number, the output of the output unit is "1," or it is "0". The neural network training model of the engine structure is established as follows: choose the single hidden layer 30-60-32 network, where 30 is corresponding to the characteristics number of fault sample, 32 is corresponding to the faults species number; If a fault occurs, the target output of the output unit corresponding this kind of fault is "1," or it is "0;" a large number of trials are conducted for the training process. Due to the network size is too big (the training sample is 244), it is difficult to achieve the global optimal convergence point, and local minimum point and concussion are easy to appear. Finally the better training parameters are chosen as: the vector is $\alpha = 0.4$, the inertia coefficient is $\eta = 0.4$, the total error is 0.01, the single error is 0.001. When the eventual iteration number is 10,000 times, the overall error is 0.17913. Training results are in the file, and fault samples can be classified and diagnosed in the use of the network forward calculation (Tables 5.1 and 5.2).

When the already trained network is tested using the test sample, the output of the network should be able to reflect the desired fault mode. Adding 5 % noise in a typical training sample randomly, the test sample can be got as shown in Table 5.3, and diagnostic test results are as shown in Table 5.4. The more close to the "1" the data in table are, the greater the degree of the fault samples belonging to the corresponding fault mode is. The data in table can also be thought as the membership degree the fault samples corresponding to fault mode.

As shown in Tables 5.3 and 5.4, the classification ability of the BP network is strong, which can correctly diagnose the corresponding fault in the case of noisy, but there are also false diagnosis, which is mainly for too big network size, complicated structure, and similar pattern characteristics under the condition of noise. In the application of BP network, the characteristic of fault mode samples should be analyzed and chosen, so that the network size can be reduced.

5.5 Fault Diagnosis Method Based on RBF ANN

At present in the study of function approximation, space mapping, the forward ANN is in extremely extensive application, where BP network and radial basis function (RBF) network are two kinds of the most effective network. In theory, RBF ANN and BP network can approximate any continuous nonlinear functions, of which the main difference lies in the use of different excitation function. The incentive function of hidden layer nodes of BP network is of overall quality, such as Sigmoid function, of which the function value is a nonzero value in the infinite range of input space. But the excitation function of RBF network, such as gaussian function, is local. Therefore, BP network used for function approximation belongs to a kind of global approximation network, of which the weight adjustment uses the gradient descent method. For each input and output data, each connection weight of the network must be adjusted, which results in the slow learning speed of network,

Table 5.1 Fault samples

Sample No.	Pressure in the oxidant tank	Pressure in the fuel tank	Flow of the oxidant				Flow of the fuel				Pressure of the oxidant pump outlet			
			1	2	3	4	1	2	3	4	1	2	3	4
1	0.963806	0.001194	0.000000	0.065720	0.019193	0.023695	0.000000	0.000000	0.040920	0.000000	0.000000	0.012111	0.043726	0.000000
2	0.000000	0.973364	0.000000	0.041716	0.020156	0.000000	0.009543	0.061251	0.066840	0.023606	0.040195	0.000000	0.000000	0.001047
3	1.000000	0.992792	0.038422	0.038422	0.000000	0.023201	0.068920	0.036773	0.070779	0.004000	0.043831	0.048353	0.011154	0.000000
4	1.000000	0.980850	0.014159	0.052804	0.059508		0.068381	0.043221	0.041040	0.004092	0.289047	0.203748	0.030518	0.000000
5	0.970477	0.218691	0.289859	0.244477	0.269010	0.233297	0.131262	0.120764	0.090435	0.121199	0.954144	1.000000	0.231807	0.290622
6	0.648496	0.967215	1.000000	0.961249	1.000000	1.000000	0.690109	0.608305	0.609434	0.669682	0.669682	0.466921	0.962753	1.000000
7	0.970667	0.110695	0.339846	0.120606	0.128464	0.141981	0.121116	0.044990	0.048559	0.026075	0.377085	0.964072	0.437667	0.394495
8	0.322856	0.931691	1.000000	0.011730	0.038432	0.000000	0.696892	0.062626	0.088386	0.102397	1.000000	1.000000	1.000000	1.000000
9	0.000000	1.000000	0.055732	0.000000	0.000000		0.313426	0.064550	0.045750	0.123231	0.077111		0.000000	0.000000
10	0.726686	1.000000	0.990970	0.083376	0.024257	0.030397	0.997824	0.125139	0.076806	0.065216	0.989832	0.259177	0.178703	0.226544
11	0.994265	1.000000	1.000000	0.388328	0.456199	0.375204	1.000000	0.699242	0.699043	0.636926	0.982800	0.422020	0.448871	0.409318
12	0.195704	0.189740	0.977591	0.051235	0.102379	0.086729	1.000000	0.038597	0.016958	0.023104	0.992651	0.101114	0.092930	0.060718
13	0.197333	0.265103	0.987747	0.062948	0.000000	0.057674	1.000000	0.030886	0.018289	0.000000	1.000000	0.096856	0.070877	0.072209
14	0.953751	0.675882	1.000000	0.174628	0.137584	0.142818	0.988205	0.101958	0.068498	0.050510	1.000000	0.249849	0.304664	0.244718
15	0.596728	1.000000	0.990829	0.047081	0.031937	0.030760	1.000000	0.108313	0.110918	0.085444	1.000000	0.044991	0.048617	0.064979
16	0.992783	0.962473	1.000000	0.115818	0.040339	0.044129	1.000000	0.065254	0.103491	0.048576	0.977414	0.289989	0.280309	0.246309
17	0.769068	1.000000	0.995847	0.066950	0.042397	0.100492	0.987112	0.038653	0.021891	0.114104	0.950989	0.261327	0.249955	0.269528
18	0.213293	0.240988	1.000000	0.011092	0.058501	0.047264	0.982217	0.000074	0.049649	0.012404	1.000000	0.063402	0.055206	0.033943
19	0.067815	0.095994	0.155501	0.010749	0.009480	0.019218	0.645287	0.000000	0.043146	0.016294	0.058609	0.037244	0.016634	0.034266
20	0.093866	0.256280	1.000000	0.058830	0.051005	0.026724	0.810179	0.000000	0.000000	0.007746	1.000000	0.053130	0.010560	0.012838
21	0.165927	0.995572	1.000000	0.015891	0.000000	0.013075	0.987561	0.031102	0.051931	0.086982	1.000000	0.097352	0.055422	0.106828
22	0.802662	0.952414	1.000000	0.456193	0.448871	0.447865	1.000000	0.611627	0.616768	0.681173	1.000000	0.459624	0.484993	0.426410
23	0.227220	0.955588	0.989493	0.000000	0.064031	0.065767	1.000000	0.000000	0.112441	0.019326	0.959192	0.044778	0.055326	0.031821
24	0.768921	1.000000	1.000000	0.468768	0.338015	0.459023	0.963815	0.631813	0.519303	0.637480	0.963763	0.480329	0.404801	0.466572
25	0.000000	0.970184	0.078982	0.078981	0.007547	0.023917	0.083841	0.153551	0.102395	0.098513	0.007929	0.047272	0.000000	0.000000
26	0.967374	0.017073	0.039025	0.126231	0.098311	0.040291	0.080038	0.029366	0.000000	0.054195	0.069754	0.040175	0.060257	0.044447

(continued)

Table 5.1 (continued)

Top section (Samples 27–32)

Features / Sample No.	Pressure in the oxidant tank	Pressure in the fuel tank (tank)	Flow of the oxidant 1	2	3	4	Flow of the fuel 1	2	3	4	Pressure of the oxidant pump outlet 1	2	3	4
27	0.000000	0.987953	0.052580	0.040301	0.000000	0.000000	0.016813	0.029069	0.072907	0.000000	0.000000	0.044924	0.035877	0.000000
28	0.225564	0.000000	0.000000	0.047386	0.012953	0.052453	0.029605	0.042064	0.000000	0.000000	0.046871	0.000000	0.019321	0.016530
29	0.998883	0.130359	1.000000	0.034666	0.123170	0.105097	0.951495	0.062127	0.076443	0.006637	1.000000	0.202132	0.152636	0.177030
30	0.605978	1.000000	0.973428	0.081036	0.074025	0.012996	0.970773	0.035486	0.063398	0.122204	0.997598	0.161675	0.171449	0.204849
31	0.951962	1.000000	1.000000	0.000000	0.056556	0.009806	0.678059	0.057744	0.000000	0.036053	0.950204	0.003409	0.078762	0.051479
32	0.052246	0.050221	0.146829	0.012346	0.009242	0.035345	0.128969	0.052751	0.000000	0.000000	0.128502	0.000000	0.014367	0.000000

Bottom section (Samples 1–17)

Features / Sample No.	Pressure of the fuel pump outlet 1	2	3	4	Speed of the turbine pump 1	2	3	4	Oxidant pressure before the injection 1	2	3	4	Fuel pressure at the branch of main pipe 1	2	3	4
1	0.045452	0.044987	0.000000	0.000000	0.016753	0.021067	0.000000	0.006630	0.000000	0.000000	0.027815	0.000000	0.029497	0.019333	0.000000	0.000000
2	0.071342	0.000000	0.026139	0.031269	0.000000	0.000000	0.000000	0.012932	0.046373	0.035282	0.000000	0.035385	0.075558	0.000000	0.057435	0.000000
3	0.004439	0.010740	0.028536	0.027394	0.000000	0.000000	0.047769	0.031900	0.000000	0.058897	0.044849	0.056268	0.002115	0.013570	0.031583	0.011564
4	0.064695	0.072311	0.007033	0.039831	0.000000	0.046862	0.039288	0.017430	0.032205	0.024445	0.000000	0.017687	0.003846	0.028141	0.065874	0.071034
5	0.063063	0.051906	0.002802	0.026488	0.000000	0.002840	0.015021	0.034039	0.166785	0.188251	0.173823	0.220075	0.044379	0.086895	0.057523	0.011054
6	0.713987	0.779886	0.763608	0.711673	0.250109	0.216194	0.292091	0.225481	1.000000	1.000000	0.936574	0.987080	0.802926	0.747340	0.839499	0.819619
7	0.028737	0.040839	0.034217	0.041476	0.000000	0.006478	0.000000	0.000000	0.246652	0.090137	0.062252	0.067961	0.048711	0.025149	0.000000	0.000000
8	1.000000	0.127103	0.237198	0.178904	0.799958	0.077899	0.083550	0.025042	0.955331	0.032369	0.056874	0.017592	1.000000	0.188120	0.179822	0.217466
9	0.219115	0.331837	0.326895	0.300085	0.045330	0.000000	0.000000	0.034896	0.023485	0.000000	0.000000	0.029148	0.229141	0.085030	0.038539	0.003712
10	0.986560	0.372614	0.321380	0.377742	1.000000	0.005792	0.000000	0.024719	0.958508	0.049007	0.000000	0.029853	1.000000	0.011154	0.000000	0.000000
11	1.000000	0.606063	0.580993	0.551491	0.957501	0.191856	0.196900	0.181653	0.965292	0.211667	0.182311	0.156588	0.962164	0.346567	0.417447	0.387536
12	0.967386	0.175119	0.116098	0.204311	1.000000	0.032696	0.047391	0.015009	0.982220	0.084097	0.053458	0.030816	1.000000	0.050240	0.135057	0.066728
13	1.000000	0.145068	0.065176	0.174426	0.983874	0.080247	0.020153	0.005863	1.000000	0.076033	0.000000	0.014035	0.979156	0.141220	0.068480	0.113527
14	1.000000	0.126613	0.144547	0.124052	0.979932	0.025350	0.064189	0.105602	1.000000	0.152141	0.166914	0.096540	1.000000	0.229276	0.154007	0.215703
15	1.000000	0.094888	0.061443	0.152756	0.951080	0.000000	0.000000	0.000000	1.000000	0.005239	0.011343	0.016594	0.956363	0.029374	0.000576	0.027462
16	1.000000	0.253491	0.285726	0.293250	0.994949	0.000000	0.000000	0.017422	0.953253	0.057266	0.023197	0.042927	0.970096	0.000000	0.000525	0.053499
17	0.981647	0.158344	0.198286	0.165425	0.953806	0.000000	0.000000	0.002960	1.000000	0.019624	0.061425	0.072577	0.966040	0.054199	0.016159	0.093434

(continued)

Table 5.1 (continued)

Features Sample No.	Pressure of the fuel pump outlet				Speed of the turbine pump				Oxidant pressure before the injection				Fuel pressure at the branch of main pipe			
	1	2	3	4	1	2	3	4	1	2	3	4	1	2	3	4
18	0.162949	0.000000	0.000000	0.022024	0.291495	0.000000	0.040341	0.049547	1.000000	0.053799	0.000000	0.031447	0.105735	0.000000	0.000000	0.000000
19	0.162120	0.000000	0.000000	0.036620	0.079726	0.034448	0.024800	0.000000	0.032684	0.038971	0.021321	0.046080	0.352619	0.032454	0.012685	0.050242
20	0.400471	0.063846	0.040211	0.058677	0.267067	0.000000	0.037730	0.027127	0.959805	0.002577	0.016574	0.000000	0.274694	0.039959	0.048089	0.000000
21	0.964664	0.034534	0.062116	0.061566	1.000000	0.026835	0.022054	0.021822	1.000000	0.027000	0.040905	0.007475	0.975748	0.015344	0.065220	0.015639
22	0.956055	0.517879	0.568507	0.551619	1.000000	0.176328	0.246852	0.192542	1.000000	0.221558	0.182894	0.192693	0.974509	0.427619	0.403311	0.407469
23	0.975674	0.022359	0.153562	0.120849	0.972494	0.057615	0.060364	0.000000	1.000000	0.012491	0.055976	0.006791	1.000000	0.000000	0.088276	0.067228
24	0.976193	0.574749	0.500750	0.537198	0.997830	0.233250	0.154040	0.230763	0.961719	0.199393	0.229823	0.200632	1.000000	0.423063	0.386122	0.466579
25	0.080894	0.118837	0.057907	0.044677	0.000000	0.000000	0.009358	0.013291	0.000000	0.030544	0.037014	0.000000	0.093022	0.056192	0.106257	0.047602
26	0.055674	0.000000	0.011057	0.045657	0.026020	0.038415	0.000000	0.000000	0.104803	0.091801	0.037613	0.102421	0.000000	0.051160	0.026239	0.059185
27	0.055416	0.061983	0.061860	0.011567	0.020430	0.048174	0.008492	0.033289	0.008867	0.000000	0.000000	0.049906	0.057851	0.040553	0.000000	0.003914
28	0.024007	0.000000	0.005499	0.021007	0.034166	0.000000	0.000000	0.000000	0.033882	0.000000	0.016312	0.000000	0.000000	0.034386	0.000000	0.038653
29	0.985507	0.011021	0.039223	0.000000	1.000000	0.000000	0.039727	0.003593	1.000000	0.045198	0.117301	0.116235	0.953769	0.073535	0.112764	0.038653
30	1.000000	0.131991	0.147286	0.109525	0.982352	0.000000	0.031930	0.000000	0.981885	0.007708	0.000000	0.000000	0.972617	0.010096	0.044880	0.028261
31	1.000000	0.007617	0.054397	0.059785	0.983673	0.038951	0.035600	0.000000	0.986694	0.025162	0.025774	0.010477	0.951447	0.013443	0.057699	0.014174
32	0.242642	0.000000	0.000000	0.000000	0.090577	0.026162	0.000000	0.046559	0.080627	0.038203	0.047359	0.000000	0.249713	0.028243	0.047837	0.017712

Table 5.2 Diagnosis results of the basic BP ANN

S/F	0	1	2	3	4	5	6	7	8	9	10	11	12	13	14	15
0	0.8877	0.0000	0.0042	0.0003	0.0000	0.0000	0.0006	0.0000	0.0001	0.0000	0.0001	0.0000	0.0011	0.0012	0.0006	0.0000
1	0.0000	0.3731	0.0004	0.0001	0.0001	0.0010	0.0000	0.0002	0.0000	0.0000	0.0001	0.0000	0.0000	0.0005	0.0000	0.0000
2	0.0000	0.0059	0.4338	0.6240	0.0001	0.0017	0.0006	0.0009	0.0024	0.0002	0.0023	0.0001	0.0000	0.0017	0.0000	0.0028
3	0.0000	0.0039	0.4061	0.6585	0.0000	0.0018	0.0007	0.0011	0.0046	0.0003	0.0015	0.0002	0.0000	0.0026	0.0000	0.0037
4	0.0000	0.0000	0.0000	0.0000	1.0000	0.0013	0.0000	0.0000	0.0000	0.0000	0.0009	0.0000	0.0000	0.0000	0.0000	0.0000
5	0.0000	0.0000	0.0000	0.0000	0.0000	0.9994	0.0000	0.0000	0.0000	0.0000	0.0000	0.0000	0.0000	0.0000	0.0000	0.0000
6	0.0000	0.0000	0.0000	0.0000	0.0000	0.0000	0.9976	0.0012	0.0000	0.0000	0.0000	0.0000	0.0000	0.0000	0.0000	0.0000
7	0.0000	0.0000	0.0000	0.0000	0.0000	0.0001	0.0005	0.9999	0.0000	0.0000	0.0000	0.0000	0.0000	0.0000	0.0000	0.0000
8	0.0000	0.0000	0.0000	0.0000	0.0000	0.0000	0.0000	0.0000	0.9993	0.0028	0.0000	0.0000	0.0000	0.0000	0.0001	0.0000
9	0.0000	0.0000	0.0000	0.0000	0.0000	0.0000	0.0000	0.0050	0.0000	0.9940	0.0000	0.0008	0.0000	0.0000	0.0000	0.0056
10	0.0000	0.0000	0.0000	0.0000	0.0000	0.0000	0.0000	0.0000	0.0000	0.0000	0.9985	0.0000	0.0000	0.0000	0.0000	0.0034
11	0.0000	0.0000	0.0000	0.0000	0.0000	0.0000	0.0000	0.0001	0.0000	0.0000	0.0000	0.9960	0.0000	0.0000	0.0000	0.0000
12	0.0000	0.0000	0.0000	0.0000	0.0000	0.0000	0.0000	0.0000	0.0000	0.0000	0.0000	0.0003	0.9884	0.0000	0.0000	0.0000
13	0.0000	0.0000	0.0000	0.0000	0.0000	0.0000	0.0000	0.0000	0.0000	0.0000	0.0000	0.0000	0.0000	1.0000	0.0000	0.0000
14	0.0000	0.0000	0.0003	0.0000	0.0000	0.0000	0.0004	0.0000	0.0000	0.0000	0.0000	0.0000	0.0001	0.0000	0.9656	0.0000
15	0.0000	0.0000	0.0000	0.0000	0.0000	0.0000	0.0000	0.0000	0.0000	0.0144	0.0000	0.0000	0.0001	0.0000	0.0000	0.9978
16	0.0000	0.0000	0.0000	0.0001	0.0000	0.0000	0.0131	0.0125	0.0000	0.0014	0.0000	0.0000	0.0000	0.0005	0.0000	0.0001
17	0.0000	0.0000	0.0000	0.0000	0.0000	0.0000	0.0000	0.0000	0.0000	0.0000	0.0000	0.0000	0.0000	0.0000	0.0000	0.0000
18	0.0000	0.0000	0.0000	0.0000	0.0000	0.0000	0.0000	0.0000	0.0000	0.0762	0.0000	0.0000	0.0000	0.0000	0.0000	0.0000
19	0.0000	0.0000	0.0000	0.0000	0.0000	0.0000	0.0000	0.0002	0.0000	0.0000	0.0000	0.0000	0.0000	0.0000	0.0000	0.0000
20	0.0000	0.0001	0.0000	0.0000	0.0000	0.0004	0.0000	0.0012	0.0000	0.0000	0.0000	0.0000	0.0000	0.0000	0.0000	0.0000
21	0.0000	0.0000	0.0000	0.0000	0.0000	0.0000	0.0000	0.0000	0.0000	0.0000	0.0077	0.0000	0.0000	0.0000	0.0000	0.0000
22	0.0000	0.0000	0.0000	0.0000	0.0000	0.0000	0.0000	0.0001	0.0005	0.0001	0.0000	0.0001	0.0000	0.0000	0.0017	0.0000
23	0.0000	0.0002	0.0000	0.0000	0.0000	0.0000	0.0000	0.0000	0.0000	0.0000	0.0000	0.0000	0.0000	0.0000	0.0000	0.0000
24	0.0000	0.6694	0.0010	0.0005	0.0000	0.0040	0.0000	0.0003	0.0001	0.0001	0.0003	0.0000	0.0000	0.0000	0.0000	0.0000

(continued)

Table 5.2 (continued)

S.\F.	0	1	2	3	4	5	6	7	8	9	10	11	12	13	14	15
25	0.1377	0.0000	0.0024	0.0002	0.0000	0.0000	0.0044	0.0000	0.0000	0.0000	0.0003	0.0000	0.0001	0.0012	0.0000	0.0000
26	0.0000	0.5089	0.0015	0.0027	0.0003	0.0049	0.0002	0.0042	0.0015	0.0000	0.0001	0.0000	0.0000	0.0005	0.0000	0.0000
27	0.0000	0.0002	0.0009	0.0022	0.0000	0.0015	0.0027	0.0081	0.0000	0.0006	0.0001	0.0000	0.0000	0.0000	0.0000	0.0000
28	0.0000	0.0000	0.0000	0.0000	0.0000	0.0000	0.0000	0.0000	0.0000	0.0000	0.0000	0.0000	0.0007	0.0065	0.0000	0.0000
29	0.0000	0.0000	0.0001	0.0000	0.0000	0.0000	0.0002	0.0000	0.0000	0.0000	0.0000	0.0000	0.0000	0.0021	0.0000	0.0000
30	0.0000	0.0000	0.0000	0.0000	0.0000	0.0000	0.0000	0.0000	0.0000	0.0000	0.0000	0.0000	0.0000	0.0000	0.0000	0.0000
31	0.0000	0.0000	0.0000	0.0000	0.0000	0.0033	0.0000	0.0000	0.0000	0.0000	0.0000	0.0000	0.0000	0.0000	0.0000	0.0000

S.\F.	16	17	18	19	20	21	22	23	24	25	26	27	28	29	30	31
0	0.0000	0.0000	0.0000	0.0000	0.0000	0.0018	0.0000	0.0000	0.0000	0.1421	0.0000	0.0000	0.0015	0.0000	0.0000	0.0000
1	0.0000	0.0000	0.0000	0.3230	0.0000	0.0000	0.0000	0.0000	0.0022	0.0000	0.7384	0.0003	0.0000	0.0000	0.0028	0.0014
2	0.0021	0.0000	0.0000	0.0000	0.0000	0.0000	0.0005	0.0000	0.0264	0.0050	0.0169	0.0001	0.0003	0.0000	0.0015	0.0010
3	0.0020	0.0000	0.0000	0.0000	0.0000	0.0000	0.0005	0.0000	0.0172	0.0084	0.0068	0.0001	0.0002	0.0001	0.0020	0.0016
4	0.0000	0.0000	0.0000	0.0000	0.0000	0.0000	0.0000	0.0000	0.0000	0.0000	0.0000	0.0000	0.0000	0.0000	0.0000	0.0000
5	0.0000	0.0000	0.0000	0.0000	0.0000	0.0000	0.0000	0.0000	0.0000	0.0000	0.0000	0.0000	0.0000	0.0000	0.0000	0.0000
6	0.0000	0.0000	0.0000	0.0000	0.0000	0.0000	0.0000	0.0000	0.0000	0.0000	0.0000	0.0000	0.0000	0.0000	0.0000	0.0000
7	0.0000	0.0000	0.0000	0.0000	0.0048	0.0000	0.0000	0.0000	0.0000	0.0000	0.0000	0.0000	0.0000	0.0001	0.0000	0.0000
8	0.0000	0.0000	0.0000	0.0000	0.0000	0.0135	0.0000	0.0000	0.0000	0.0000	0.0000	0.0000	0.0000	0.0001	0.0000	0.0000
9	0.0119	0.0000	0.0000	0.0000	0.0000	0.0001	0.0000	0.0000	0.0000	0.0000	0.0000	0.0000	0.0000	0.0012	0.0000	0.0000
10	0.0000	0.0000	0.0000	0.0000	0.0000	0.0019	0.0000	0.0000	0.0000	0.0244	0.0000	0.0000	0.0000	0.0000	0.0000	0.0000
11	0.0000	0.0072	0.0000	0.0002	0.0000	0.0000	0.0000	0.0000	0.0000	0.0000	0.0000	0.0000	0.0000	0.0000	0.0000	0.0000
12	0.0000	0.0000	0.0000	0.0003	0.0000	0.0000	0.0000	0.0000	0.0000	0.0000	0.0000	0.0000	0.0000	0.0000	0.0000	0.0000
13	0.0000	0.0000	0.0000	0.0000	0.0000	0.0000	0.0000	0.0000	0.0000	0.0000	0.0000	0.0000	0.0003	0.0000	0.0000	0.0000
14	0.0000	0.0000	0.0000	0.0000	0.0000	0.0000	0.0000	0.4307	0.0000	0.0000	0.0000	0.0000	0.0000	0.0000	0.0000	0.0000
15	0.0001	0.0000	0.0000	0.0000	0.0000	0.0000	0.0000	0.0000	0.0000	0.0080	0.0000	0.0000	0.0000	0.0000	0.0000	0.0000
16	0.9998	0.0000	0.0000	0.0000	0.0000	0.0003	0.0000	0.0000	0.0000	0.0000	0.0000	0.0000	0.0000	0.0003	0.0000	0.0000

(continued)

Table 5.2 (continued)

S\F.	16	17	18	19	20	21	22	23	24	25	26	27	28	29	30	31
17	0.0000	0.9963	0.0000	0.0000	0.0000	0.0000	0.0000	0.0000	0.0000	0.0000	0.0000	0.0000	0.0345	0.0000	0.0000	0.0000
18	0.0000	0.0000	0.9997	0.0000	0.0000	0.0000	0.0000	0.0000	0.0000	0.0000	0.0000	0.0000	0.0000	0.0000	0.0000	0.0000
19	0.0000	0.0001	0.0000	0.9819	0.0000	0.0000	0.0000	0.0000	0.0000	0.0000	0.0000	0.0000	0.0014	0.0000	0.0000	0.0000
20	0.0000	0.0000	0.0000	0.0000	0.9841	0.0000	0.5499	0.0000	0.0000	0.0000	0.0058	0.0000	0.0000	0.0488	0.0000	0.0001
21	0.0000	0.0000	0.0000	0.0000	0.0000	0.9752	0.0000	0.3289	0.0000	0.0000	0.0000	0.0000	0.0000	0.0000	0.0000	0.0000
22	0.0000	0.0000	0.0000	0.0000	0.0000	0.0000	0.5085	0.0000	0.0000	0.0000	0.0000	0.0000	0.0000	0.7439	0.0000	0.0002
23	0.0000	0.0000	0.0000	0.0000	0.0000	0.0001	0.0000	0.9998	0.0177	0.0000	0.0000	0.0000	0.0000	0.0000	0.0000	0.0000
24	0.0000	0.0000	0.0000	0.0000	0.0011	0.0000	0.0001	0.0000	0.9899	0.0001	0.0002	0.0000	0.0000	0.0000	0.0001	0.0001
25	0.0000	0.0000	0.0000	0.0000	0.0000	0.0016	0.0000	0.0000	0.0000	0.7629	0.0000	0.0000	0.0020	0.0000	0.0000	0.0000
26	0.0000	0.0000	0.0000	0.0003	0.0003	0.0000	0.0056	0.0000	0.4960	0.0000	0.8797	0.0000	0.0000	0.0000	0.0003	0.0024
27	0.0000	0.0003	0.0000	0.0088	0.0000	0.0000	0.0000	0.0000	0.0000	0.0000	0.4837	0.8533	0.0000	0.0000	0.0000	0.0131
28	0.0000	0.0000	0.0000	0.0000	0.0000	0.0000	0.0000	0.0000	0.0000	0.0000	0.0000	0.0000	0.9887	0.0000	0.0000	0.0000
29	0.0615	0.0000	0.0000	0.0007	0.0000	0.0000	0.0002	0.0000	0.0000	0.0000	0.0000	0.0000	0.0000	0.5877	0.0000	0.0001
30	0.0000	0.0000	0.0000	0.0000	0.0000	0.0000	0.0000	0.0000	0.0000	0.0000	0.0000	0.0000	0.0000	0.0000	0.9990	0.0000
31	0.0000	0.0000	0.0188	0.0013	0.0000	0.0000	0.0023	0.0000	0.0000	0.0000	0.0000	0.0000	0.0000	0.0000	0.0000	0.9999

NOTE: 1.S. denotes the fault samples, F. denotes the fault modes. 2. The data in Table 5.2 denotes the degree of the sample belonging to each mode. The closer the degree to unit, the more possible the sample belongs to the mode

Table 5.3 Oxidant main valve fault (Model 1)

Mode number	Parameter mode (%)					Fault mode
	$\dot{m}_o$	$\dot{m}_f$	poc	p_f^l	n	
1	−0.72	0.46	−0.51	0	0.05	Mild
2	−3.15	2.06	−2.25	0.07	0.25	Medium
3	−5.81	5.24	−5.49	0.3	0.78	Serious

Table 5.4 Oxidant main valve fault (Model 2)

Mode number	Parameter mode (%)					Fault mode
	$\dot{m}_o$	$\dot{m}_f$	Poc	p_f^l	n	
4	0.21	−0.74	0.01	−0.42	0.07	Mild
5	0.62	−2.04	0.05	−1.145	0.2	Medium
6	2.73	−5.48	0.43	−3.9	1.02	Serious

and the easiness to fall into the local minimum point. Although, with the development of the neural network technology, a lot of the improved BP network learning algorithm appears, its bottleneck problems caused by the algorithm itself is still unable to be solved well.

Radial basis function network (RBFN) is a local approximation network. The orthogonal least squares (OLS) method is used to select RBF center, and the method of solving linear equations is adopted to obtain weights. Therefore, there are advantages of fast learning speed, high precision, and a strong ability of function approximation and pattern classification. Based on this, the OLS method is used to design RBF network, and applied to the engine fault pattern recognition.

5.5.1   RBF ANN

In 1985, Powell proposed RBF method of multivariate interpolation. In 1988, Broomhead and Lowe firstly applied RBF to artificial ANN design, making up the RBF ANN [87].

RBF network is a kind of three layer forward network. The basic idea of RBF ANN is to use RBF as the "basis" of the hidden layer, thus constituting the hidden layer space, so that the input vector can be directly mapped to the hidden layer space without the connection of the weight, in other words, the weight is fixed to 1 from the input to the hidden layer unit. On the whole the mapping of network from the input layer space to the output layer space is nonlinear, but it is linear from network output to adjustable parameters. So that the weight of the network can be solved directly by the linear equation group, which greatly accelerates the learning speed and avoids the local minimum problem, and overcomes the intrinsic defects of BP ANN based on gradient descent method. RBF ANN is a kind of typical local approximation network, which has a certain advantage in approximation capacity, pattern classification ability, and learning speed contrast to BP network

(1) RBF network structure and algorithm

RBF network is a kind of three layer static forward network, and its structure is as shown in Fig. 5.3. Input layer nodes just pass the input signal to the hidden layer; Hidden layer nodes are made up of role functions (kernel functions), and the unit number is determined by the described problem; But the output layer nodes are usually simple linear function, which responds to the effect of input mode. The transformation from the input layer space to hidden layer space is nonlinear, but from hidden layer to output layer space it is linear. The transformation function of neurons in hidden layer is RBF, which is a kind of nonlinear function of local distribution and center radial symmetry.

The role function of hidden layer nodes will influence the input signal in the local, in other words, when the input signal is near the central range of the effect function, hidden layer nodes will produce larger output. From this, this kind of network is with local approximation ability, so RBF network is also called the local perception field network [89].

As the form of the basis function, there are the following kinds:

$$f(x) = \exp^{-(x/\sigma)^2} \tag{5.24}$$

$$f(x) = \frac{1}{(\sigma^2 + x^2)^\alpha}, \quad \alpha > 0 \tag{5.25}$$

$$f(x) = (\sigma^2 + x^2)^\beta, \quad \alpha < \beta < 1 \tag{5.26}$$

These functions above are radial symmetry, although there are all kinds of kernel functions, what the most commonly used is Gaussian kernel function, as follows:

$$R_j(x) = \exp[-\frac{\|x - c_j\|^2}{2\sigma_j^2}], \quad j = 1, 2, \ldots, l \tag{5.27}$$

Fig. 5.3 RBF ANN structure

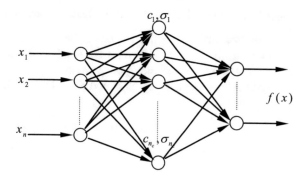

Set the input as $X = (x_1, x_2, ..., x_n)$, n denotes the dimension of input sample; The output is $Y = (y_1, y_2, ..., y_m)$, m denotes the node number of the output layer; On the equation, $R_j(x)$ denotes the output of the first j hidden layer node; x denotes a n dimension vector, c_j denotes the center of the first j basis function, which has the same dimension with x, namely $c_j = (c_{j1}, c_{j2}, ..., c_{jn})T, j = 1, 2, ..., l$; σ_j denotes the first j variables of perception (the gaussian function width vector), which determines the width of the basis function around the central point, namely the size of the field of perception, input vector only near the center c_j has a larger output; l denotes the number of sensing unit, $\|\cdot\|$ denotes the vector norm, $\|x - c_j\|$ denotes the distance between x and c_j, $R_j(x)$ has the only maximum in the c_j, as the increase of $\|x - c_j\|$, $R_j(x)$ rapidly attenuates to zero, for a given input $x \in R^n$, only a small part of centers near x is activated.

The input layer implements nonlinear mapping of $x \rightarrow R_j(x)$, and the output layer implements linear mapping of $R_j(x) \rightarrow y_k$:

$$y_k = \sum_{j=1}^{p} w_{jk}R_j(x), \quad k = 1, 2, ..., m \tag{5.28}$$

where m denotes the output layer node number; w_{jk} denotes the weights between hidden layer and output layer; j denotes the hidden layer nodes; k denotes the output layer nodes.

BP network is nonzero in the limited range of input space and of high order nonlinear in the parameter, so that only the nonlinear parameter adjustment technology can be used. However, the RBF network is nonzero in only a partial scope of the input space, that is only when the input is in a small local scope of the input space, the basis function can generate a valid nonzero response, and its parameters adjustment can adopt linear technology, thus has the characteristics of high learning.

RBF network learning process includes hidden layer unit study and output layer unit study. When studying, hidden layer unit adopts the clustering algorithm of unsupervised learning (often using the K-value algorithm). Output neurons are a linear summation, so when studying, supervised learning is used, such as the least square method used to complete the study.

Two stages of learning algorithms are illustrated below, that is the clustering algorithm of the hidden layer and the least squares training of output layer:

1) Initialize each clustering center c_j, $j = 1, 2, ..., p$, p denotes the clustering number, the value of p denotes the number of training samples;
2) All the sample models are grouped at the most recent basis functions centers c_j, and the input signals are classified to q classes. The condition is:

$$\|x_i - c_q\| = \min_{j} \|x_i - c_j\| \tag{5.29}$$

3) Get the new center of each category after the completion of classification, that is:

$$c_j = \frac{1}{m_j} \sum_j x_j \qquad (5.30)$$

where m_j denotes the sample number of the input signal which belongs to the j class.

When the center of the basis function c_j does not change, stop training. The stable value of the basis function center c_j is requested; otherwise, turn to 3 and recalculate c_j until it is stable.

4) After the training of the value of the basis function center c_j is finished, the width σ_j of the basis function can be determined.

The above four steps complete the clustering, so that the output of the hidden layer unit is determined. The clustering algorithm is an unsupervised learning method, and next is the training of the least square method of the output unit.

OLS is derived from the linear regression model. In order to be general, assume that there is only one unit in the output layer. Set the network training samples as $\{X_n, d(n)\}$, $n = 1, 2, \ldots, N$. Where N denotes the training sample number; $X_n \in R^n$ denotes the network's input data vector; $d(n) \in R^1$ denotes the expected output response of the network. According to the linear regression model, the expectation output of network is correspondingly represented as:

$$d(n) = \sum_{i=1}^{M} p_i(n)w_i + e(n), \quad n = 1, 2, \ldots, N; \quad i = 1, 2, \ldots, M \qquad (5.31)$$

where M denotes the unit number of the hidden layer, $M < N$; $p_i(n)$ denotes the return operator; w_i denotes the model parameters, which is actually the weights between the hidden layer and the output layer; $e(n)$ denotes the residual.

One subset of the input sample data vector set $\{X_n | n = 1, 2, \ldots, N\}$ is generally chose as the center t_i ($1 \le i \le M$) of RBF. A regression matrix P can be got corresponding to the input samples when a set of t_i ($1 \le i \le M$) is determined. What is to note here is that the residual error "e" of the regression model is related to the choice of the changes and the choice of the number M of the return operator. The contribution to reduce the residual error for each return operator is different, and the prominent operator should be chosen.

The task of OLS method is by learning to choose the appropriate return operator vector P_i ($1 \le i \le M$) and its number M, so that the network output can meet the quadratic performance index. The basic idea of OLS method is: through the orthogonalization of P_i ($1 \le i \le M$), the contribution of P_i to reduce the residual error is analyzed, suitable return operator is chosen, and M is determined according to the performance indicators. The steps of OLS method is introduced as follows.

1) Select a hidden layer unit number M in advance.
2) Select a set of RBF center vector t_i $(1 \leq i \leq M)$ in advance;
3) According to the center of RBF selected on the last step, use input sample vector $X_n(n = 1, 2, \ldots, N)$, and calculate $p_i(n) = R(\|X_n - t_i\|)$ $(n = 1, 2, \ldots, N;\ i = 1, 2, \ldots, M)$, so that matrix P can be got.
4) Calculate each columns of the orthogonalization regression matrix by the following equations:

$$u_i = p_i \qquad\qquad \alpha_{ik} = \frac{u_i^T p_k}{u_i^T u_i}$$

$$u_k = p_k - \sum_{i=1}^{k-1} \alpha_{ik} u_i \quad (1 \leq i \leq k;\ \ k = 2, \ldots, M)$$

5) Compute $g_i = \frac{u_i^T d}{u_i^T u_i}$ $(1 \leq i \leq M)$ and $\varepsilon_i = \frac{g_i^2 u_i^T u_i}{d^T d}$ $(1 \leq i \leq M)$;

6) Use $A = \begin{bmatrix} 1 & \alpha_{12} & \alpha_{13} & \cdots & \alpha_{1M} \\ 0 & 1 & \alpha_{23} & \cdots & \alpha_{2M} \\ 0 & 0 & 1 & \cdots & \vdots \\ \vdots & & & 1 & \alpha_{M-1,M} \\ 0 & 0 & 0 & \cdots & 1 \end{bmatrix}$ to solve the connection weight vector

W by $g = AW$, Where $g = [g_1, g_2, \ldots, g_M]T$;

7) Check whether it meets $1 - \sum_{i=1}^{M} \varepsilon_i < \rho$, where $0 < \rho < 1$ denotes the selected tolerance. If it meets the equation above, then stop counting, otherwise, turn to 2, that is to say, select the RBF center again.

From the above process, at the end of the online learning, the number of the hidden layer neurons, and the weights of the network output are determined at the same time. Thus, RBF network is with the characteristics of structure self-adaptive determination, non-relation between output and the initial weights, etc.

5.5.2 Application Examples

The fault model of liquid missile motor is set up based on certain physical phenomena and the parameter change of each assembly. When the liquid missile motor is at work, every fault will cause the transition process of different characteristics. Therefore, the motor model established should not only be able to reflect the normal work of the motor, but be able to reflect the working characteristics of the motor under fault condition. Document [87] carries out extensive researches on the fault simulation of the large pump pressure liquid rocket engine, and sets up the fault simulation data from Tables 5.5, 5.6, 5.7, 5.8, 5.9, 5.10, 5.11, 5.12, 5.13 and 5.14.

Table 5.5 Oxidant main valve fault (Model 3)

Mode number	Parameter mode (%)					Fault mode
	$\dot{m}_o$	$\dot{m}_f$	Poc	p_f^l	n	
7	−0.67	0.31	−0.52	−0.08	0.06	Mild
8	−2.97	1.33	−2.29	−0.38	0.31	Medium
9	−5.43	3.2	−5.71	−1.01	0.86	Serious

Table 5.6 Oxidant main valve fault (Model 4)

Mode number	Parameter mode (%)					Fault mode
	$\dot{m}_o$	$\dot{m}_f$	Poc	p_f^l	n	
10	−0.17	−0.22	−0.2	−0.23	0.13	Mild
11	−0.49	−0.64	−0.59	−0.66	−0.37	Medium
12	−0.79	−1.05	−0.96	−1.07	0.59	Serious

Table 5.7 Oxidant main valve fault (Model 5)

Mode number	Parameter mode (%)					Fault mode
	$\dot{m}_o$	$\dot{m}_f$	Poc	p_f^l	n	
13	−2.8	−3.91	−3.58	−3.95	−2.2	Mild
14	−5.75	−8.12	−5.27	−8.05	−4.5	Medium
15	−8.69	−12.7	−11.06	−12.3	−6.87	Serious

Table 5.8 Oxidant main valve fault (Model 6)

Mode number	Parameter mode (%)					Fault mode
	$\dot{m}_o$	$\dot{m}_f$	Poc	p_f^l	n	
16	−6.78	3.57	5.96	−0.39	0.27	Mild
17	−9.61	5.53	11.36	−0.18	0.66	Medium
18	−12.09	5.29	14.47	0.05	1.07	Serious

Table 5.9 Oxidant main valve fault (Model 7)

Mode number	Parameter mode (%)					Fault mode
	$\dot{m}_o$	$\dot{m}_f$	Poc	p_f^l	n	
19	1.57	−0.21	−6.4	−5.2	−2.41	Mild
20	2.74	0.55	−11.75	−9.04	−4.04	Medium
21	3.86	1.18	−16.3	−12.33	−5.4	Serious

Table 5.10 Oxidant main valve fault (Model 8)

Mode number	Parameter mode (%)					Fault mode
	$\dot{m}_o$	$\dot{m}_f$	Poc	p_f^l	n	
22	−1.86	−2.54	−2.34	−2.58	−1.44	Mild
23	−5.85	−8.27	−5.39	−8.2	−4.08	Medium
24	−5.98	−11.57	−10.13	−11.28	−6.29	Serious

Table 5.11 Oxidant main valve fault (Model 9)

Mode number	Parameter mode (%)					Fault mode
	$\dot{m}_o$	$\dot{m}_f$	Poc	p_f^l	n	
25	−1.1	−3.15	−1.83	−2.69	−1.56	Mild
26	−2.6	−5.21	−3.72	−4.27	−2.73	Medium
27	−5.23	−11.97	−9.59	−11.27	−6.37	Serious

Table 5.12 Testing samples

Mode number	Parameter mode (%)					Fault mode
	$\dot{m}_o$	$\dot{m}_f$	Poc	p_f^l	n	
28	−15.49	10.83	−10.61	1.065	2.07	1
29	−20.9	15.06	−14.02	1.936	3.31	1
30	1.63	−4.85	0.15	−2.63	0.57	2
31	4.87	−11.78	1.2	−5.73	1.98	2
32	−14.92	5.9	−11.45	2.36	2.09	3
33	−20.25	5.29	−15.6	−3.68	3.18	3
34	−1.07	−1.43	−1.31	−1.59	−0.81	4
35	−1.35	−1.79	−−1.64	−1.82	−1.02	4
36	−11.5	−15.8	−14.9	−16.78	−9.3	5
37	−14.47	−24.19	−19.09	−21.8	−11.87	5
38	−14.29	8.9	15.35	0.003	1.47	6
39	−16.28	10.39	20.02	0.006	1.87	6
40	4.93	1.72	−20.5	−15.18	−6.55	7
41	5.93	2.2	−23.58	−15.68	−5.54	7
42	−10.18	−15.25	−13.03	−14.57	−8.1	8
43	−12.43	−19.47	−16.08	−18.14	−10.02	8
44	−11.07	−18.44	−14.66	−15.04	−9.54	9
45	−14.57	−26.27	−19.75	−23.2	−12.66	9

Table 5.13 Relationship between network training precision ρ and diagnosis result of RBF ANN

Mode number	Network output 1	Network output 2	Separation result 1	Separation result 2	Mode number	Network output 1	Network output 2	Separation result 1	Separation result 2
1	1.2331832	1	Right	Right	24	8	8	Right	Right
2	1.155342	1	Right	Right	25	9	9	Right	Right
3	1.155342	1	Right	Right	26	9	9	Right	Right
4	1.155342	2	Right	Right	27	9	9	Right	Right
5	2	2	Right	Right	28	1.155342	1	Right	Right
6	2	2	Right	Right	29	1.155342	1	Right	Right
7	2.990107	3	Right	Right	30	2	2	Right	Right
8	3	3	Right	Right	31	2	2	Right	Right
9	3	3	Right	Right	32	3	3	Right	Right
10	4	4	Right	Right	33	3	3	Right	Right
11	4	4	Right	Right	34	4	4	Right	Right
12	4	4	Right	Right	35	4	4	Right	Right
13	5	5	Right	Right	36	5	5	Right	Right
14	5	5	Right	Right	37	5	5	Right	Right
15	5	5	Right	Right	38	6	6	Right	Right
16	6	6	Right	Right	39	6	6	Right	Right
17	6	6	Right	Right	40	7	7	Right	Right
18	6	6	Right	Right	41	7	7	Right	Right
19	7	7	Right	Right	42	8	8	Right	Right
20	7	7	Right	Right	43	8	8	Right	Right
21	7	7	Right	Right	44	9	9	Right	Right
22	8	8	Right	Right	45	9	9	Right	Right
23	8	8	Right	Right					

Table 5.14 Fault diagnosis result

Mode number	Network output							Result
	y_1	y_2	y_3	y_4	y_5	y_6	y_7	
1	1							1 (Right)
2	1							1 (Right)
3	1							1 (Right)
4	0	1						2 (Right)
5	0	1						2 (Right)
6	0	1						2 (Right)
7	0	0	1					3 (Right)
8	0	0	1					3 (Right)
9	0	0	1					3 (Right)
10	0	0	0	1				4 (Right)
11	0	0	0	0	1			5 (False)

Mode number	Network output							Result
	y_1	y_2	y_3	y_4	y_5	y_6	y_7	
12	0	0	0	1	0			4 (Right)
13	0	0	0	0	1			5 (Right)
14	0	0	0	0	1			5 (Right)
15	0	0	0	0	1			5 (Right)
16	0	0	0	0	0	1		6 (Right)
17	0	0	0	0	0	1		6 (Right)
18	0	0	0	0	0	1		6 (Right)
19	0	0	0	0	0	0	1	7 (Right)
20	0	0	0	0	0	0	1	7 (Right)
21	0	0	0	0	0	0	1	7 (Right)

Table 5.15 Fault diagnosis results (part)

Mode No.	1	2	7	8
Node No.				
c1	0.94619280537298	0.94619280537298	0.94556731013865	0.93173642775157
c2	0.47502081252106	0.47502081252106	0.47502081252106	0.47502081252106
c3	0.94556731013865	0.93173642775157	0.94619280537298	0.94619280537298
c4	0.47502081252106	0.47502081252106	0.47502081252106	0.47502081252106
c5	0.47502081252106	0.47502081252106	0.47502081252106	0.47502081252106
c6	0.47502081252106	0.47502081252106	0.47502081252106	0.47502081252106
c7	0.47502081252106	0.47502081252106	0.47502081252106	0.47502081252106
c8	0.47502081252106	0.47502081252106	0.47502081252106	0.47502081252106
c9	0.47502081252106	0.47502081252106	0.47502081252106	0.47502081252106
Winning node	1	1	3	3
Result	Right	Right	Right	Right

5.5.3 Calculation Results and Analysis

Set all the samples as training samples, the characteristic parameters are used for the input of RBF ANN and the fault mode category is the output. When RBF width σ_i is set as 0.01, and the network training accuracy is set as 0.991, the number of neurons in hidden layer is 39. When in validation, there are 44 fault modes separated correctly; When RBF width σ_i is 0.01, and network training accuracy is 0.9991, the number of neurons in hidden layer is 44, when in validation, all 45 fault modes are divided correctly, as shown in Table 5.15.

Table 5.15 shows that when the network training accuracy ρ is set high enough, RBF network can make completely correct diagnosis to the existing fault modes.

In the simulation, it is also found that if the fault data from Tables 5.5 to 5.13 are used as training samples, and the fault data in Table 5.14 are used as test samples, the extrapolating ability of RBF ANN can be tested, because the fault in Table 5.14 is beyond the scope of training samples. But in the results only 29 samples get right separations, and no matter how to adjust the network parameters, good results cannot be got. Thus, the extrapolation ability of RBF ANN is poor.

In conclusion, when using RBF ANN diagnoses the motor fault, its effectiveness mainly depends on whether there is enough, certain typical fault data. RBF ANN is the same as all other forward network, very difficult to make correct diagnosis to new fault.

5.6 Fault Diagnosis Method Based on Improved ART2 ANN

BP, RBF, and other mentor learning networks are widely applied in fault diagnosis field because there are advantages of high accuracy and they can realize fault quantitative. However, mentor learning network generally cannot make correct diagnosis for new fault. A fully learned mentor network, when it encounters a new model, will make the existing connection weight disrupted, and leads to the learned schema information disappear. To make the learned schema information not disappear, it is needed to train the network for the original learning mode along with the new pattern. Therefore, it is not suitable for intelligent diagnosis of the online learning and fault. However, no mentor learning network can find the main characteristic of input mode and make adaptive clustering of data set without a priori knowledge classification, through the study of its inherent laws of the change of environment by system itself, so as to realize pattern recognition. Therefore, no mentor learning network is suitable for motor fault intelligent diagnosis.

5.6.1 Selection of no Mentor Learning ANN Model

Competitive learning is a kind of unsupervised learning (no mentor study). Competitive learning refers to that each neurons at the same level that compete, and the victory neuron modifies the weights connected to itself. In unsupervised learning, only some learning samples are provided to the network, but ideal output is not provided. Network makes self-organization according to input samples, and classifies them into corresponding model classes. Because it is not needed to provide the ideal output samples, supervision pattern classification method is promoted. This mechanism can be used for pattern classification.

Basic competitive learning network, such as Hamming network, WTA (Winner Take-All) network, is composed of two levels, which refer to the input level and competition level. In competitive level, neurons compete with each other, which eventually leads to that only one or a few neurons are active (activated, that is to say, there is the output value), so as to adapt to the current input samples. Neurons that win the competition represent the pattern classes that the current input samples belong to. Competition can also be realized by the inhibition of neurons.

The basic competitive learning network structure is shown in Fig. 5.4.

Figure 5.5 shows the single neuron of the competitive learning network model. The weight summation is calculated as follows:

$$S_j = \sum x_i \omega_{ji} \qquad (5.32)$$

where ω_{ji} denotes the connection weights from input neurons i to competition layer neurons j, x_i denotes the first i element of the input sample vector. In the WTA mechanism, neuron j with maximum weight summation in competition level wins, the output is:

$$a_j = \begin{cases} 1 & s_j > s_i, \forall i, i \neq j \\ 0 & 其它 \end{cases} \qquad (5.33)$$

After competition, modify the network connection weights according to certain rules.

Fig. 5.4 Basic two layers structure of competitive learning

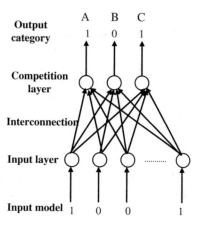

Fig. 5.5 Process neurons by competition

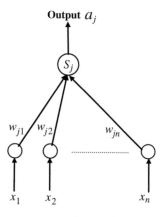

At present, the competition learning networks widely used in the pattern classification are Kohonen self-organizing feature map model (SOFM) and Stephen self-adaptive resonance theory model (ART). Self-adaptive resonance theory is a kind of network system achieving the self-stability and the self-organization to arbitrarily complex environment input mode that Stephen Grossberg who is in the Boston university and his colleagues put forward in the study of human cognitive process, of which memory patterns are similar to biological memory form, and memory capacity can be increased with the increase of learning mode. It can not only make real-time online learning, but also learn in a dynamic environment. Although the self-organizing feature map network can play a role similar to the cluster, it still cannot be directly used for classification and recognition. In order to make it used to identification, some supervised learning is needed, for example, if it is used for the clustering center, all the data that the center is on behalf of can be put to the category that the center belongs to. Sometimes in order to improve the accuracy of recognition, the training samples can be used to finely adjust weights.

Therefore, Kohonen proposed the learning vector quantization (LVQ). SOFM and LVQ cannot always get reasonable clustering result because there is a large relationship between final convergence results and initial value. For this purpose, the document [90] proposed an improved algorithm-generalized vector quantization (GLVQ). GLVQ is not sensitive to initial value, and reasonable results can be got in different conditions. However, classification category number is needed to be known in advance for this algorithm. However, because self-stability and self-organization mechanism are introduced, ART ANN has the ability to quickly identify the patterns, and can quickly adapt to the mode not studied; Do real-time learning, and adapt to the non-balance environment; Have the unitary ability, and according to some characteristics in the proportion of all the features, sometimes use it as key characteristics, sometimes treat it as a noise; When the system responds falsely to the environment, it can quickly identify new objects by improving vigilance of the system. Therefore, ART ANN is more fit for motor fault intelligent diagnosis.

5.6.2   Basic Structure and Theory of ART ANN

The development of ART has experienced three stages. ART1 is the most basic, which can only handle binary signal; ART2 can handle any analog input; ART3 is the most complex, which adopts multilayer structure, and includes human neurons bioelectricity-chemical reaction mechanism in the operation model of the neurons. Combined with the motor fault diagnosis practice, adopt ART2 ANN which can handle arbitrary input. Figure 5.6 is the structure of the first j processing unit of ART-2 network.

ART2 network system can be divided into attention subsystem and adjustment subsystem. Among them, what the attention subsystem does is the vector competition selection from bottom to up and the comparison of vector similarity. The adjustment subsystem is to check whether the similarity can achieve a satisfactory standard, and make the corresponding action: success or reset. The attention subsystem is composed of F1 field and F2 filed. F1 and F2 both belong to short-term-memory (STM). The connection between F1 and F2 is interconnection, which acts as the role of long-term memory, so called the Long-Term-memory (LTM) layer.

F1 filed is called feature presentation field, which consists of the bottom, middle, and top, as shown in Fig. 5.6. The bottom and middle make a close positive feedback loop. In F1 field there are two kinds of neurons in different functions. One kind is a hollow circle, completing the contrast of two different inputs (incentives of excitement and inhibition); The other kind is a solid circle, completing the modulus calculation. As a nonlinear operation function, the purpose of $f(\cdot)$ is to suppress noise and extrude useful signal. a, b, c, d, and e are the network parameters of the neural circuits. The specific algorithm in F1 filed is as follows:

Fig. 5.6 ART2 ANN
structure

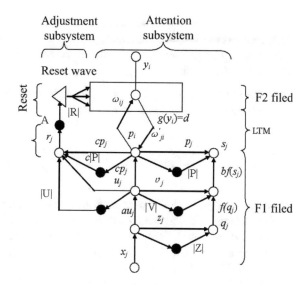

(1) When the input mode $x = (x_1, x_2, \ldots, x_n)T$ is handled by underlying neurons z_j
     and q_j, the output is as follows:

$$z_j = x_j + au_j \quad q_j = z_j/(e + |Z|) \tag{5.34}$$

where $Z = (z_1, z_2, \ldots, z_n)$, $|\cdot|$ denotes its norm. At first F2 field has not yet been
incented, so there is no effect from F2 filed to F1 filed, namely $u_j = 0$. e is a very
small constant, usually taken as $e = 0$. Equation (5.34) completed the normalized
task of the input sequence, which is very important for processing simulation data
input, but it will also create some problems [91].

(2) The output of the middle neurons v_j and u_j is:

$$v_j = f(q_j) + bf(s_j) \quad u_j = v_j/(e + |V|) \tag{5.35}$$

where b the same as a in Eq. (5.28), is the network parameter, reflecting the internal
feedback size of the F1 layer, and affecting the speed of the middle model $U = (u_1,
u_2, \ldots, u_n)T$ closing to input mode. When these two parameters are obtained too
small or too large, input mode x and the network factor are mixed together slowly in
F1 layer and cannot convergence fast, destabilizing the instability of the classification results. When F2 is not activated, $s_j = 0$.
   Take transformation function $f(\cdot)$ as

$$f(x) = \begin{cases} x & x > \theta \\ 0 & 其它 \end{cases} \tag{5.36}$$

where θ is called threshold, which is in order to remove the basement noise of each component of input vector, so all should be taken different values according to the practice. The greater θ is, the smaller the noise effect is, but useful signal will also be removed more.

(3)   The middle and top in F1 field make a close positive feedback loop, of which the output of p_j and s_j is as follows:

$$p_j = u_j + \sum_{i=1}^{M} \omega'_{ji} g(y_i) \quad g(y_i) = \begin{cases} d & i = I \\ 0 & i \neq I \end{cases} \quad s_j = p_j/(e + |P_j|) \tag{5.37}$$

where ω_{ji} denotes the LTM coefficient from top to down, M denotes the maximum category number, d denotes feedback constant from F2 to F1 field (the choice of parameter d sees later), I denotes the selected category, namely the winning neuron. At first, as a result that F2 field has not been activated, there is no feedback signal from F2 to F1 filed, namely $p_j = u_j$.

What can be seen from the above equations, F1 field plays the role of the pretreatment to input data. Stable output $P = (p_1, p_2, ..., p_n)T$ can be got after input mode x is handled by F1 field, and be sent into the F2 field by connection weight.

F2 field is called category representation field, of which the role is to improve the contrast of the filter wave input mode from F1 to F2 and send out replaced wave. Among them, the contrast enhancement is implemented through competition. The signal is input from F1 to F2 field, where the input of the first i neuron sent to the F2 field is:

$$t_i = \sum_{j=1}^{n} \omega_{ij} p_j \quad (i = 1, 2, ..., M) \tag{5.38}$$

where ω_{ji} denotes the LTM coefficient from bottom to up.

According to the competitive learning mechanism, if the first i node in F2 field is activated (won), then other nodes are in the state of inhibition, then

$$I = \arg(\max\{t_i, \quad i = 1 \sim M\}) \tag{5.39}$$

The output vector of F2 field is $Y = (y_1, y_2, ..., y_n)$, and at this time $y_i = \begin{cases} 1 & i = I \\ 0 & i \neq I \end{cases}$.

In the orientation subsystem of Fig. 5.6, define similarity r_i and similar measure $|R|$ between STM signal $U = (u_1, u_2, ..., u_n)T$ after being processed in the F1 and activated LTM signal $P = (p_1, p_2, ..., p_n)T$ in F2 filed:

$$r_i = \frac{u_i + cP_i}{e + |U| + c|P|}$$

$$|R| = \left| \sum_{i=1}^{M} r_i^2 \right| \tag{5.40}$$

If $|R| + e \leq \rho$ (ρ denotes vigilance parameters, of which the choice directly determines the classification accuracy. The greater ρ is, the more classification category number is, even fills all possible classes, and the higher the accuracy is; On the contrary, the smaller ρ, the rougher the classification is, and classification results are even instable), it represents the similarity between the input sample and the first I class in F2 field is not enough, and sends replaced wave [when at the moment of program design, cycle is not over, recover to Eq. (5.34)], finds a model class again [by Eqs. (5.34)–(5.37) calculate similarity measure again)] or defines a new class (that is, in the F2 add a neurons again); Otherwise the match is successful, and the network comes in the process of unsupervised learning, namely modifying weight according to the following equations.

$$\omega_{ij}(k+1) = \omega_{ij}(k) + g(y_i)(p_i(k) - \omega_{ij}(k))$$

$$\omega_{ji}'(k+1) = \omega_{ji}'(k) + g(y_i)(p_i(k) - \omega_{ji}'(k)) \tag{5.41}$$

Set initial weights again and make iterative calculation according to the above equations until $\omega_{ji}(k+1) = \omega_{ji}(k)$, $\omega_{ij}(k+1) = \omega_{ij}(k)$. Among them, $g(y_i) = \begin{cases} d & i = J \\ 0 & i \neq J \end{cases}$. What can be seen from the above equations, the modification of the winning node weight is not only related to the current input mode, but also related to the previous model. Do not modify the weight of not winning node. That is to say, if the current input mode belongs to some class of the network, modify the center value of the class model, otherwise do not make any modification.

In addition, the document [84] also points out that in order to avoid the false reset of coefficient before and after adjustment, it is needed to choose parameters d, making $cd/(1 - d) < 1$, and $cd/(1 - d)$ up to 1.

5.6.3   Improved ART2 Algorithm

(1) The adjustment of the initial weights of network LTM layer

Document [92] points out that in order to avoid the possibility of transposition the winner makes in the process of learning, bottom–up weight vector $WI = (\omega_{I1}, \omega_{I2}, \ldots, \omega_{In})$ should satisfy

$$0 < |WI| \leq 1/(1-d) \tag{5.42}$$

This new started output may be able to occupy, and it ensures that in the process of learning, *WI* has been always rising. When using ART2 algorithm for pattern recognition, a certain amount of calculations is needed to meet the above condition for *WI*. To reduce the calculations, and improve the real time of the pattern recognition, take the initial weights ω_{ij} from bottom to up as the same value, then by Eq. (5.42)

$$\omega_{ij} = \frac{1}{2(1-d)\sqrt{M}} \tag{5.43}$$

(2) The problems and solutions of the ART2 pretreatment signal distortion and the same phase inseparable

By nonlinear transformation function $f(x) = \begin{cases} x & x > \theta \\ 0 & 其它 \end{cases}$, the term less than the threshold θ contained in the input mode vector x is set to 0 after process treatment, which makes the input signal all lose some information and distort. Namely, at the same time of signal feature enhancement, noise suppression, characteristic information is reduced wrongly, which causes the signal characteristics distort in the process of pretreatment. This phenomenon is called "pre-signal distortion problem".

Individual neuron model in Fig. 5.6 is as shown in Fig. 5.7.

In figure, the hollow arrows indicate the excited incentive to neurons, and the solid arrows indicate inhibited incentive to neurons. The output *V* satisfies the following equation

$$\varepsilon \frac{dv}{dt} = -A + (1 - Bv)J^+ - (C + Dv)J^- \tag{5.44}$$
$$0 < \varepsilon \ll A \ll 1, \quad B \ll 1, \quad C \ll D \approx 1$$

A, *B*, *C*, and *D* are the parameter in neurophysiology. Excited incentive J^+ denotes a component of the input vector, the inhibition incentive J^- denotes the mould of the input vector. Approximate solution for this style (5.43) is

$$v = J^+/(A + DJ^-) \approx J^+/J^- \tag{5.45}$$

What can be seen is that the basic function for neurons is the normalization of the input vector. As a result, the input mode became a normalized vector after the treatment of F1 field, which is used for pattern classification, of which extend

Fig. 5.7 Individual neuron model

information is lost, causing the inaccurate of classification results, so it is not appropriate clearly to use this structure directly for the design of classifier. For example, on the plane coordinates, point $A(1,1)$ represents some clustering center, point $B(10,10)$ is another clustering center. After this classifier, A and B will reach 100 % of the similarity because of the completely same phase information, and will be returned to the same class. In fact, the two samples are far away. Therefore, wrong classification will be caused if the amplitude information is not considered. This is the same inseparable phase problem.

For the distortion problem of the preprocessing signal, document [93] proposes that nonlinear transformation function should be changed to:

$$f(x) = \begin{cases} x & x > \theta, \ x < -\theta \\ 0 & 其它 \end{cases} \tag{5.46}$$

But the change will cause new problems in the F1 layer. In the competitive learning mechanism of the F2 layer, negative items of the middle layer model U, P cannot win in the process of the competition learning. Equation (5.38) is amended as:

$$t_{max} = \max\{|t_i|, i = 1 \sim M\} \quad I = \arg(t_{max}) \tag{5.47}$$

Adopt new pretreatment function $f(x)$ and new competition learning machine Eq. (5.47), and the problem of signal distortion can be solved, so that the network can not only receive normal negative input data, but also will not affect their normal classification.

The same phase inseparable problem can be solved by the following coordinate translation transformation processing.

$$x'_{ji} = x_{ji} - \frac{1}{m} \sum_{j=1}^{n} x_{ji} \tag{5.48}$$

After the data conversion, the sample group in the same phase in the first quadrant can be separated around the origin, making them have different phase.

In addition, in order to achieve the orderly output of ART2 ANN, the known class of sample order can be input. For unknown class of samples, they can be put after the known classes of samples.

5.6.4 Implementation of Improved ART2 Algorithm

According to the improvement and adjustment of ART2 ANN algorithm, the following ART2 improved algorithm procedures can be concluded. Assuming that the total number of input patterns is M, and the neuron for which output neurons value is 1 which represents the category the current input mode belongs to.

(1) The initialization
   According to the network parameter selection method in Sect. 5.6.2, select network parameters a, b, c, d, e, vigilance parameter ρ, and thresholds θ.

(2) The formation of initial ART2 network
   Set n as the number of input neurons, namely the characteristic number of the input mode, m denotes the number of the output neurons, set $m = 1$. Set the initial weight as: $\omega_{mj} = 1/(2(1 - d)\sqrt{n}), \omega'_{jm} = 1$ $(j = 1, 2, ..., n)$ and set the cycle count $k = 1$.

(3) The input sample vector x_k
   Pretreat the sample according to Eq. (5.48) in order to solve the same phase inseparable problem. Make cycle calculation for the Eqs. (5.34)–(5.38) and (5.47) in order to get stable output vector P, U of F1 field. Set the flag $flag = 0$ and the inner cycle count $l = 1$.

(4) Competition choice
   The input P of F1 field is sent to F2 field by the weight. Calculate the input of F2 field according to Eq. (5.47). The output neurons of F2 field make competition in the equation of WTA according to Eq. (5.47). Assume that the winning neuron is I, and the input activated value that the neurons receive is maximum. According to the equation $y_i = \begin{cases} 1 & i = I \\ 0 & i \neq I \end{cases}$, calculate the network output vector Y. According to Eq. (5.31), recalculate the output vector P of F1 field.

(5) Vigilance test
   Calculate similarity measure $|R|$ according to Eq. (5.40).
   If $|R| + e \geq \rho$, set the logo $flag = 1$, indicating that the matching pattern has been identified, and modify the weight according to Eq. (5.41). Set $k = k + 1$, and if $k \leq M$, turn to (3), or turn to (6).
   If $|R| + e < \rho$, shield current winning nodes (make $tI = 0$), making it no longer participate in the competition, and set the logo $flag = 0$, indicating no matched mode is found. Set the cycle count $l = l + 1$, if $l \leq m$, return to (4), and find the pattern matching the input sample again; Otherwise it indicates that all existing patterns could not match the input sample, so a new category should be opened up, namely set $m = m + 1$ (add a new output neurons in F2 filed), and make the neuron as the winning node. After the weight is initialized, modify the weight according to Eq. (5.41). Set the logo $flag = 1$. Set $k = k + 1$, and if $k \leq m$ (indicating there are also samples needed for classification), turn to (3); Or turn to (6).

(6) The end

5.6.5 Fault Diagnosis Examples

In order to verify the application effect of the improved ART2 algorithm in fault diagnosis, the fault data in Tables 5.5 and 5.11 are as the input mode of ART2 ANN. Take $a = 8$; $b = 8$; $c = 0.1$; $d = 0.9$; $e = 0$; Vigilance parameters $\rho = 0.999$; Nonlinear function threshold $\theta = 0.1$, the calculation results are shown in Table 5.14.

From the above results, there is only one model diagnostic error in 21 modes, which shows that ART2 algorithm can be used basically in fault diagnosis. But when the data in Tables 5.12 and 5.13 are also as the diagnosis mode, some problems arise.

(1) Because ART2 algorithm adopts no overseers learning mechanism, when used for fault classification, although the classification results are correct, what kind of fault this class belongs to cannot be known. While in a sequential input method, however, if all samples of some a class are classified as a former class, the back sample will be wrongly judged. Therefore, it is necessary to add supervision learning mechanism for ART2 ANN.

(2) The same as all clustering algorithms, ART2 algorithm cannot separate two fault samples which are in high distance similarity and angle similarity. This is mainly due to that there are only four characteristic parameters of the fault mode, and there is inseparable property between some samples themselves. No matter in the distance similarity of the pattern space or in the angle similarity, the similarity degree between the two samples is bigger than the corresponding samples of the same kind. So, it is not enough that the clustering method is used to separate such faults. But, the function mapping method can be used to solve the problem.

In fact, if the classification accuracy is adjusted higher, ART can be very good to be used for the diagnosis of separable fault pattern. The establishment of the separable fault mode depends on the choice of the motor characteristic parameters, and document [94] makes a lot of researches in this aspect. Due to the limitation of the practical situation, borrow the fault simulation data obtained from document [94] to verify the application of the ART algorithm in the diagnosis of separable fault pattern.

Table 5.16 denotes the training sample. Take $a = 8$; $b = 8$; $c = 0.1$; $d = 0.9$; $e = 0$; Vigilance parameters $\rho = 1$; Nonlinear function threshold $\theta = 0.1$, the result is as Table 5.17.

In brief, ART2 ANN has strong nonlinear approximation ability, which not only can be used for system modeling, fault diagnosis, and can also be used for parameter prediction, system identification, parameter optimization, and motor health monitoring technology. There are very good clustering features when it is used for pattern clustering, not only making diagnosis for existing faults, but also recognizing new faults. But, it is not the best method to use the ART2 algorithm separately for fault diagnosis. It is needed to be used with other expert system methods to make its application space wider in fault diagnosis field.

Table 5.16 Training samples

Fault mode	Input samples							
1	−1	−1	−1	0	1	1	1	−1
2	−1	−1	−1	1	−1	1	1	1
3	1	−1	1	−1	1	−1	−1	0
4	1	1	1	−1	−1	1	−1	1
5	−1	−1	−1	−1	−1	1	1	0
6	−1	−1	−1	1	−1	−1	1	1
7	−1	−1	−1	−1	−1	1	1	−1
8	0	−1	−1	1	−1	−1	1	1

Table 5.17 Diagnosis result

Mode number	Network output								Result
	y_1	y_2	y_3	y_4	y_5	y_6	y_7	y_8	
1	1								1 (Right)
2	0	1							2 (Right)
3	0	0	1						3 (Right)
4	0	0	0	1					4 (Right)
5	0	0	0	0	1				5 (Right)
6	0	0	0	0	0	1			6 (Right)
7	0	0	0	0	0	0	1		7 (Right)
8	0	0	0	0	0	0	0	1	8 (Right)

5.7 FTART ANN-based Fault Diagnosis Method

ART algorithm is an unsupervised learning algorithm, and its classification accuracy needs artificial determination. Once identified precision is too low, learning again is needed, and incremental learning is hard to conduct. So, the application of ART algorithm in the fault diagnosis field is by certain restriction, needing the combination of other supervision learning algorithm or expert system to conduct clustering analysis. Document [94] proposes FTART algorithm (Field Theory-based ART). The algorithm combines the advantages of Field Theory, ARTC, and ARTMAP algorithm and adopts the response function similar to the Field Theory, the basic algorithm of ART and the organization structure of ARTMAP, making it have incremental learning ability; Use super ellipsoid to divide sample space, and classification accuracy is high; Expand decision domain dynamically, and make self-adaptive adjustment network topology structure; Have the ability of online learning, strong summing up ability, and high efficiency.

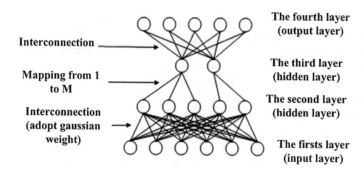

The fourth layer
(output layer)

The third layer
(hidden layer)

The second layer
(hidden layer)

The firsts layer
(input layer)

Interconnection

Mapping from 1
to M

Interconnection
(adopt gaussian
weight)

Fig. 5.8 FRART network structure

5.7.1 FTART Structure and Basic Theory

FTART network is composed of four layer neurons, of which the structure is as shown in Fig. 5.8. In figure, the input layer interconnects the second layer through the Gaussian connection weight. Gaussian function is as follows:

$$f(x) = e^{-\left(\frac{x-\mu}{\sigma}\right)^2} \tag{5.49}$$

Hidden layer neurons adopt Sigmoid response function. The second layer neurons are connected to the third layer neurons in groups, in which a group of neurons linked to the same neuron in the third layer belongs to the same class. The two layer neurons classify the input and output, respectively, and make the connection between them in order to realize the supervision learning. Thus a competitive three layers neural network classifier is formed, which can achieve arbitrary division to the sample set, having strong applicability. The connection weight between the third layer neurons and the fourth represents the characteristics component of the classification. What the whole network outputs is the center vector of a certain model.

Field theory gets its name from the CPM (coulomb potential model) model that Bachmann puts forward, which is a kind of relaxation model, of which the study result is to make an energy function minimization. In the domain theory, the response function is the formula similar to coulomb's law:

$$E_u = -\sum_{i=1}^{m} \|u - x_i\|^{-l} \frac{(u - x_i)}{\|u - x_i\|} \tag{5.50}$$

where u denotes "the space position vector of the test charge", E_u denotes "field vector" at the point u, x_i denotes the first i sample vector, m denotes the total number of samples, $\|\cdot\|$ denotes the Euclidean distance between two space vector, l denotes a constant ($l = 2$ represents ordinary electrostatic field). Domain theory algorithm views test pattern as a positive charge in hyperspace, namely the test charge, and then determines the movement of the charge on the negative electric field caused by the training pattern according to Eq. (5.50). When the test charge becomes still in a minimum point of energy in the end, the test model gets the same classification as the point. Domain theory model is the only neural network model for which one time of learning is only needed. The learning speed is so fast that it can conduct the real-time monitoring learning, and it has large memory capacity, so that there will not be a strange phenomenon, which is a very good kind of strange association pattern classifier. But, the response function that the Field Theory clustering algorithm adopts is similar to the form of coulomb's law, when u is approximate to x_i, E_u tends to infinity, which easily leads to overflow, and is difficult to guarantee implementation precision in the computer. But FTART adopted Gaussian function which overcomes the problem.

FTART ANN adopts the basic algorithm of ART, but realizes the supervised learning between the second and the third layer of the network by establishing a link, and solves the problem that ART algorithm cannot output as required.

5.7.2 Improvement and Its Mathematical Description of FTART ANN

Set the input mode that the first layer neuron of the network receives as $X_s = (x_1^s, x_2^s, \ldots, x_n^s)$ $(s = 1, 2, \ldots, M)$, where M denotes the total number of the input mode.

The input value that the second layer neurons j receives from the first layer neurons i is

$$B_{ji} = \exp\left(-\left(\frac{x_i^s - \omega_{ji}}{v_{ji}}\right)^2\right) \tag{5.51}$$

where ω_{ji} denotes the response center value of the Gaussian connection right between the input neurons i and the second layer input neurons j, v_{ji} denotes the response characteristics width of the connection weight. The response center value and response characteristics width of each Gaussian connection weight uniquely identifies a domain in geometric sense. That is to say the location of the attraction domain is determined by the response center, and its size is determined by the response characteristics width. Obviously, by Eq. (5.51), the following equations are established

$$x_i^s = \omega_{ji} \Rightarrow B_{ji} = 1$$
$$x_i^s \neq \omega_{ji} \Rightarrow B_{ji} \to 0 \tag{5.52}$$

The output of the second layer neurons j can be determined by the following equation:

$$b_j = f\left(\sum_{i=1}^{n} B_{ji} - \theta_j\right) \tag{5.53}$$

where θ_j denotes the neuron threshold, $f(\cdot)$ denotes the Sigmoid function, and the form is as follows:

$$f(u) = 1/(1 + e - gu) \tag{5.54}$$

where u denotes the independent variable, g denotes the gain of S function.

In all the second layer neurons, neurons belonging to the same group conduct competition in the equation that the winner is king (WTA), and the winning neuron (that is, the neurons of the maximum output) makes output according to the original value, and neuron not winning does not make output). The same set of neurons refers to all the neurons linked to the same third layer neurons. Equation (5.52) shows that the closer the components x_i^s of input mode X_s parts from attraction domain center, the greater the value of B_{ji}. If the attraction domain is viewed as a model class (a collection of all similar patterns), attraction domain center is equivalent to the clustering center in the clustering problem. Therefore, what can be seen from Eq. (5.51) is that the size of b_j reflects the degree input model which belongs to the model class. The greater the value is, the greater the membership degree is. But the second layer neurons are only the initial classification to the pattern space, of which the output value represents the membership degree the current input mode belonging to the subtypes. Each set of neurons in the second layer neurons presents a subtypes (called subtypes neurons). To determine the classification that the input mode belongs to, WTA competition algorithm is needed to determine the output value of the third layer neurons.

The output value of the third layer neurons can be calculated as the following equation:

$$c_k = f\left(\sum_{j=1}^{m} W(k,j)b_j - \beta_k\right) \tag{5.55}$$

where j denotes the victory neurons in the second layer, $W(k, j) = 1$ indicates that the connection exists between the neurons and the third layer neurons k. $W(k, j) = 0$ indicates that there is no connection between them. $f(\cdot)$ denotes the Sigmoid function similar to Eq. (5.54), and β_k denotes the threshold of neurons k. Competition between all the third layer neurons is conducted, and the winning neurons (the neurons of the largest output value) are the category that the input mode belongs to.

In the process of network learning, when the input mode is located in the attraction domain determined by ω_{ji} and v_{ji}, ω_{ji} and v_{ji} do not make any change. Otherwise, attraction domain will be adjusted based on the location of the input mode and the original attraction domain to make input mode as the typical attractor of the attraction domain. Adjustment equation is as follows:

$$\omega'_{ji} = \begin{cases} (\omega_{ji} + 0.3\upsilon_{ji} + x_i^s)/2, & x_i^s \in (-\infty, \omega_{ji} - 0.3\upsilon_{ji}) \\ \omega_{ji}, & x_i^s \in (\omega_{ji} - 0.3\upsilon_{ji}, \omega_{ji} + 0.3\upsilon_{ji}) \\ (\omega_{ji} - 0.3\upsilon_{ji} + x_i^s)/2, & x_i^s \in (\omega_{ji} + 0.3\upsilon_{ji}, +\infty) \end{cases} \qquad (5.56)$$

$$\upsilon'_{ji} = \begin{cases} (3\upsilon_{ji} + 10\omega_{ji} - 10x_i^s)/6, & x_i^s \in (-\infty, \omega_{ji} - 0.3\upsilon_{ji}) \\ \upsilon_{ji}, & x_i^s \in (\omega_{ji} - 0.3\upsilon_{ji}, \omega_{ji} + 0.3\upsilon_{ji}) \\ (3\upsilon_{ji} - 10\omega_{ji} + 10x_i^s)/6, & x_i^s \in (\omega_{ji} + 0.3\upsilon_{ji}, +\infty) \end{cases} \qquad (5.57)$$

When an existing attraction domain diverges from its typical attractor, the attraction domain will move to the typical attractor as the following equation, where δ is the adjustment step length of gaussian connection response center value.

$$\omega'_{ji} = \omega_{ji} + \delta(x_i^s - \omega_{ji}), \quad 0 < \delta < 1 \qquad (5.58)$$

5.7.3 Design of FTART ANN

(1) The learning algorithm of FTART ANN

If the output in the fourth layer of FTART network is used in clustering analysis, it is actually the clustering center, namely the center value ω_{Ji} $(i = 1, 2, ..., n)$ of the response function in Eq. (5.52), where J denotes the winning neuron number in the third layer, n denotes the characteristic vector dimension numbers of input pattern; But if used in pattern recognition, it is the corresponding input mode. Considering the FTART algorithm which is used for fault diagnosis, rather than the clustering analysis, thus remove the fourth layer neurons, and use only competitive neurons in the front three layers. Thus, the adopted competition rules and the activation function in the second and third layer neurons are slightly different from that the original FTART adopts.

Set the input mode as $X_s = (x_1^s, x_2^s, ..., x_n^s)$ $(s = 1, 2, ..., M)$, then the number of input layer neurons is n. The category number that input mode X_s is corresponding to is $y(s)$. The FTART network design steps are as follows:

1) Load fault samples;
2) Initialize the network parameters υ_{ji}, θ_j, β_k, gain g and output error limits Err;
3) Form the initial network;

Input layer neurons are signified as i, and the total number is n. The second neurons are signified as j, and at this time there is only one neuron, namely the total number is $m = 1$; The third neurons are signified as k, and there is also only one neuron, namely total number is $l = 1$. Connect all the first neurons and the first neurons in the second layer with gaussian rights, namely $\omega_{mi} = x_i^s$

($m = 1; s = 1; i = 1, 2, ..., n$), set the input mode as the network initial value of the response function center in the second layer neuron. Connect the first neurons in the second layer and the third layer again. If use $W(k, j) = 1$ to indicate the second layer neurons j is connected with the third layer neuron k, and use $W(k, j) = 0$ to mean there is no connection between the two neurons, then $W(l, m) = 1$ ($l = 1; m = 1$). The response characteristic width of the second layer neurons is $v_{mi} = 1(m = 1; i = 1, 2, ..., n)$.

4) Input sample $X(s)(s = 1, 2, ..., M)$ and repeat the following steps:

① According to Eqs. (5.52) and (5.53) to calculate the output b_j of the second layer neurons, where $j = 1, 2, ..., m$.

② In all the second layer neurons, the same set of neurons conduct competition in the type of WTA, and the output value of neurons that wins in the competition is sent into the third layer neurons linked.

③ Calculate the output value of the third layer neurons according to Eq. (5.56).

④ All the third layer neurons conduct competition in the type of WTA, and L denotes the neurons winning the competition. That is to say, L represents the category number that the input model belongs to calculated by FTART network.

a. If there is no classification error, the neurons of the class that input mode belongs to would win in the competition (i.e., the practical output L is equal to category number $y(s)$ that the current input model belongs to), and the output error (the output error = $1 - c_L$) is less than the error limit. For ρ, find out the second layer neurons (set as neuron J) connected to the neuron L which has the largest output. According to Eqs. (5.57) and (5.58), modify the gaussian connection weight ω_{Ji} and response characteristics width v_{Ji} connected to the neurons;

b. If there is no classification error, but the output error is beyond the error limit, one neuron should be added in the second layer (namely set $m = m + 1$), and be made connected with the third layer neurons L (namely, set $W(L, m) = 1$) and connected with the first layer neurons (namely, set $\omega_{mi} = x_i^s$). Set characteristic width value v_{mi} as default values. Other network parameter values are also set as the default value.

c. If there is a classification error (namely, $L \neq y(s)$), but the model class that the current input mode belongs to has been in existence in the network (namely, $1 \geq y(s)$), calculate the error between the practical output $cy(s)$ of the third layer neurons and the desired output 1. If the error is greater than the limit, add a subtype neuron in the second layer, (set $m = m + 1$) and make it connected with the third layer neurons $y(s)$ (namely, $W(y(s), m) = 1$) and connected with the first layer neurons (set $\omega_{mi} = x_i^s$). Set characteristic width v_{mi} as default values. Other network parameter values are also set as the default value. If it is not beyond the error limit, do nothing.

d. If the classification error arises, and the model class that the current input mode belongs to does not exist in the network (namely $1 < y(s)$), then add a neuron between the second and third layer (namely, $m = m + 1, l = l + 1$). Connect all the newly added neurons in first layer and the second layer with gaussian weight; Namely, set $\omega_{mi} = x_i^s$ ($m = 1; s = 1; i = 1, 2, \ldots, n$). Then connect the second layer neuron m and the third layer neuron l; Namely, set $W(l, m) = 1$. Other network parameters are set as default values.

(2) FTART network authentication

After the network training is completed, it begins to discriminate the input vector. This method is quite simple and consistent with the training method except the learning modification process after the network discrimination.

Once a mode is input, discriminate whether it belongs to a new class or not:

1) If yes, as the method described in point 4 of Step 4 above, a neuron should be added in the second and third layer. Set up the initial value of the network parameters
2) If the model does not belong to a new class, calculate all the outputs b_j ($j = 1, 2, \ldots, m$) of the second layer neurons according to Eqs. (5.52) and (5.54); In all the second layer neurons, the same set of neurons conducts competition in the type of WTA, and then all the outputs of those won in the competition are sent into the third layer neurons linked to it; Calculated the outputs of the third layer neurons according to Eq. (5.56); All the neurons in the third layer conduct competitions in the type of WTA, and then the winning neuron represents the category number of the current input mode.

5.7.4   Diagnosis Examples and Analysis for FRART ANN

In order to verify the application of FTART algorithm in fault diagnosis, FTART ANN input mode still adopts the fault data in Tables 5.5 and 5.13. All the response characteristic widths are 0.3, namely $v_{ji} = 0.3$; $\theta_j = 0.1, \beta_k = 0.1$, the gain is $g = 2.4$, and output error limits $Err = 0.1$.

The calculation results are shown in Table 5.17. Due to the limitation of space, only partial results are listed there.

Calculation results show that all 27 fault modes are correctly diagnosed.

In conclusion, the self-adaptive resonance ANN model FTART based on the domain theory combines the advantages of the domain theory and self-adaptive resonance theory. The learning speed is fast, the induction ability is strong, and the efficiency is high. It can also generate hidden layer neurons self-adaptively according to the training samples. It overcomes the problem that the traditional forward ANN is difficult to set up the hidden layer neurons. FTART is a kind of preferable ANN classifier because it is better than the BP algorithm in learning precision and speed. It means that the fault diagnosis based on FTART ANN method is effective.

Chapter 6
Fault Diagnosis Method Based on Wavelet Transform

6.1 Theory of Wavelet Transform

6.1.1 Introduction

In 1981, Jean Morlet, a geology physical scientist of France, proposed the concept of "wavelet analysis" in the data processing of geology the first time based on the comparison of Fourier transform and Gabor transform. Overcoming the limitation of Fourier transform, wavelet transform can decompose the signal in the time-frequency domain together and had rapidly attracted a great many attentions in the application area. In the following, the characters of wavelet transform are introduced by the usage of abrupt signal, frequency modulation signal, and amplitude modulation signal [104].

1. Abrupt signal

For a sine signal $f(t) = \sin(2\pi \times 2 \times t)$, sampled by a frequency of 256 Hz with 512 sampling point, set an abrupt point in every period. Transform the signal with continuous wavelet transform by using the Haar wavelet and the result is shown in Fig. 6.1. It can be seen that the wavelet coefficients of small scales (1–7) around the time of abrupt points are bigger than the others located in the similar scales. Such phenomenon indicates that the abrupt signal can be detected and located on the image of wavelet coefficient image. Additionally, it is worth noting that the bright lines located at big scales (20–37) are matched with the peak of the sine wave, while the dark lines are corresponding to the zero points.

© Springer-Verlag Berlin Heidelberg and National Defense Industry Press 2016
W. Zhang, *Failure Characteristics Analysis and Fault Diagnosis
for Liquid Rocket Engines*, DOI 10.1007/978-3-662-49254-3_6

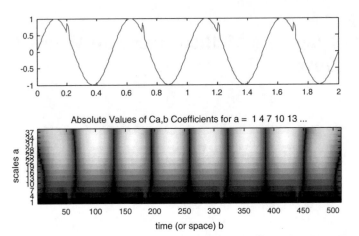

Fig. 6.1 Abrupt signal and the wavelet coefficient image

2. Frequency modulation signal

For a frequency modulation signal

$$f(t) = \begin{cases} \sin(2\pi \times 6 \times t), & t \in [0, 1] \\ \sin(2\pi \times 12 \times t), & t \in [1, 2] \end{cases} \tag{6.1}$$

whose sampling frequency $f_s = 256$ Hz and sampling points $N = 512$. The wavelet coefficient image is shown in Fig. 6.2. It can be seen that the big coefficients before the 256th point are located around the scale of 11, while the others are located around the scale of 5.

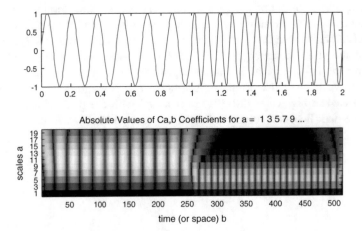

Fig. 6.2 Frequency modulation signal and corresponding wavelet coefficient image

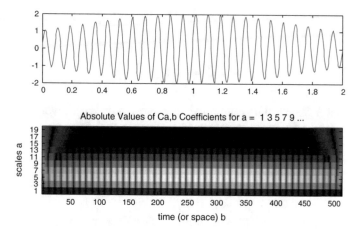

Fig. 6.3 Amplitude modulation signal and corresponding wavelet coefficient image

3. Amplitude modulation signal

Figure 6.3 illustrates an amplitude modulation signal and the corresponding wavelet coefficient image. It can be seen that the big coefficients of the signal are concentrated around the scales of 3–7 and the values indicated by the color are decreased from the center to both end points. Similarly, effect of the end points can be clearly noticed around both edges of the image.

In summary, Fourier transform cannot locate the abrupt and frequency modulation signals in the time domain and indicated the amplitude variation of amplitude modulation signal. However, the wavelet transform, by decomposing the signal in the time-frequency domain together, can successfully illustrate the character which Fourier transform cannot and is more powerful in the signal processing.

6.1.2 Basic Theories of Wavelet Analysis [104]

Oscillating with finite length and mean value of zero, wavelet can be scaled and moved to construct a series of base functions. Wavelet analysis is equal to projecting a function on these base functions.

1. Continuous wavelet analysis

Generally, for a function $f(t)$ with finite energy, namely $f(t) \in L^2(R)$, its continuous wavelet transform is defined by the inner product with a series of determined functions

$$W_f(a,b) = \langle f, \psi_{a,b} \rangle = |a|^{-1/2} \int_R f(t) \overline{\psi \left(\frac{t-b}{a} \right)} \mathrm{d}t \tag{6.2}$$

where the series of functions are obtained by the scaling and moving of the basic wavelet

$$\psi_{a,b}(t) = \frac{1}{\sqrt{a}} \psi\left(\frac{t-b}{a}\right)$$
(6.3)

where a and b denote the scaling and shift parameter, respectively, and $a \in R_{+}$, $b \in R$. Correspondingly, the reconstruction function is defined by

$$f(t) = \frac{1}{C_{\psi}} \int_{-\infty}^{\infty} \int_{-\infty}^{\infty} \frac{1}{a^2} W_f(a,b) \psi\left(\frac{t-b}{a}\right) dadb$$
(6.4)

Suppose that $\psi(t) \in L^2(R)$ and corresponding Fourier transform is denoted by $\hat{\psi}(\omega)$. If $\hat{\psi}(\omega)$ satisfies the allowed condition, namely,

$$C_{\psi} = \int_{R} \frac{\left|\hat{\psi}(\omega)\right|^2}{|\omega|} d\omega < \infty$$
(6.5)

Then, $\psi(t)$ can be seen as a basic wavelet or mother wavelet.

Additionally, science $\psi_{a,b}(t)$ can be treated as observing windows at different scales and different moments, the basic wavelet $\psi(t)$ must satisfy the constraint condition of general window functions, namely,

$$\int_{-\infty}^{\infty} |\psi(t)| dt < \infty$$
(6.6)

Based on the above analysis, it can be seen that $\hat{\psi}(\omega)$ must be a continuous function. It means that $\hat{\psi}(0) = \int_{-\infty}^{\infty} \psi(t) dt = 0$ to satisfy the reconstruction condition.

2. Dyadic wavelet transform

For the sake of calculation, wavelet transform can be performed on discrete sampling with binary mode. To reconstruct $f(t)$ on the discrete scales of

$$\{s_j = 2^{-j} | j \in Z\}$$
(6.7)

the basic wavelet $\psi(t)$ of the wavelet transform $Wf_\psi(s,b)$ can be restricted by a stable condition, namely,

$$A \le \sum_{j\in z} \left|\hat{\psi}(2^{-j}\omega)\right|^2 \le B \tag{6.8}$$

where A and B denote two constants and $0 < A \le B < \infty$. Wavelets, satisfying Eq. (6.8), can be called binary wavelet; and corresponding wavelet transform, named Binary wavelet transform, is determined by

$$W_{-j}f_\psi(b) = Wf_\psi(2^{-j},b) = f * \psi_{2^{-j}}(b) = \frac{1}{2^{-j}} \int_{-\infty}^{+\infty} f(t)\psi\left(\frac{b-t}{2^{-j}}\right)dt \quad j \in Z \tag{6.9}$$

According to the convolution theorem, the following equation exists

$$\hat{W}_{-j}f_\psi(\omega) = \hat{f}(\omega)\hat{\psi}(2^{-j}\omega) \tag{6.10}$$

Moreover, based on *Parseval equality*, Eq. (6.8) is equal to

$$A\|f\|_2^2 \le \sum_{j\in z} \left\|W_{-j}f_\psi\right\|_2^2 \le B\|f\|_2^2 \tag{6.11}$$

Additionally, Eq. (6.10) can be derived by Eq. (6.8) and the binary wavelet is one of the basic wavelet. Correspondingly, the reconstructed equation is determined by

$$f(t) = \sum_{j\in z} W_{-j}f_\psi * \psi^*_{2^{-j}}(t) = \sum_{j\in z} \int_{-\infty}^{\infty} W_{-j}f_\psi(b)\left[\frac{1}{2^{-j}}\psi^*\left(\frac{t-b}{2^{-j}}\right)\right]db \tag{6.12}$$

where $\psi * (t)$ is a binary wavelet and named as the dual wavelet of $\psi(t)$. Corresponding Fourier transform is denoted by

$$\hat{\psi}^*(\omega) = \frac{\overline{\hat{\psi}}(\omega)}{\sum_{j\in z}\left|\hat{\psi}(2^{-j}\omega)\right|^2} \tag{6.13}$$

and $\psi * (t)$ satisfies the stable condition of

$$\frac{1}{B} \le \sum_{j\in z} \left|\hat{\psi}^*(2^{-j}\omega)\right|^2 \le \frac{1}{A} \tag{6.14}$$

Similarly, the decompose and reconstruction equations can be calculated by the inner product form of

$$\tilde{f}_\psi(2^{-j}, b) = \int_{-\infty}^{\infty} f(t)\left(2^{j/2}\bar{\psi}\left(\frac{t-b}{2^{-j}}\right)\right) dt \quad j \in z \tag{6.15}$$

$$f(t) = \sum_{j \in z} 2^{j/2} \int_{-\infty}^{\infty} \tilde{f}_\psi(2^{-j}, b)\psi^*\left(\frac{t-b}{2^{-j}}\right) db \tag{6.16}$$

In applications, the scale function φ and wavelet function ψ are transformed to the filter forms in order to improve the calculation efficiency. For example, φ is transformed to a low pass filter of $H_0(\omega)$, while ψ is a high pass filter of $G_0(\omega)$.

Suppose that $D = S_{2^0}^d f(n)(n \in Z)$ denoting the sampling series of the original signal $f(x)$, $W_{2^j}^d f(n)(n \in Z)$ denoting the wavelet coefficients, and $S_{2^j}^d f(n)(n \in Z)$ denoting the estimations of D. In this instance, the binary wavelet transform of D is constructed by $\{S_{2^j}^d f, W_{2^j}^d f\}_{1 \le j \le J}$. Figure 6.4 illustrates the filter structure of the discrete binary wavelet transform. It can be seen that the idea of discrete binary wavelet transform is decomposing $S_{2^j}^d f$ to be $S_{2^{j+1}}^d f$ and $W_{2^{j+1}}^d f$ at every scale of j.

Suppose that H_j and G_j denote the discrete filters constructed by H_0 and G_0 by adding 2^{j-1} samplings of zeros, they can be obtained by the following:

$$j = 0$$
$$\text{While} \quad j < J$$
$$W_{2^{j+1}}^d f = S_{2^j}^d f * G_j$$
$$S_{2^{j+1}}^d f = S_{2^j}^d f * H_j$$
$$j = j + 1$$
$$\text{End of While}$$

If the amount of nonzero samplings of original signal $\{S_{2^0}^d f(n)\}_{n \in z}$ is N, then both $S_{2^j}^d f$ and $W_{2^j}^d f$ have N numbers of nonzero samplings and the super boundary of the scale is log2N. In this instance, the computing amount of the algorithm is

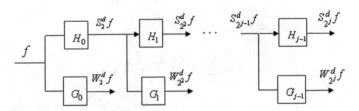

Fig. 6.4 Filter structure of the discrete binary wavelet transform

$O(N\log 2N) = C \cdot N\log 2N$, where C is determined by the number of the nonzero samplings in H_0 and G_0.

Corresponding reconstruct algorithm is listed as follows

$$j = J$$
$$\text{While} \quad j > 0$$
$$S_{2^{j-1}}^{d} f = S_{2^{j}}^{d} f * \bar{H}_{j-1} + W_{2^{j+1}}^{d} f * \bar{G}_{j-1}$$
$$j = j - 1$$
$$\text{End} \quad \text{of} \quad \text{While}$$

3. Multi-resolution analysis and Mallat algorithm

Founded on the function space, Multi-Resolution Analysis (MRA) or named as Multi-Scale Analysis (Multi-Resolution Analysis) is actually rooted in the image processing area by Mallat. In those days, the image was commonly and severally decomposed at different scales in order to obtain the useful information. In this instance, MRA provides a simple way for the construction of the orthogonal wavelet basis and theoretical evidence for the fast algorithm of the orthogonal wavelet transform.

In the program of MRA, the continuous function should be projected on a series of little orthogonal spaces, which are constructed by the shift and stretch the scale function $\phi(t)$, presenting as low pass filter, and the wavelet function $\psi(t)$, presenting as band pass filter. The relationship between the scale space and wavelet space is illustrated as Fig. 6.5.

It can be seen that the wavelet space is determined by the difference of the adjacent scale space, namely

$$W_j = V_j - 1 - V_j \tag{6.17}$$

and $L^2(R) = \underset{j \in z}{\oplus} W_j$.

Fig. 6.5 Scale space and wavelet space

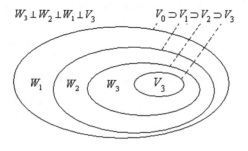

Definition A MRA in the space of $L^2(R)$ denotes a series of close sub-spaces $\{V_m\}_{m \in z}$ satisfying the following conditions.

1. Consistent monotonicity: $V_m \subset V_{m+1}, \forall m \in Z$;
2. Progressive completeness: $\lim_{m \to +\infty} V_m = \bigcap_{m \in z} V_m = \{0\}$, $\lim_{m \to -\infty} V_m = \bigcup_{m \in z} V_m = L^2(R)$;
3. If $f(x) \in V_m$, then $f(x + k2^{-j}) \in V_m, k, j \in Z$;
4. Flexibility: If $f(x) \in V_m$, then $f(2x) \in V_{m+1}$;
5. Existence of the Riesz basis: There exists $\varphi \in V_0$ to make $\{\varphi(x - n)\}m \in z$ is a Riesz basis of V_0.

Based on the definition, the following equation exists.

$$V_0 = V_1 \oplus W_1 = V_2 \oplus W_2 \oplus W_1 = V_3 \oplus W_3 \oplus W_2 \oplus W_1 = \cdots \qquad (6.18)$$

In this instance, an arbitrary signal $f(t) \in V_0$ can be decomposed to be the detail part W_1 and the estimation part V_1. Moreover, the estimation part V_1 can further be decomposed and both detailed part and estimation part can be obtained at more little scale. As the iteration goes on, the signal can be checked at different scales with different resolutions.

As one of the most interested wavelets, the orthogonal wavelet can be constructed by the MRA. For a MRA approximation in the space of $L^2(R)$, there exists a scale function $\phi(t) \in L^2(R)$ to make

$$\phi_{j,k}(t) = 2^{-j/2} \phi\left(\frac{t}{2^j} - k\right), \quad k \in Z \qquad (6.19)$$

To be the standard orthogonal basis in the subspace of V_j. Suppose that W_{j+1} denotes the orthogonal supplementary subspace of V_{j+1} in V_j, namely,

$$V_j = V_{j+1} \oplus W_{j+1} \qquad (6.20)$$

Then,

$$L^2(R) = \underset{j \in Z}{\oplus} W_j \qquad (6.21)$$

And there exists a wavelet function $\psi(t) \in W_0$, which can make

$$\psi_{j,k}(t) = 2^{-j/2} \psi\left(\frac{t}{2^j} - k\right), \quad k \in Z \qquad (6.22)$$

Construct the standard orthogonal basis in W_j then,

$$f(t) = \sum_{j,k \in Z} \langle f, \psi_{j,k} \rangle \psi_{j,k} \qquad (6.23)$$

It is known that every scale function is determined by the discrete filter named the conjugate image filer, which can be calculated by

$$h[k] = \left\langle 2^{-1/2}\phi(t/2), \phi(t-k) \right\rangle \tag{6.24}$$

Theorem *Suppose that ϕ is a scale function, and h denotes corresponding conjugate image filter. For the function of ψ, corresponding Fourier transform is*

$$\hat{\psi}(\omega) = \frac{1}{\sqrt{2}}\hat{g}\left(\frac{\omega}{2}\right)\hat{\phi}\left(\frac{\omega}{2}\right) \tag{6.25}$$

where $\hat{g}(\omega) = e^{-i\omega}\hat{h}^*(\omega + \pi)$.
    Let

$$\psi_{j,k}(t) = 2^{-j/2}\psi\left(\frac{t}{2^j} - k\right), \quad k \in Z \tag{6.26}$$

Then, for the arbitrary scale groups of 2^j, $\{\psi_j, k\}$ $k \in Z$ is one of the standard orthogonal bases in W_j.

From the theorem, it can be concluded that $g[k] = (-1)1 - kh[1 - k]$ and a set of standard orthogonal basis can be constructed by a group of arbitrary conjugate image filters.

Because $\phi(t) \in V_0 \subset V_{-1}$, $\psi(t) \in W_0 \subset V_{-1}$, therefore, the following bi-scale functions exist

$$\phi(x) = \sqrt{2}\sum_k h_k\phi(2t-k), \quad k \in Z \tag{6.27}$$

$$\psi(x) = \sqrt{2}\sum_k g_k\phi(2t-k), \quad k \in Z \tag{6.28}$$

where both series of $\{h(k)\}$ and $\{g(k)\} \in l^2$. In this instance, the fast calculating algorithm of the orthogonal wavelet decomposition, namely Mallat algorithm, can be derived by the bi-scales functions. Let the jth layer approximate and detailed coefficients of the signal f denoted by $a_{j(n)} = \langle f, \phi_{j,n} \rangle$ and $d_{j(n)} = \langle f, \psi_{j,n} \rangle$, respectively. Then the wavelet decomposition can be represented by

$$a_{j+1}[p] = \sum_{n=-\infty}^{+\infty} h[n-2p]a_j[n] = a_j * \bar{h}[2p] \tag{6.29}$$

$$d_{j+1}[p] = \sum_{n=-\infty}^{+\infty} g[n-2p]a_j[n] = a_j * \bar{g}[2p] \tag{6.30}$$

and corresponding reconstruction algorithm is determined by

$$a_j[p] = \sum_{n=-\infty}^{+\infty} h[p - 2n]a_{j+1}[n] + \sum_{n=-\infty}^{+\infty} g[p - 2n]d_{j+1}[n]$$

$$= \breve{a}_{j+1} * h[p] + \breve{d}_{j+1} * g[p]$$

(6.31)

With the computing complexity of $O(n)$, the Mallat algorithm is called fast wavelet decomposition algorithm. In Eqs. (6.30) and (6.31), h_k denotes the low pass filter and g_k denotes the high pass filter. The decomposition algorithm of the orthogonal wavelet transform is illustrated in Fig. 6.6, the reversal transform can be performed by similar way and is illustrated in Fig. 6.7. The frequency segmentation sketch map of the orthogonal wavelet transform is illustrated in Fig. 6.8, where S denotes the original signal, A denotes the approximate sub-signal and D denotes the detailed sub-signal.

4. Wavelet packet decompose theory

From Fig. 6.8, it can be seen that the scale of the MRA is changed in the manner of binary style and the spectral resolution of the detailed sub-signal and the

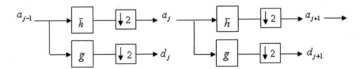

Fig. 6.6 Decomposition algorithm of the orthogonal wavelet transform

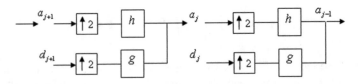

Fig. 6.7 Reconstruction algorithm of the orthogonal wavelet transform

Fig. 6.8 Frequency segmentation sketch map of the orthogonal wavelet transform

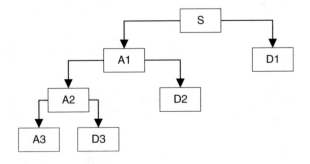

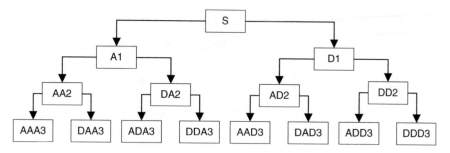

Fig. 6.9 Frequency segmentation sketch map of the wavelet packet analysis

temporal resolution of the approximate sub-signal are not high enough. Comparatively, the wavelet packet analysis can provide a decomposition manner with higher resolution by decomposing the detailed sub-signals. Figure 6.9 illustrates the spectral segmentation sketch map of the wavelet packet analysis of three layers. In the figure, A denotes the approximate sub-signal and D denotes the detailed sub-signal. Finally, the original signal S = AAA3 + DAA3 + ADA3 + DDA3 + AAD3 + DAD3 + ADD3 + DDD3.

By referring to the space structure of the MRA, it is known that $L^2(R) = \bigoplus_{j \in Z} W_j$. Namely, the MRA divides the Hilbert space $L^2(R)$ into the orthogonal summation of a series of subspace $W_j (j \in Z)$, wavelet subspace of the wavelet function $\psi(t)$, at different scales. In order to enhance the resolution, the wavelet subspace W_j is intended to be subdivided. Naturally, both scale subspace V_j and wavelet subspace W_j can be presented by U_j^n. Let

$$\begin{cases} U_j^0 = V_j \\ U_j^1 = W_j \end{cases} j \in Z \tag{6.32}$$

where V_j and W_j denote the scale and wavelet spaces. Then, the orthogonal decomposition of the Hilbert space of $V_{j+1} = V_j \oplus W_j$ can be re-written by

$$U_{j+1}^0 = U_j^0 \oplus U_j^1 \quad j \in Z \tag{6.33}$$

where U_j^n and U_j^{2n} denote the close spaces of functions $u_n(t)$ and $u_{2n}(t)$. Let $u_n(t)$ satisfy the following bi-scale equations of

$$\begin{cases} u_{2n}(t) = \sqrt{2} \sum_{k \in Z} h(k) u_n(2t - k) \\ u_{2n+1}(t) = \sqrt{2} \sum_{k \in Z} g(k) u_n(2t - k) \end{cases} \tag{6.34}$$

where $g(k) = (-1)kh(1 - k)$. Then, when $n = 0$, the following equation exists

$$\begin{cases} u_0(t) = \sum_{k \in Z} h_k u_0(2t - k) \\ u_1(t) = \sum_{k \in Z} g_k u_0(2t - k) \end{cases} \tag{6.35}$$

In MRA, $\phi(t)$ and $\psi(t)$ satisfies the bi-scale equations of

$$\begin{cases} \phi(t) = \sum_{k \in Z} h_k \phi(2t - k) & \{h_k\}_{k \in Z} \in l^2 \\ \psi(t) = \sum_{k \in Z} g_k \phi(2t - k) & \{g_k\}_{k \in Z} \in l^2 \end{cases} \tag{6.36}$$

Apparently, Eqs. (6.36) and (6.34) are equivalent. Expand $n \in Z_+$, Eq. (6.35) can be re-written to be

$$U_{j+1}^n = U_j^n \oplus U_j^{2n+1} \quad j \in Z, \quad n \in Z_+ \tag{6.37}$$

Definition The series of $\{u_n(t)\}(n \in Z_+)$ constructed by Eq. (6.35) is named as the Orthogonal Wavelet Packet determined by the basis function of $u_0(t) = \phi(t)$.

Since $\phi(t)$ is only determined by hk, $\{u_n(t)\}$ $n \in Z$ is also called as the orthogonal wavelet packet of the series of $\{hk\}$.

6.2   Fault Diagnosis Based on Wavelet Analysis for LRE

6.2.1   Wavelet Packet Decomposition and Feature Extraction

Similar to the spectrum, the coefficient of wavelet transform is the function of scale and time shift and the corresponding modulus reflects the energy distribution of the signal. Therefore, the energy along with the scale is

$$E(a) = \sum_b WT_x^2(a, b)$$

And the energy along with time is

$$E(b) = \sum_a WT_x^2(a, b)$$

The wavelet packet transform provides a more precise analysis method by decomposing the detailed sub-signal of the wavelet transform. For example, if the signal is decomposed to be m layers, then there are $2m$ windows and the energy of

$2m$ frequency bands can be used to be feature. Of course, only parts of these bands can be used for the statistic index of peak value, root mean square, kurtosis, shape factor, etc. Combining with the previous knowledge, the wavelet packet transform can be used for the weak feature extraction by selecting those bands with the high signal-to-noise ratio (SNR).

6.2.2   Time-Series Analysis Method and Its Application

Obtained by the wavelet transform or wavelet packet transform, the sub-signals can be analyzed by the time-series model, which was usually used of the stationary signal processing. If the sub-signal was presented by a time-series model, the model parameter, error variance, Kullback–Leibler information and power density spectral can be calculated as the feature and used for fault diagnosis.

Generally, the time-series model of a signal is determined by

$$X(n) = -\sum_{k=1}^{p} a_k x(n-k) + \sum_{k=0}^{q} b_k u(n-k) \quad b_0 = 1$$

and is named as autoregressive moving average (ARMA) model. If $b_k = 0$, $k = 0$, ..., q, then it is called AR model (Fig. 6.10).

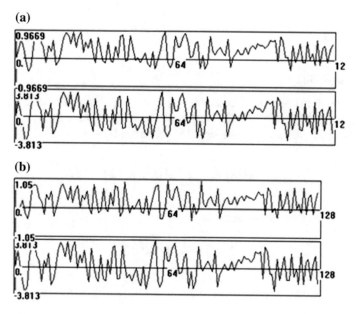

Fig. 6.10 AR models of the raw signal. **a** Raw signal and the Levinson-AR model. **b** Raw signal of the Burg-AR model

1. Levinson algorithm

Calculations of the parameters of AR model are performed by Yule–Walker equation

$$
\begin{bmatrix} \rho_1 \\ \rho_2 \\ \rho_3 \\ \cdots \\ \rho_k \end{bmatrix} = \begin{bmatrix} 1 & \rho_1 & \rho_2 & \cdots & \rho_{k-1} \\ \rho_1 & 1 & \rho_1 & \cdots & \cdots \\ \rho_2 & \rho_1 & 1 & \cdots & \rho_2 \\ \cdots & \cdots & \cdots & \cdots & \rho_1 \\ \rho_{k-1} & \cdots & \rho_2 & \rho_1 & 1 \end{bmatrix} \cdot \begin{bmatrix} \phi_{k1} \\ \phi_{k2} \\ \cdots \\ \cdots \\ \phi_{kk} \end{bmatrix}
$$

where $\rho_i = \gamma_i / \gamma_0$, $i = 1, \ldots, k$ $(k < n)$ and $\gamma_k = E(x_t x_{t+k})$ denoting the correlation function. It can be seen that the matrix with dimension of $k \times k$ is with the type of Toeplitz matrix. Therefore, the reserve can be obtained by iteration calculation of

$$
\phi_{11} = \rho_1
$$

$$
\phi_{k+1,k+1} = \left[\rho_{k+1} - \sum_{j=1}^{k} \rho_{k+1-j} \phi_{kj} \right] \left[1 - \sum_{j=1}^{k} \rho_j \phi_{kj} \right]^{-1}
$$

$$
\phi_{k+1,j} = \phi_{kj} - \phi_{k+1,k+1} \phi_{k,k-(j-1)}, \quad j = 1, 2, \ldots, k
$$

The error variance is estimated by

$$
\sigma_k^2 = \gamma_0 - \sum_{j=1}^{k} \phi_j \gamma_j
$$

2. Burg algorithm

Based on the linear prediction theory, Burg algorithm is resulted from the combination of forward prediction method and Levinson algorithm.

The forward prediction of the signal is presented by

$$
\hat{x}^f(n) = -\sum_{k=1}^{p} a^f(k) x(n-k)
$$

Corresponding prediction error is

$$
\hat{e}^f(n) = x(n) - \hat{x}^f(n)
$$

And the power density function is

$$\rho^f = E\{|e^f(n)|^2\}$$

Comparatively, in the backward prediction condition, these equations are re-written as

$$\hat{x}^b(n-p) = -\sum_{k=1}^{p} a^b(k)x(n-p+k)$$

$$e^b(n) = x(n-p) - \hat{x}^b(n-p)$$

$$\rho^b = E\{|e^b(n)|^2\}$$

Suppose that the partial differential of ρ^b, ρ^f on the coefficients of a_i, $i = 1, 2, ...,$ p is zero, then

$$\rho^b_{min} = \rho^f_{min}$$

$$a^b(k) = a^f(k) \quad k = 1, 2, ..., p$$

If both $a^b(k)$ and $a^f(k)$ is complex, then

$$a^b(k) = \left[a^f(k)\right]^*$$

Namely, the minimum squared errors of the forward and backward prediction method are equivalent and the coefficients of the predictors are conjugate. The prediction errors of different orders are determined by the following equations.

$$e^f_m(n) = e^f_{m-1}(n) + k_m e^b_{m-1}(n-1) \tag{6.38}$$

$$e^b_m(n) = e^b_{m-1}(n-1) + k_m^* e^f_{m-1}(n)$$
$$\hat{\rho}_m = \left(1 - |k_m|^2\right)\hat{\rho}_{m-1} \tag{6.39}$$

where $m = 1, 2, ..., p$, and the initial condition is

$$e^f_0(n) = e^b_0(n) = x(n)$$

$k_m = \hat{a}_m(m)$ denotes the reflection coefficient and is determined by

$$k_m = \frac{-2\sum_{n=m}^{N-1} e^f_{m-1}(n)e^b_{m-1}(n-1)}{\sum_{n=m}^{N-1} \left|e^f_{m-1}(n)\right|^2 + \sum_{n=m}^{N-1} \left|e^b_{m-1}(n-1)\right|^2} \quad m = 1, 2, 3, ..., p \tag{6.40}$$

Coefficients of the AR model are calculated by Levinson algorithm iteratively.

$$\hat{a}_m(k) = \hat{a}_{m-1}(k) + k_m \hat{a}_{m-1}^*(m-k) \quad k = 1, 2, 3, \ldots, m-1 \tag{6.41}$$

The iteration steps of Burg algorithm is as follows:

Step 1   Based on the initial condition of $e_0^f(n) = e_0^b(n) = x(n)$, calculate k_1 by
         Eq. (6.41);

Step 2   According to the equation of $r(0) = \frac{1}{N}\sum_{n=0}^{N}|x(n)|^2$, calculate $\hat{a}_1 = k_1, \rho_1 = \left(1 - |k_1|^2\right)r(0)$ in the condition of $m = 1$;

Step 3   Combing k_1 and Eq. (6.39), calculate $e_1^f(n)$ and $e_1^b(n)$, then estimate k_2 by
         Eq. (6.41);

Step 4   By the relationship of Eqs. (6.42) and (6.40), calculate $\hat{a}_2(1), \hat{a}(2)$ and ρ_2
         of the condition of $m = 2$;

Repeat Step 1 to Step 4 until $m = p$, then parameters of AR model at all orders
can be obtained.

3. Power density function estimation of AR parameter and the order selection

For the Gauss white noise, corresponding estimations of maximum entropy
spectrum and the AR spectrum is equivalent and is determined by

$$p_{AR}\left(e^{jw}\right) = \frac{\sigma^2}{\left|1 + \sum_{k=1}^{p} a_k e^{-jwk}\right|^2}$$

where σ^2 denotes the error variance. Let $a_0 = 1$, then the denominator can be
obtained by the DFT of the series of a_i, $i = 1, 2, 3, \ldots, p$.

The order of AR model can be determined by several criteria listed as follows.
Final prediction error criteria:

$$FPE(k) = \rho_k \frac{N + (K+1)}{N - (K+1)}$$

Information theory criteria:

$$AIC(k) = N\ln(\rho_k) + 2k$$

where ρ_k denotes the error variance at the order of k. As the increasing of the order
k from unit, both FPE(k) and AIC(k) will reach the minimum value at the most
suitable order of the AR model illustrates the reconstructed signals obtained by
different AR models. In the Levinson model condition, the minimum value of FPE
is 3.59836, the order is 1 and the coefficient is −0.190172. Comparatively, in the
Burg model condition the minimum FPE value is 2.00604, the order is 3 and the

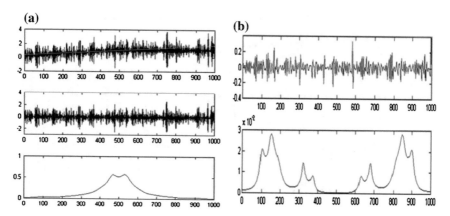

Fig. 6.11 Identification of the signal spectrum. **a** Raw signal and the power spectrum. **b** AR spectrum of the signal spectrum

coefficients are −0.19848, 0.0380587 and 0.0376586. It can be seen that the Burg algorithm is more precise with less error than that obtained by Levinson algorithm.

Figure 6.11 illustrates the sub-signal obtained by the wavelet decomposition of three layers and corresponding spectrum of the AR model. It can be seen that the reconstructed sub-signal of the last seven frequency bands presents to be stationary signal and the energy is concentrated in the upper boundary of the AR spectrum. Combing wavelet decomposition and the AR modeling, the spectral peak of the signal can be obtained. Figure 6.11b illustrates the sub-signal of the second frequency band and corresponding AR spectrum. Several peaks can be noticed. It is included that the combination of wavelet decomposition and AR spectrum is more powerful than the traditional power density spectrum.

6.2.3   Harmonic Wavelet and Its Application

1. Wavelets and affection on the time-frequency analysis

Figure 6.12 illustrates the temporal waveforms of db2, db6, db10, db16 wavelet and corresponding spectra. It can be seen that the compact supporting property of Daubechies wavelet increases with the increasing of N and the spectrum is close to a box. In the vibration signal processing and feature extraction, the spectrum with box shape is very important. Figure 6.13 illustrates the Morlet wavelet, Mexican hat wavelet, Meyer wavelet, Gauss wavelet and their spectra. It can be seen that the Meyer wavelet has good box shape and can plays an important role in the engineering application.

The wavelets of db$N$ series have N zero-order origin moment and are limited support. The bigger the value of N, the regular the wavelet is. Here, the regulation of the wavelet function $\phi(x)$ is defined by

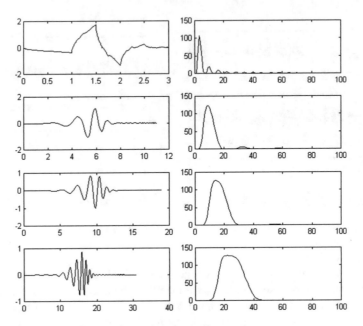

Fig. 6.12 Daubechies series wavelets and corresponding spectra

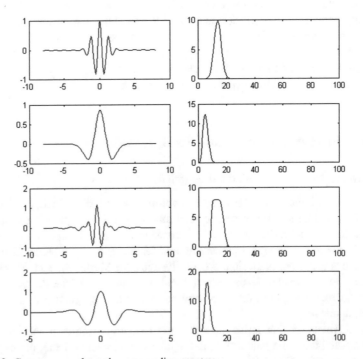

Fig. 6.13 Common wavelet and corresponding spectra

$$\int t^p \phi(t)dt = 0 \quad p = 1, 2, 3, \ldots, n$$

In the frequency domain, it can be written as

$$\hat{\phi}(\varpi) = \varpi^{n+1}\hat{\phi}_0(\varpi) \quad \text{and} \quad \hat{\phi}_0(\varpi = 0) \neq 0$$

In this instance, the wavelet function $\phi(x)$ has a vanishing moment of n order and the bigger n, the more regular of the wavelet is.

2. Harmonic wavelet transform

Proposed by Newland [113], the harmonic wavelet holds good box shape in the frequency domain and is very simple in the time domain. The temporal equation is

$$w(x) = \frac{e^{i4\pi x} - e^{i2\pi x}}{i2\pi x}$$

and corresponding Fourier transform $\hat{w}(\omega)$ is

$$\hat{w}(x) = \begin{cases} \frac{1}{2\pi} & 2\pi \leq w < 4\pi \\ 0 & \text{else} \quad w \end{cases}$$

where $w(x)$ is a complex function, whose real and imaginary parts are shown in Fig. 6.14.

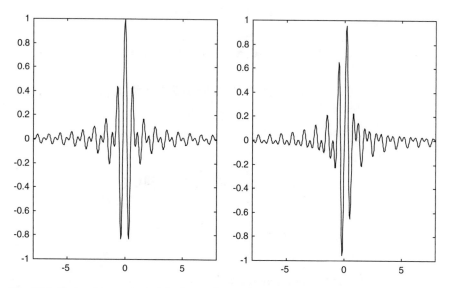

Fig. 6.14 Real and image part of the harmonic wavelet

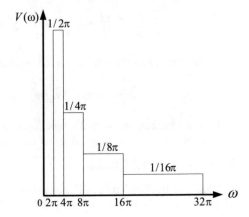

Used for the signal decomposition, the harmonic wavelet should expand to be a series of orthogonal wavelet basis and the wavelets with different degree are corresponding with different frequency ranges. Figure 6.15 illustrates the spectral distribution of different harmonic wavelets. By combining Fig. 6.14, it can be seen that the harmonic wavelet locally supports in the time domain and sharply cuts-off in the frequency domain.

$$w\left(2^j - k\right) = \frac{\exp(i4\pi(2^j - k)) - \exp(i2\pi(2^j - k))}{i2\pi(2^j x - k)} \tag{6.42}$$

Denote $v(\omega)$ as the Fourier transform of $w(2^j - k)$, then

$$v(w) = \int_{-\infty}^{\infty} w(2^j x - k)e^{-iwx}dx$$

Let $z = 2^j x - k$, the above equation can be re-written as

$$v(w) = \int_{-\infty}^{\infty} w(z)e^{-iw(z+k)2^{-j}}2^{-j}dz = 2^{-j}e^{-iwk/2^j}W(w/2^j) \tag{6.43}$$

Furthermore, define the complex wavelet coefficient as the following

$$a_{j,k} = 2^j \int_{-\infty}^{\infty} f(x)w^*\left(2^j x - k\right)dx$$

$$\tilde{a}_{j,k} = 2^j \int_{-\infty}^{\infty} f(x)w\left(2^j x - k\right)dx$$

And when $f(x)$ is real function, $a_{j,k}^* = \tilde{a}_{j,k}$.

Then, the function $f(x)$ can be expanded by the harmonic wavelet.

$$f(x) = \sum_{j=-\infty}^{\infty} \sum_{k=-\infty}^{\infty} \left[a_{j,k} w(2^j x - k) + \tilde{a}_{j,k} w^*(2^j x - k) \right]$$

It is very easy to prove that $\sum_{j=-\infty}^{\infty} 2^{-j}(|a_{j,k}|^2 + |\tilde{a}_{j,k}|^2) = 1/2\pi \int_{-\infty}^{\infty} |F(\omega)|^2 d\omega = \int_{-\infty}^{\infty} f^2(x) dx$ and included that the harmonic series defined by Eq. (6.43) is a complete series of basis function.

For the harmonic expansion of the real series of fr, $r = 0, 1, \ldots, N - 1$, where $N = 2n$. Perform the DFT on fr, we can obtain the Fourier coefficient F_r, $r = 0, 1, 2, \ldots, N - 1$ and $F_{N-1-r} = \bar{F}_r$. Correspondingly, the wavelet coefficient $a_{N-1-s} = \bar{a}_s$. Table 6.1 illustrates the wavelet coefficients along with the degree increase. It can be seen that there are 2^j coefficients of a_{2+k}^j at the degree of j, where $k = 0, 1, 2, \ldots, 2^j - 1$. Every coefficient presents the amplitude of the wavelet transform and the Fourier transform of the wavelet is determined by the coefficient of F_{2+k}^j. Considering contributions of all a_{2+k}^j on F_m, the following equation exists

$$F_m = 2^{-j} \sum_{k=0}^{2^j - 1} a_{2^j + k} e^{\left(-\frac{i 2\pi m k}{2^j} \right)}$$

It is indicated that a_{2+k}^j can be obtained by the reverse Fourier transform of the series of $F_m (m = 2^j + k, k = 0, 1, 2, \ldots, 2^j - 1)$.

Figure 6.16 illustrates the algorithm of the real series of F_r, $r = 0, 1, 2, \ldots, 15$. It can be seen that the wavelet coefficient as is obtained by the IFFT of the segments divided by Fs.

$$as = \text{IFFT}(Fs) \quad s = 2j, 2j+1, \ldots, 2j+1-1 \ldots, \quad j = -1, 0, \ldots, \log 2(N/4)$$

$[x] = \max\{n \leq x | n \in Z\}$, When j is the maximum, as is corresponding to the high semi-spectrum of the signal, namely, $fs/4 \sim fs/2$, where fs is the sampling frequency. Figure 6.17 illustrates the harmonic transform of a simulation signal.

Table 6.1 Coefficients of the wavelet transform	Degree of the wavelet	Coefficients
	−1	a_0
	0	a_1
	1	a_2, a_3
	2	a_4, a_5, a_6, a_7
	…	…
	j	$a_2^j, \ldots, a^{2j+1} - 1$
	…	…
	$n - 2$	$aN/4, \ldots, aN/2$
	$n - 1$	$aN/2$

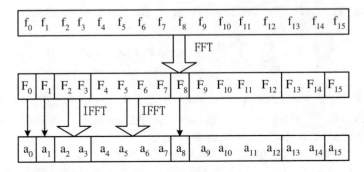

Fig. 6.16 Flow chart of the harmonic wavelet transform

(a)   **(b)**

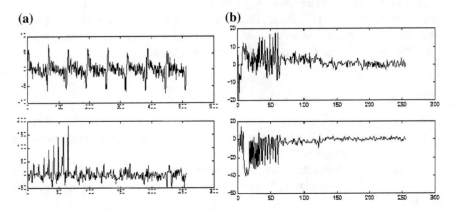

Fig. 6.17 harmonic wavelet transform of the simulation signal. **a** Harmonic signal and corresponding spectrum. **b** Real and image part of the signal in sub-figure (**a**)

3. Harmonic wavelet packet transform

Based on the expansion of the wavelet basis function, the spectral definition of the wavelet packet transform can be presented as

$$\hat{\varphi}_{m,n}(\omega) = \begin{cases} \frac{1}{2\pi(n-m)}, & \omega \in [2\pi m, 2\pi n] \\ 0, & \text{else} \end{cases} \tag{6.44}$$

Corresponding temporal expression can be obtained by the reverse Fourier transform and presented as

$$\phi_{m,n}(t) = \frac{\exp(j2\pi nt) - \exp(j2\pi mt)}{j2\pi(n-m)t} \tag{6.45}$$

where $m, n \in R$ denotes the layers of the wavelet packet transform. In this instance, the wavelet packet transform is defined by

$$W_f(m,n,k) = \int\limits_{-\infty}^{\infty} f(t)\phi_{m,n}^*(t-k)\mathrm{d}t$$

corresponding spectral Fourier can be denoted as

$$\hat{W}_f(m,n,\varpi) = \hat{f}(\varpi)\hat{\phi}_{m,n}^*(\varpi)$$

and the sketch of the wavelet packet transform is illustrated in Fig. 6.18.

In application, the harmonic wavelet packet transform can be used in the fault diagnosis. Figure 6.19 illustrates an example of the harmonic wavelet packet transform.

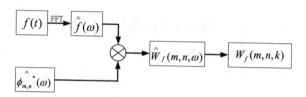

Fig. 6.18 Harmonic wavelet packet transform

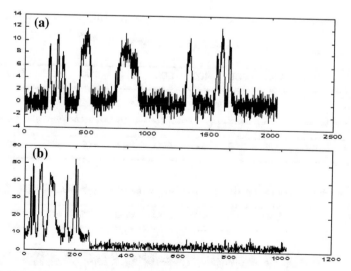

Fig. 6.19 Harmonic wavelet packet transform of the signal. **a** Original signal. **b** Harmonic wavelet transform of the signal

6.2.4 Abnormal Vibration Monitoring and Diagnosis of Turbo Pump

Generally, the fault of LMR is gradually developed in the application and an accurate detection of the initial fault may play an important role for the diagnosis, prognosis, and control of the LMR fault.

In fact, the abnormal trend detection of the turbo pump can be performed by decomposing the vibration to be the parts of high and low frequency and the vibration trend is corresponding to the part of low frequency.

Case study: In a test, the idler wheel of turbo pump looses and wears to break. The thrust disappears after several seconds of the beginning and the vibration increases heavily. Figure 6.20a illustrates the temporal waveform and spectrum of

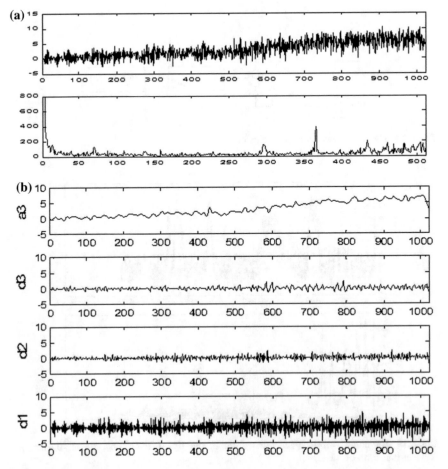

Fig. 6.20 Case study by the wavelet transform. **a** Raw signal and spectrum. **b** Wavelet transform

the turbo pump vibration, collected in a test and Fig. 6.20a illustrates the sub-signals obtained by the harmonic wavelet packet transform. It can be seen that the trend development of the vibration is clearly shown by the sub-signal of a3. Furthermore, by combing the red line alarm analysis, the abnormal trend analysis may have a good performance on the condition monitoring of the turbo pump.

6.2.5   Sub-synchronous Precession Analysis of Turbo Pump Based on Wavelet Analysis

Figure 6.21 illustrates the vibration collected in the test when the turbo pump worked with the speed of 9800 rpm. The first critical frequency of the system is around 100 Hz. It can be seen that the vibration protrudes at the frequencies of the first three working frequency and the first critical frequency. It is indicated that sub-synchronous precession happens in the system.

6.2.6   Fault Diagnosis of LRE Based on Wavelet-ANN

With lots of different expressions, the wavelet-ANN used for the nonlinear approximation is denoted by

$$y(x) = \sum_m \sum_n w_{m,n} \varphi \left(\frac{x - n}{2^m} \right)$$

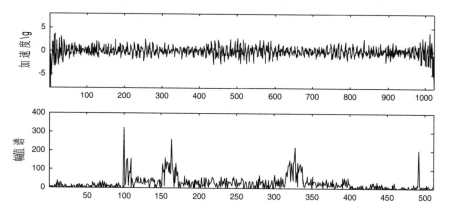

Fig. 6.21 Sub-synchronous precession of the turbo-pump

When used for the adaptive decomposition, it is

$$y(t) = \sum_k \sum_j w_k \varphi\left(\frac{t - b_{k,j}}{a_k}\right) \quad t = 1, 2, 3, \ldots$$

In the nonlinear modeling and prediction, the wavelet-ANN is described as

$$y(t) = \sum_{k=1}^{P} w_k y(t - k) \varphi\left(\frac{t - b_k}{a_k}\right) + \varepsilon(t) \quad t = 1, 2, 3, \ldots$$

When used for the classification, the wavelet-ANN is expressed by

$$y_n(m) = f(u_n(m)) = f\left(\sum_{k=1}^{K} w_{k,m} \sum_{t=1}^{T} s_n(t) \varphi\left(\frac{t - b_k}{a_k}\right)\right)$$

where $s_n(t)$, $t = 1, 2, 3, \ldots, T$ the nth input signal; $y_n(m)$, $m = 1, 2, 3, \ldots, M$ denotes the output vector, $w_{k,m}$ denotes the weight, $\phi(\frac{t-b_k}{a_k})$, $k = 1, 2, 3, \ldots, K$ denotes the wavelet basis function and a_k denotes the scale $f(x)$ is sigmoid function. It is worth noting that different scales of the wavelet is only corresponding to different time shift b_k. In this instance, inspired by the wavelet reconstruct equation, the wavelet-ANN is designed as the following

$$v_p(k) = f\left(\sum_{j=1}^{N} \sum_{i=1}^{R} w_{i,j,k} \sum_{t=1}^{M} x_p(t) \varphi\left(\frac{t - b_{i,j}}{a_j}\right)\right)$$

It can be seen that every scale is corresponding to several time shifts and every combination of the scale a_j and time shift $b_{i,j}$ has a weight $w_{i,j,k}$ supposed that the Morlet wavelet function $\phi(t) = \cos(\omega t) \exp\left(-\frac{t^2}{2}\right)$, $f(x) = \frac{1}{1+e^{-x}}$ and $\varphi_2(t) = \sin(\omega t) \exp\left(-\frac{t^2}{2}\right)$ is used, the gradient vector of parameter can be calculated by

$$\varepsilon = \frac{1}{2} \sum_{p=1}^{size} \sum_{k=1}^{S} (d_{p,k} - v_{p,k})^2$$

and the total error is

$$t' = \frac{t - b_{i,j}}{a_j}$$

Let

$$\frac{\partial \varepsilon}{\partial w_{i,j,k}} = -\sum_{p=1}^{size} \left(d_{p,k} - v_{p,k}\right) \frac{\partial v_{p,k}}{\partial w_{i,j,k}} = -\sum_{p=1}^{size} \left(d_{p,k} - v_{p,k}\right) v_{p,k} \left(1 - v_{p,k}\right) \sum_{t=1}^{M} \varphi(t') x_p(t)$$

$$\frac{\partial \varepsilon}{\partial a_j} = -\frac{1}{a_j} \sum_{p=1}^{size} \sum_{k=1}^{S} \left(d_{p,k} - v_{p,k}\right) v_{p,k} \left(1 - v_{p,k}\right) \sum_{i=1}^{R} w_{i,j,k} \sum_{t=1}^{M} x_p(t) [\varphi_2(t')\omega + t'\varphi(t')] t'$$

$$\frac{\partial \varepsilon}{\partial b_{i,j}} = -\frac{1}{a_j} \sum_{p=1}^{size} \sum_{k=1}^{S} \left(d_{p,k} - v_{p,k}\right) v_{p,k} \left(1 - v_{p,k}\right) w_{i,j,k} \sum_{t=1}^{M} x_p(t) [\varphi_2(t')\omega + t'\varphi(t')]$$

where size denotes the batch of the samples, S, R, M, N denote the dimension of the output vector, time shift amount, dimension of the input vector, and the scale amount. Minimize the total error by conjugate gradient method and construct the search directions of $S(a)$, $S(b)$, $S(w)$ for a_j, $b_{i,j}$, $w_{i,j,k}$. For example, for the calculation of a_j, $j = 1, 2, 3, \ldots, N$

$$g(a)_j = \frac{\partial \varepsilon}{\partial a_j} \quad j = 1, 2, 3, \ldots, N$$

Let $s(a)^0 = -g(a)^0$ in the beginning. Then, perform the iteration as the following

Step 1 Calculate the output vector $v_{p,k}$. If the total error is smaller than the threshold, terminate the iteration. Otherwise, go to Step 2.

Step 2 Calculate the gradient vector $g(a)^i$

Step 3 Calculate the search direction

$$s(a)^i = -g(a)^i + \frac{g(a)^{iT} g(a)^i}{g(a)^{i-1} g(a)^{i-1}} s(a)^{i-1}$$

Step 4 Update the coefficient

$$a = a + \alpha s(a)^i, g(a)^{i-1} = g(a)^i$$

where α is determined by the linear search methods, such as second order interpolation, three order interpolation, etc. Obtain α when minimizing the total error ε and then return to Step 1 to calculate w and b similarly.

Table 6.2 lists several parameters of four different fault modes, where *mo* denotes the flux of the oxidant, *mf* denotes the flux of the fuel, *poc* denotes the pressure before the oxidant spray nozzle, *pf* denotes the branch pressure of the main fuel pipeline and *n* denotes the rotating speed of the turbo pump. The four fault modes are (1) main leaflet fault of the oxidant, (2) main leaflet fault of the fuel, (3) faults of main leaflet on both oxidant and fuel pipeline and (4) filter screen choke of the oxidant. It is worth noting that the parameter value in the Table is the

Table 6.2 Fault data of the LMR

Sample	mo	mf	poc	pf	n	Fault mode
1	−0.72	0.46	−0.51	0.00	0.05	1
2	−3.15	2.06	−2.25	0.07	0.25	1
3	−7.81	5.24	−5.49	0.30	0.78	1
4	0.21	−0.74	0.01	−0.42	0.07	2
5	0.62	−2.04	0.05	−1.145	0.20	2
6	2.73	−7.48	0.43	−3.90	1.02	2
7	−0.67	0.31	−0.52	−0.08	0.06	3
8	−2.97	1.33	−2.29	−0.38	0.31	3
9	−7.43	3.20	−5.71	−1.01	0.86	3
10	−0.17	−0.22	−0.20	−0.23	0.13	4
11	−0.49	−0.64	−0.59	−0.66	−0.37	4
12	−0.79	−1.05	−0.96	−1.07	0.59	4

Table 6.3 Diagnosis result of the wavelet-ANN

	Ideal output				Actual output			
1	0.9	0.1	0.1	0.1	0.74279	0.249777	0.255837	0.263427
2	0.9	0.1	0.1	0.1	0.98843	0.0098240	0.011223	0.013218
3	0.9	0.1	0.1	0.1	0.99996	2.34e−005	3.19e−005	4.68e−005
4	0.1	0.9	0.1	0.1	0.29682	0.693035	0.166294	0.16887
5	0.1	0.9	0.1	0.1	0.09154	0.897238	0.013523	0.014194
6	0.1	0.9	0.1	0.1	0.00079	0.998818	1.60e−006	1.87e−006
7	0.1	0.1	0.9	0.1	0.17708	0.207224	0.638831	0.141411
8	0.1	0.1	0.9	0.1	0.00108	0.0025648	0.926535	0.000330
9	0.1	0.1	0.9	0.1	3.3e−008	2.93e−007	0.998331	1.65e−009
10	0.1	0.1	0.1	0.9	0.17734	0.185276	0.227966	0.643295
11	0.1	0.1	0.1	0.9	0.00220	0.0027250	0.007678	0.912953
12	0.1	0.1	0.1	0.9	0.00067	0.0008752	0.003022	0.942848

ratio between the difference of collected value minus the normal value and the normal value. Namely, all collected parameter are normalized by the normal values.

In the fault diagnosis, morlet wavelet was used with eight scales and four time shifts at every scale. The center frequency ω of morlet is set to be 1.75 and the conjugate learn length of the ANN is 0.7. The result is shown in Table 6.3. It can be seen that 75.8 % of the fault modes are correctly diagnosed by the wavelet-ANN method.

Chapter 7
Fault Diagnosis Method Based on Artificial Immune System

Inspired by the nature immune system, artificial immune system (AIS) is a kind of intelligent algorithm and provides learning mechanisms of noise endurance, self-organization, memory, etc. Combining the advantages of classifier, ANN, and machine learning, AIS provides a new way for the fault diagnosis.

7.1 Artificial Immune System

It is known that one can resist invading of many diseases and heal by oneself sometimes, since the existence of the natural immune system in the body. From the 1940s, with the development of the iatrology, the recognition and understanding of the natural immune system is gradually deepening and consummating and the immunology is developed. Aiming at analyzing the structure and function of the natural immune system, the immunology is devoted to the investigation of the rooting, differentiation, character, and function of the antibody with the ways of cytology, biochemistry, etc. and reveal the mechanism of recognition and answer of the antibody, repelling non-self and preserving oneself [114]. Inspired by the advantage of the natural immune system, the AIS is developed and provides a new way for the engineering application.

7.1.1 Natural Immune System

1. Composing of the natural immune system

With the development of the modern immunology, it is proved that the natural immune system is integrated and can perform the immune function of the body. As the basis of the immune response, the natural immune system is composed of

© Springer-Verlag Berlin Heidelberg and National Defense Industry Press 2016
W. Zhang, *Failure Characteristics Analysis and Fault Diagnosis for Liquid Rocket Engines*, DOI 10.1007/978-3-662-49254-3_7

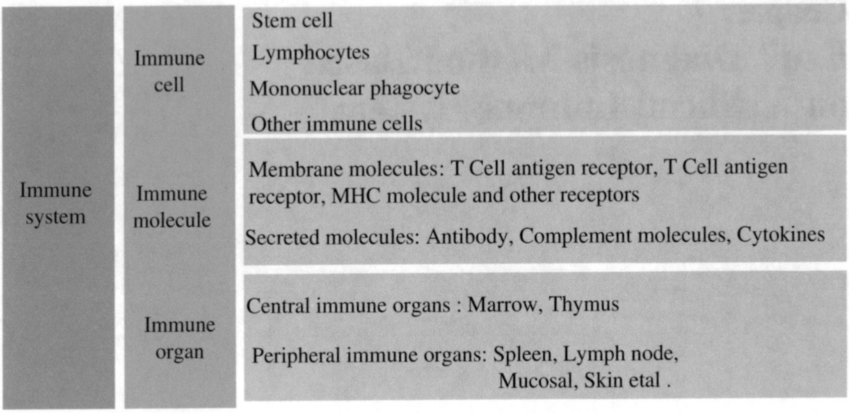

Immune system	**Immune cell**	Stem cell
		Lymphocytes
		Mononuclear phagocyte
		Other immune cells
	Immune molecule	Membrane molecules: T Cell antigen receptor, T Cell antigen receptor, MHC molecule and other receptors
		Secreted molecules: Antibody, Complement molecules, Cytokines
	Immune organ	Central immune organs : Marrow, Thymus
		Peripheral immune organs: Spleen, Lymph node, Mucosal, Skin etal .

Fig. 7.1 Tree structure of the natural immune system

immune apparatus, immune cell, and immune molecule. Figure 7.1 illustrates the tree structure of the natural immune system.

As the body's main blood-forming organs, the bone marrow is a place for the production and differentiation of various immune cells. With powerful capability of differentiation, the Pluripotent stem cells of bone marrow can perform the prolif-eration and differentiation operation to produce the non-lymphoid cells and lym-phoid stem cells. The non-lymphoid cells grow to be red blood cells, myeloid cells, monocytes–macrophage cell lines, and megakaryocytic cells, while the lymphoid cells develop to be prolymphocytic. Lymphoid stem cells differentiate into T cells in the thymus and then differentiate into B cells in the bone marrow. In the thymus, T lymphocyte develops, differentiates, and mellows and the main ingredient is immature T cell precursors. The pro-T cells in the bone marrow enter the thymus via the blood and become the T lymphocytes in the differentiation and maturation in thymus.

The immune cells involved in the cells performing immune responses, such as the lymphocytes and the monocytes–macrophages. Including B cells and T cells, the lymphocytes are the most important immune cells and play a central role in the immune response. Including immune cell membrane molecule, immune molecules play an important role in development of immune system, activation of immune cell, immune response inducing, and regulation.

2. Main function of the immune system

Being a specific physiological response of the body, the immunization maintains a stable internal environment by identifying and excluding antigenic foreign bodies. The general defense process of the immune system is illustrated in Fig. 7.2.

The main function of the immune system is listed in Table 7.1.

When foreign bodies (antigens) invade the body, the macrophages uptake the foreign bodies at first and pass the information to the immune memory antigen

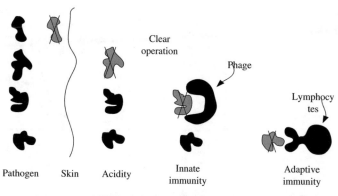

Multi-layered protection system

Fig. 7.2 Defense progress of the immune system

Table 7.1 Classification, physiological, and pathological reactions of immune function

Functions	Normalities	Abnormalities
Immune defense	Resist and clear invasion of pathogenic microorganisms and other antigens	Hypersensitivity (excessive) Immunodeficiency disease (less)
Immune homeostasis	Remove damaged or aging cells, modulating the immune response	Autoimmune diseases
Immune surveillance	Prevent, remove malignant cells with mutations or aberrations or persistent viral infection	Tumorigenesis, persistent viral infection

which is the equivalent of computer devices. The agent information is obtained by the spin-off process. If it is judged to be "non-self," the immune cells, including B type cell and T type cell, are activated and proliferated to attack the "non-self." Immune response is the defense mechanism against foreign substances harmful to the body and can be functionally divided into immune recognition, immune learning, immune memory, immune tolerance, adaptive immunity, and so on.

(1) Immune recognition is to identify the "self" and "non-self" by the combination of the receptor and the agent. The strength of the combination is called affinity.
(2) Immune response is the whole process including the recognition, activation, differentiation, and immune response generation of the antigen molecules. The immune response, generally caused by antigen, is a series of reactions by a variety of immune cells participate. First, the antigens are presented to the appropriate T cells by the antigen-presenting cells. Then the activated cells further activate the other immune cells to produce a corresponding immune response through molecules or cells secrete interleukin. The antigens and target cells are further engulfed, which is activated by the inflammatory

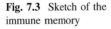

Fig. 7.3 Sketch of the immune memory

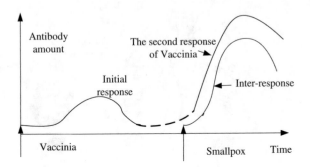

response mediated in cytokines through the complement component. Finally, the antigens are removed and the interval balance and stability are maintained.

(3) Immune tolerance is a specific unresponsiveness when the immunocompetent contacts the antigenic substances. It is caused by the function loss or death of the specific lymphocyte induced by the antigens. Normally, the body's own tissue antigens are tolerated, which is called self-immune tolerance. The autoimmune disease may happen when the self-tolerance is destroyed.

(4) Immune memory. When the specific antigen appeared in the body again, the response interval of antibody is much shorter than the first time. The amount of the antibody increases quickly and the maintaining time also increases. Such phenomenon is called immune memory. If the antigen is similar to that invading the body at first time, the associative memory will be inspired and the response speed will greatly be enhanced. Figure 7.3 illustrates the immune memory sketch for the cowpox and smallpox. It can be seen that the response is very slow when the body is invaded by the cowpox at the first time. When the smallpox, similar to the cowpox in the protein structure, invades the body, both response speed and strength are enhanced owing to the associative memory mechanism. Such phenomenon is the basis of the artificial cultivation vaccine.

(5) Immune regulation denotes the interaction between the internal cells, immune cells and the molecules, immune system and the other systems, neural or endocrine system, etc. In this instance, a net structure with mutual assistance and mutual restraint function will be constructed and will be used for the stability maintaining of the internal environment by maintaining a suitable strength of the immune response.

3. Basic character of the immune system

The most important feature of biological immune system is immune memory, self-identification of antibodies, and immune diversity. In the view of information processing, the immune system holds the characters of tolerance, learning and understanding, robustness and adaptability, distribution and self-organization, etc. [115]

(1) Tolerance, which is presented by the immune tolerance.
(2) Learning and understanding, which are presented by the immune response and the immune memory. Namely, the body can learn from the new antigen and memorize the information of the antigen to be the own knowledge.
(3) Robustness and adaptability. The cross response inspired by the associative memory is the presentation of the adaptability. The powerful capability of the immune system makes the body develop to be an adaptive robust evolution system.
(4) Distribution. It indicates that the immune system is composed of many basis localized parts and is not the centralized control. Such distribution makes the function of the immune system not influenced by the damage of the local organization.
(5) Diversity. Generally, the amount of antibody is much less than that of antigen outside the body. However, the metabolism of the body is very fast. New antibodies will be carried by the new cells into the body and the kind of the antibody changes with the variation of the antigens. The numerous types and numbers of antibodies make the immune system show diversity.
(6) Self-organization. For the immune regulation, the immune system is dynamic and the condition is transformed from one to another. The self-organization indicates that the system is a cohesive and self-maintenance orderly system constructed by the variation of itself.

Artificial immune system

The diversity, immune memory, distribution processing, adaption, robustness, etc. of the natural immune system have inspired the interest of people to explore the mechanism and the AIS has been developed from the 1990s. Advanced characters of the natural immune system have been simulated by the AIS and applied in the engineering application. In 1974, the first mathematical model of the immune system has been proposed by Jerne and has constructed the foundation of the immune calculation. In 1986, Farmer proposed the mathematical description of the immune net and provides a way for the development of the immune mechanism-based calculation system. In 1989, Varela discussed the convergence performance of the immune net and the adaptive ability of different antibody mutations and made tremendous contributions for the application of AIS in engineering. In 1994, Forrest proposed the negative selection algorithm and has simulated the tolerance process of the self-cell. Many novel concepts on the computing immune system were developed. In 1999, the clone selection mechanism was developed by Hunt based on the investigation of Burnet and the hyper-mutation doctrine was proposed. All the above investigations have established a theoretical foundation for the AIS applications.

1. Immune model

From the last of 1980s to the beginning of 1990s, the immune system was introduced to be a novel computing intelligence and many artificial immune models have been developed at the middle of 1990s. In these models, the computing

Fig. 7.4 Basic sketch of the immune algorithm

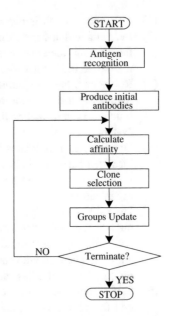

mechanism of the immune system was intended to be extracted out in order to obtain the ability of distribution, self-learning, and adaption abilities and to solve the problems in applications.

Generally, the AIS can be divided into two classes. The first class is the model based on the immune system theory. The representative models include the ARTIS model proposed by Hofmeyr in 1999 [116] and the multi-agent model proposed by Ballet [117], Dipankar in 1998 [118]. The other class is the net model developed on the immune net theory. The representative models include the dynamic immune net model developed by Ishida in 1996 [119], the aiNet model developed by Castro in 2000 [120], and the RLAIS model proposed by Timmis in 1995 [121].

2. Immune algorithm

Immune algorithm is a computing process inspired by the immune mechanism [122]. The functions of T cell, B cell, and antibodies are integrated to be a detector to simulate the immune response of the natural immune system, including the production of antibodies, self-tolerance, clone proliferation, immune memory, etc. The basis sketch of the immune algorithm is illustrated in Fig. 7.4.

In the application, the problem should be described by the concepts of immune system and correspond to the immune mechanism at first. Then define the immune elements and design the immune algorithm. The basic sketch of the immune algorithm is the reflection of the common frame and different mechanisms may induce different immune algorithms. For example, the group-based and the net-based algorithms are the main contents of the immune algorithm. The negative selection algorithm and the clone selection algorithm are two kinds of group-based algorithm.

7.2 Application of Negative Selection Principle to Fault Detection and Diagnosis of LRE

LRE, known as the heart of the spaceflight equipment, is a kind of complicated structural liquid and heat dynamic system running with high temperature, high pressure, strong cauterization, and heavy density of energy emitting. So it is sensitive to faults which need to be detected and diagnosed quickly and effectively.

Nowadays, artificial intelligence is widely used to diagnose the fault of the LRE, such as neural network, expert system, and so forth. Neural network, applied to detect and diagnose the turbo pump fault, has very huge and intricacy structure and needs a relative longer training time. The pattern recognition method relies on a great number of data. While the expert system is difficult with knowledge express, its illation relies on the fault mechanism which is the puzzle of the fault diagnosis. During recent years, the fault diagnosis method based on the integration of qualitative and quantitative knowledge has developed rapidly and is quick but ignores the deep-seated relationship in the system.

AIS, enlightened by the biology immune system, is one of the artificial intelligences, which simulate the function of the biology system. It has the advantage of noise tolerance, self-organizing, memory, and so on. It supplies a new method to solve problems. Basing on the mechanism of lymphoid cell's negative selection, Forrest (1996) supposed the negative selection algorithm which could detect the change of a system. So according to the fact of the LRE, the negative selection algorithm was applied.

7.2.1 Negative Selection Algorithm

1. AIS and artificial recognition ball (ARB)

In the engineering and science field, AIS simulates the mechanism of the human beings' immune system, studies the technology of computing and information disposing and applies the theory in the engineering program. Timmis said "AIS, comes from the theory of biology, consults the immune function, mechanism and model and then is used to solve the complicated problems." According to the theory of immunology based on colony, antibodies exist in the form of colony. The ones with similar functions get structural resemblance. The kindred antibodies have strong affinities with each other also. The ARB, abstracted from the antibodies, is the set of the similar kind of functional antibodies. Different functional kinds of antibodies are corresponded to different kinds of ARB. The recognition ability of the antibodies to antigen is quantificationally measured by the ARB's radius. So Timmis introduced the concept of ARB with equal radius (Fig. 7.5).

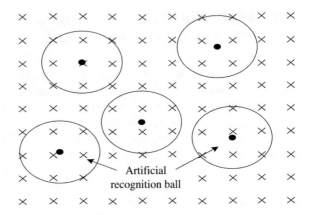

Fig. 7.5 Artificial recognition ball proposed by Timmis

But actually, the different kinds of antibodies' recognition ability are discriminating. So the ARBs with different radiuses were applied in this paper (Fig. 7.6). The stronger the recognition ability, the longer the radius is.

2. Negative selection algorithm

Introduced by Forrest, American computer professor, negative selection algorithm was used to detect the change of the computer circumstance. It came from the mechanism of differences between self and non-self, which conformed to the progress of T cells' maturation. When the detectors generated randomly, they will be deleted if they detect self, while the others will be reserved as mature detectors to detect the non-self. If the mature detectors are matched with the objects of the system, then the system has got something wrong (Fig. 7.7).

Namely, for the question field $U \in R^l$ which includes the self-set $S \subseteq U$ and non-self set $N \subset U$, and $S \cup N = U$, $S \cap N = \emptyset$. The process of detecting is the

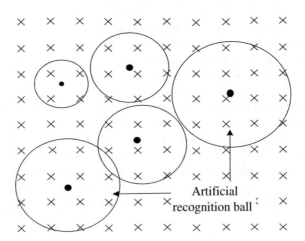

Fig. 7.6 Artificial recognition ball with variable radial

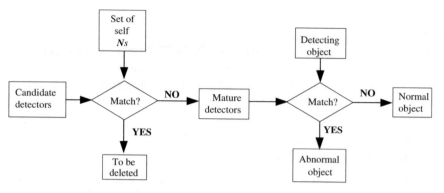

Fig. 7.7 Negative selection algorithm

classification of self and non-self. For the set of detectors (Immune cell, antibody, and so on) $D : D = \{x_1, x_2, \ldots x_i\}, x_i \in R^k, k \leq l, i \in N, N = 1, 2, 3, \ldots$. $f : f(I, x) \rightarrow \{p \in R | p \geq 0 \wedge p \leq 1\}$ is the match function, and $x \in D, R$ is the set of real number. ε is a match threshold. The classification is completed by Eq. (7.1).

$$\text{match}\,(f, \varepsilon, I, D) = \begin{cases} I \in N & 1 > f(I, x) > 1 - \varepsilon \\ I \in S & 0 < f(I, x) < 1 - \varepsilon \end{cases} \tag{7.1}$$

In this paper, the negative selection algorithm is coded by real number. The ARBs are used too. The steps of the algorithm are listed as follows:

Step 1 Define *self* and *non-self* as a multiset of real number of length l over a finite data series, which comes from the object that we wish to protect or monitor.

Step 2 Create the ARBs by the acquired data, and then compute the center and the radius of the ARBs.

Step 3 Use the created ARBs to detect the system state, and calculate the distance of the state between *self* and *non-self* ARBs. If the distance to the center of *self* ARB is less than the ARB's radius, then the state could be judged as *self*. Otherwise it is *non-self*, which means the object got some faults.

7.2.2 Case Study

1. Model of the system

Generally, the LRE with the stable condition can be described by a nonlinear equation of

$$F(Y, X, E) = 0 \tag{7.2}$$

where Y denotes the fault mode, X denotes the matrix composed by the condition parameters, and E denotes the disturbance elements. Based on Eq. (7.2), it can be seen that the fault mode Y is determined by both condition parameters and disturbance elements. When X or E is far from the designed point, the faults may happen.

2. Data acquisition and fault mode
(1) Expression of self and non-self

Based on the testing data, the simulating calculation was aimed at some fixed type of LRE. Parameters and fault modes are listed in Tables 7.2 and 7.3, where *parameters 1th* and *2th* have only one detecting point, respectively, while others have four detecting points. All data came from these detecting points were listed ordinal together, and then became a vector of 30-dimensional vector. *Eight* common fault modes are selected to validate whether this method is useful and advanced. The condition of the LRE could be fixed by the 30-dimensional vector, as follows:

Table 7.2 Parameters of monitoring

Serial number	Parameters
1	Pressure of enforcing air of the oxidant closet
2	Pressure of enforcing air of the fuel
3	Flux of oxidant
4	Flux of fuel
5	Pressure of the exit of oxidant pump
6	Pressure of the exit of fuel pump
7	Rotate speed of turbine pump
8	Pressure of oxidant before spurt
9	Pressure of fuel at the offshoot on the main pipeline

Table 7.3 Fault mode of the LRE

Mark	Fault mode
Mode1	Increase of the pressure of oxidant closet
Mode2	Increase of the pressure of fuel closet
Mode3	Increase of both closet
Mode4	Decrease of both closet
Mode5	Leakage of offshoot of oxidant pipe
Mode6	Oxidant start valve obstruct
Mode7	Fuel start valve leakage
Mode8	Gas leakage of turbine intake

$\vec{X}_i = \{x_1, x_2, \ldots x_n\}_i$, $n = 30$, $i \in N$, where N is the total amount of the known state data, and n is the amount of the detecting points. The fault mode is figured as $Y_j, j = 1, 2, \ldots m, m = 8$. So Y_j is the set of similar states' data.

Self, figured as C, is defined as normal state. All faults are *non-self* and figured as $\overline{C}$. Therefore, *non-self* is build up by the m quantitative known fault modes and the unknowns. Namely, $\overline{C} \supseteq \{Y_1, Y_2, \ldots, Y_m, \ldots\}$. The definition of *self* and *non-self* is used to monitor LRE and detect the faults.

(2) Artificial recognition ball (ARB)

The kind and recognition ability of ARB depends on its location in the fault space and its radius. Each fault mode (including the normal state one) is corresponded with an ARB. The center of the ARB is figured as $\vec{Y}_i, i = 0, 1, \ldots, m, \ldots$; when i equals 0, it is the normal state, and $r_i, i = 0, 1, \ldots, m, \ldots$ is the ARB's radius. The Euclid distance is used to denote the similarity of the data.

$$d(\vec{X}_i, \vec{X}_j) = \sqrt{\sum_{k=1}^{n} (x_{ik} - x_{jk})^2}, \quad i, j = 0, 1, \ldots, N \tag{7.3}$$

The smaller the $d(\vec{X}_i, \vec{X}_j)$, the more similar the state $\vec{X}_i$ and $\vec{X}_j$. The probability of the two belongings to part of the same fault mode is higher. For the known fault modes, the center of each ARB, $\vec{Y}_i = \{y_1, y_2, \ldots, y_n\}, n = 30$, is calculated by the average of the known data of each fault mode:

$$\vec{Y}_i = \frac{\sum_{k=1}^{n} \vec{X}_{ik}}{h}, \quad h \in N \tag{7.4}$$

$\vec{X}_{ik}$ is one of the data vectors of the fault mode Y_i, and h is the amount of the fault state data. For one fault mode, the distance of state data to the center could be calculated by

$$\overline{d}_i = \frac{\sum_{k=1}^{h} d(\vec{X}_{ik}, \vec{Y}_i)}{h} \tag{7.5}$$

Then, the radius of ARB is

$$r_i = \alpha \times \overline{d}_i \times \left[1 + \frac{h - \text{Sum}(d(\vec{X}_{ik}, \vec{Y}_i) < \overline{d}_i)}{\text{Sum}(d(\vec{X}_{ik}, \vec{Y}_i) < \overline{d}_i)} \right] \tag{7.6}$$

$\alpha > 0$, h is the total amount of the mode i. $\text{Sum}(d(\vec{X}_{ik}, \vec{Y}_i) < \overline{d}_i)$ is the amount of the state vectors whose distance to the center is smaller than the average distance of the mode i. α is an alterable parameter, which could be changed if necessary. Its value will be discussed in the result.

It could be recognized that the radius of the ARB depends on the density of the fault data, because it is not reasonable that the data whose $d(X_{ik}, Y_i) > \overline{d}_i$ will not be contained in the range of the ARB when the radius equals the average distance of $\overline{d}_i$. Therefore, let

$$\theta = \frac{h - \mathrm{Sum}(d(\overrightarrow{X}_{ik}, \overrightarrow{Y}_i) < \overline{d}_i)}{\mathrm{Sum}(d(\overrightarrow{X}_{ik}, \overrightarrow{Y}_i) < \overline{d}_i)}, \quad i = 1, 2, 3, \ldots, m \tag{7.7}$$

and $\theta_i > 0$, which is the value of modification. The more dispersive the data, the bigger the θ_i, and the less dispersive, the smaller the θ_i, and the radius is closer to the average distance.

(3) Detection and diagnosis scheme

For the condition $\overrightarrow{I} \in U$, it needs to be estimated whether it is normal or not first. If $d(\overrightarrow{I}, \overrightarrow{Y}) < r_0$, then $\overrightarrow{I} \in C$. But if $\overrightarrow{I} \in \overline{C}$, then it must be judged which fault it is.

For the condition $\overrightarrow{I} \in U$, if $T \subset X$, $T = \{\overrightarrow{t}_1, \overrightarrow{t}_2, \ldots, \overrightarrow{t}_p\}$, p is the amount of the known state data.

① Let

$$\delta = 1 - \frac{\min_{i=1,2,\ldots,p}\left\{d(\overrightarrow{I}, \overrightarrow{t}_i)\right\}}{r_i} \tag{7.8}$$

If $\overrightarrow{t}_i \in Y_j$, when $\lambda \leq \delta < 1$, then $\overrightarrow{I} \in Y_j$, $j = 1, 2, \ldots, m$. The λ is a parameter related to the congeneric similarity. This method conforms to the way of human beings' cognizance that if two things represent some enough similar characters then they are congeneric. The δ is used to estimate the similarity. It is affected by the distance of the state vector and the corresponding ARB's radius. In this paper, to relieve the computation, $\lambda = 0.95$.

② If $0 \leq \min_{i=1,2,\ldots,p} \{d(\overrightarrow{I}, \overrightarrow{Y}_i)\} < r_i$, then $\overrightarrow{I} \in Y_i$, $i = 0, 1, \ldots, m$.

④ If the mode of the state cannot be fixed by rule 1 and 2 then a new ARB denoting a new fault mode needs to be created. The center locates at $\overrightarrow{I}$, and the radius equals $\min_{i=1,2,\ldots,p} \{d(\overrightarrow{I}, \overrightarrow{t}_i)\} - \delta$. Put this new fault mode in the data storeroom, in order to make the fault storeroom more perfect.

3. Results

According to the known fault data, the simulation was carried out by programming on Visual C++ 6.0. The detected results are shown in Figs. 7.8 and 7.9.

For the ordinary faults or those aroused by little or a bit big parameter changing, this method is effective and of high exactness. The detecting result is greatly affected by ARBs' radius. The longer the radius, the bigger the ability of fault tolerance and the ratio of detecting will decrease. While on the opposite position, the ratio of miscarriage of justice will increase. So it seems to have an optimized radius to make the method more effective. Then *alpha* (α) is used to optimize the radius and control the simulation process according to the truth. When $\alpha = 1$, the ratios of exactness of testing sample and training sample are both high. So in this paper, $\alpha = 1$.

Fig. 7.8 Result of fault detecting

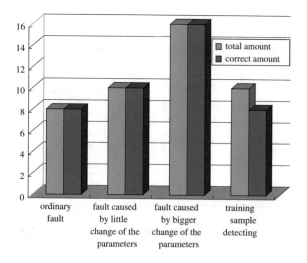

Fig. 7.9 Affect caused by the radius of ARB, α is the ratio, a regulable parameter

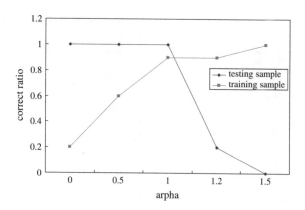

The diagnosis results are shown in Tables 7.4, 7.5, 7.6, 7.7, 7.8, and 7.9, and the normal ARB's radius is 0.127825 (data are standardized). When the degree of a fault increases, the distance between the ARB and the normal state increases correspondingly. Namely, the more visible the fault, the more removed its ARB to the normal state.

The fault ARBs surround around the normal state in the multi-dimension space (Fig. 7.10). If some faults are sensitive to the changing parameters, then they will have a longer radius (as: Mode8). For the fixed diagnosis method, the faults could be separated and both superposed partly. The states in the superposed areas could be diagnosed as different modes, which shows that some different faults could cause same change of the parameters. This is why some faults are difficult to be diagnosed. Some fault will put up another fault's characters during the faults whose parameters' changing affects each other greatly. But when the fault became serious, they could be separated soon.

Table 7.4 Distance between faults and the normal state when the faults increased by parameter changing bigger

Parameter changing	0.02	0.06	0.10	0.16	0.20
Mode1	0.136423	0.137168	0.138444	0.141342	0.143924
Mode2	0.137811	0.141787	0.146736	0.155817	0.162817
Mode3	0.137988	0.14635	0.14875	0.16029	0.1693
Mode4	0.134932	0.133648	0.134173	0.138298	0.143132
Parameter changing	0.02	0.06	0.10	0.12	0.14
Mode6	0.137082	0.14119	0.153292	0.148647	0.142532
Parameter changing	0.10	0.30	0.50	0.60	0.70
Mode5	0.138165	0.147268	0.158511	0.164177	0.169718
Mode7	0.133915	0.1305	0.128515	0.127936	0.127676
Mode8	0.119364	0.342221	0.91105	0.69195	0.7846

Table 7.5 Distance between the center of the fault modes and normal state

	Distance to the normal state	ARB's radius
Normal state (Mode0)		0.127825
Mode1	0.139996	0.0255326
Mode2	0.151802	0.0177411
Mode3	0.15516	0.0177411
Mode4	0.136112	0.019124
Mode5	0.16106	0.0385692
Mode6	0.148239	0.0318421
Mode7	0.128142	0.019124
Mode8	0.644305	0.398956

Table 7.6 Distance of a fault mode between the another

Distance	Mode2	Mode3	Mode4	Mode5	Mode6	Mode7	Mode8
Mode1	0.0518361	0.0425543	0.0736546	0.108123	0.0530702	0.0508228	0.680894
Mode2	0	0.0295684	0.0915005	0.091006	0.0723117	0.0832397	0.691117
Mode3		0	0.10481	0.117339	0.076907	0.0882757	0.691438
Mode4			0	0.064282	0.0614759	0.0318734	0.673393
Mode5				0	0.0906294	0.0881856	0.689898
Mode6					0	0.0553519	0.705565
Mode7						0	0.664926
Mode8							0

Table 7.7 Results of diagnosis

Simulation order	1	2	3
Mode1	Mode1	Mode1	Mode1
Mode2	Mode1	Mode2	Mode2
Mode3	Mode1	Mode1	Mode3
Mode4	Mode1	Mode4	Mode4
Mode5	Mode4	Mode5	Mode5
Mode6	Mode1	Mode1	Mode6
Mode7	Mode1	Mode7	Mode7
Mode8	Mode0	Mode7	Mode8

Table 7.8 Results of diagnosis with the "red-line" detection

Simulation order	1	2	3
Mode1	Mode1	Mode1	Mode1
Mode2	Mode2	Mode2	Mode2
Mode3	Mode3	Mode3	Mode3
Mode4	Mode4	Mode4	Mode4
Mode5	Mode5	Mode5	Mode5
Mode6	Mode6	Mode6	Mode6
Mode7	Mode7	Mode7	Mode7
Mode8	Mode8	Mode7	Mode8

Table 7.9 Results of diagnosis for an unknown fault

Simulation order	1	2	3	4	5	6	7
Mode	Mode1	Mode1	Mode1	Mode9	Mode9	Mode9	Mode9

This method is also not very good at deal with the superposed states alone (Table 7.7). So the red line method is used as an assistant. The results are shown in Table 7.8.

Fig. 7.10 Fault ARBs
around the normal state

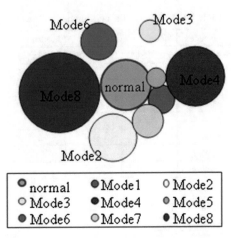

When the method confronts some unknown states which are not able to be diagnosed, a new ARB will be created, which means the system took place a new fault (Table 7.9).

The average consumed time of fault detecting is 0.015 s, while the average consumed diagnosis time is 0.02 s (Table 7.10). It is quick enough to the sampling time, and could be used as online monitor method.

Based on the above investigation, it can be concluded that

1. Performing detection within 0.02 s and diagnosis within 0.05 s, the proposed method is fast enough for the real-time condition monitoring of the LRE.
2. The fault diagnosis works well for the modes with large parameter varying and does not work well for those faults sharing the similar parameter varying. However, by combining the red line alarm detection, the diagnosis effect can be improved.
3. The proposed method is sensitive for the parameter varying. Being a fault detection method from the system vision, the detection capability can be further improved by setting a suitable radius of ARB.
4. The proposed method can learn the unknown fault mode in the program of diagnosis. It is worth noting that the unknown mode cannot be further analyzed if the mechanism of the fault is unknown.

Table 7.10 Consumed time of fault detection and diagnosis

Simulation order	1	2	3	4	5
Consumed time of detection	0.015	0.015	0.015	0.015	0.015
Consumed time of diagnosis	0.016	0.031	0.047	0.015	0.016

7.3 Application of the Clone Selection Principle in Start-up Progress Simulation of LRE

It is known that 30 % of the faults, such as pipe damage, jam, the combustor damage, etc., happen in the start-up stage. Therefore, the investigation of the start-up stage is important for the performance improvement of the LRE. With the advantage of decreasing the study period and the cost, the simulation is one of the key ways for the investigation of the LRE and provides a fast and convenient way for the design and performance optimization.

Reference [125] developed a model for the system collocate, structure optimization, and scheduling of start-up or shut off investigation based on the mechanism of LRE. However, the model is complicated and the result is sensitive to the system parameter. Comparatively, Reference [126] developed a more generalized model based on the modularization design of the LRE by Simulink and the whole model can be assembled by the module of many subsystems. However, the input and output of the module are not definitely defined and is difficult to be planted on other LRE modeling.

In this instance, a system modeling method was investigated on the basis of the AIS. With the capability of noise resistance, self-organization, and memory, the AIA can describe the system by the usage of the test data and is very convenient for the complex system modeling.

7.3.1 Clone Selection Principle and Algorithm

1. Mechanism of the clone selection principle

In the natural immune system, B cell will be activated and cloned when the lymphocyte detects the antigen. Then the cloned B cell will perform variations and produce given antibodies for the antigen. Finally, the variation antibodies are composed by memory antibodies and effect antibodies. Both types of antibodies hold good capability for the counteraction of the antigen. The mechanism of the clone selection principle is illustrated in Fig. 7.11.

It can be seen that the clone and multiplication of the antibodies are excited by the antigen and a progress of affinity maturation is corresponding to the clone, multiplication, and variation of the antibodies. With the maturation of the antibodies, the affinity increases. Correspondingly, the clone selection principle of the AIS is performed by the clone, multiplication, and variation operators, and the pullulation of the AIS is adaptive.

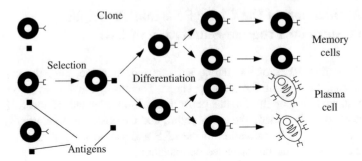

Fig. 7.11 Clone selection mechanism

2. Clone selection algorithm

In 2000, De Castro has proposed the clone selection algorithm based on the clone selection principle of the natural immune system [114], and the detailed sketch is illustrated in Fig. 7.12.

Step 1   Produce the solution candidate set of P by the memory subset of M and the reserve population Pr, namely, $P = Pr + M$;

Step 2   Select n numbers of individual Pn by the calculation of affinity.

Step 3   Clone the n individuals with biggest affinities and form a temporary group of C. The bigger the affinity, the greater the number of the antibody clones.

Fig. 7.12 Clone selection algorithm

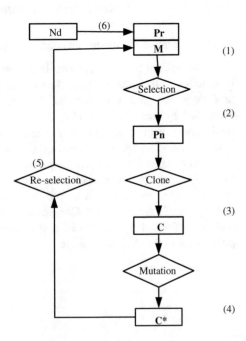

Step 4 Perform high-frequency variation in the temporary group of C based on the individual affinity and produce a group of C* with individual variation.

Step 5 Choose the individuals with highest affinities to form the memory set of M and replace some individuals of P by the improved individuals of C*.

Step 6 Replace d numbers of individuals randomly. The lower the affinity, the more possible the individual is replaced.

It can be seen that the clone selection algorithm produces a memory group of antibody with high affinities and this memory antibody group is used for the immune defense and monitoring. For the sake of modeling, a testing sketch is proposed here based on the clone selection principle (Fig. 7.13):

(1) Produce the initial group based on the given data.

(2) Compare the unknown samples with the initial group and find the antigens with the closest similarity.

(3) Clone and mutate the antibodies with high affinities to produce an antibody group C* with higher affinity.

(4) Select the antibodies with higher affinities in the group C* to construct another group C and calculate the average affinity of C. If the average affinity is bigger than the threshold, then the antibody group C is similar to the test samples. Otherwise, repeat the clone and mutation steps on group C until the algorithm reaches the convergence condition.

Fig. 7.13 Sketch of the simulation test

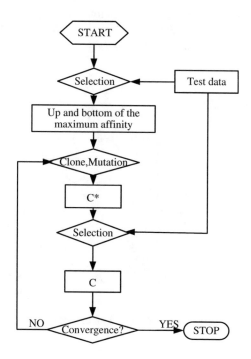

7.3.2 Case Study

1. Parameter selection and calculation

Reflecting the performance of the LRE, the thrust is one of the important parameters and should be carefully monitored in the start-up progress. Therefore, it is with important significance to calculate the thrust at arbitrary moment in the start-up progress. In a given condition of LRE, the thrust F is the function of the flux of oxidant $\dot{m}_o$, the flux of fuel $\dot{m}_f$, and the pressure of combust chamber Pc, namely,

$$F = f(\dot{m}_o, \dot{m}_f, p_c) \tag{7.9}$$

In the start-up progress, the thrust increase with the passing of time. In this instance, the time is included as one of variables of the thrust and Eq. (7.9) is rewritten as

$$F = f(t, \dot{m}_o, \dot{m}_f, p_c) \tag{7.10}$$

Collected with the interval of 0.1 s in the test progress, the data is composed of 13 training samplings and 12 testing samplings and corresponding variation of the thrust is illustrated in Fig. 7.14.

For the sake of calculation, the thrust is supposed to be invariable in a very short interval. Furthermore, the exponential of the thrust is used in the modeling in order to decrease the error, namely,

$$F' = e^F \tag{7.11}$$

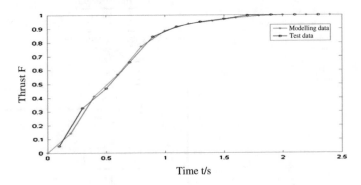

Fig. 7.14 Thrust of the start-up process

The match affinity is calculated by the Euclid distance and is given by

$$d_{ij} = |x_i - x_j| = \sqrt{\sum_{k=1}^{4} (x_{ik} - x_{jk})^2} \tag{7.12}$$

where $x = (t, \dot{m}_o, \dot{m}_f, p_c)$ denotes the input vector of the model. When $d_{ij} \to 0$, the input vectors xi and xj show very high similarities and $F'_i \to F'_j$.

The upper and bottom boundaries of the thrusts are determined at first based on the affinity of the testing data and the training data. Then the linear interpolation was applied to determine the thrust of arbitrary moment.

$$F''_i = F'_{\text{down}} + (F'_{\text{up}} - F'_{\text{down}}) \times i/\text{MAX} \tag{7.13}$$

where $i = 0, 1, 2, \ldots, \text{MAX-1}$, MAX denotes the amount of the interpolation. The bigger the value of MAX, the more the interpolation points in the estimation range. In the test, let MAX = 500.

Based on the linear interpolation, a nonlinear function is used to calculate the thrust. Suppose that the slopes of the linear functions around the upper point are denoted by $k2$ and $k1$, then let

$$\text{Kup} = k2 - k1 \tag{7.14}$$

when Kup > 0, it is indicated that the thrust increases more and more faster and the bigger than the value of Kup, the more apparent the nonlinear variation of the thrust in the start-up progress. Let

$$\Delta_i = \text{Kup} \times (F'_{\text{up}} - F'_{\text{down}}) \times (\sqrt{(t_i - t_{\text{down}}) \times (t_i - t_{\text{up}})}/m\_\text{dDIS})/\text{MAX} \tag{7.15}$$

$$F^*_i = F''_i - \Delta_i \times N \tag{7.16}$$

where $i = 0, 1, 2, \ldots, \text{MAX-1}$, m_dDIS denotes the distance between the upper and bottom boundary, and N denotes the control variable of the offset. When Kup > 0, $\Delta > 0$, the thrust F^*_i is lower than the linear interpolation value of F''_i. Comparatively, when Kup < 0, $\Delta < 0$, F^*_i is bigger than the linear interpolation value of F''_i.

The error is the most important parameter to evaluate the accuracy of the model. In this test, the single point error is given by

$$\varepsilon_n = (F^*_n - F'_n)/F'_n, \quad n = \{1, 2, \ldots, 12\} \tag{7.17}$$

The total error of the test is determined by the mean value of the signal errors:

$$E = \sum_{i=1}^{12} \varepsilon_i / 12 \tag{7.18}$$

2. Results

In the test, all codes were compiled by the software of Visuals C++ 6.0. The settings of the parameters are listed as follows: the amount of the antibody group $C*$ is MAX = 500, the antibody group C is Num = 50, $N = 100$, and the convergence criterion is $d_{ij} < 0.001$. The result is illustrated in Fig. 7.15.

(1) It can be seen that the exponential transform of the trust plays an important role in the decrease of the error. When $t < 1.1$, the single error decreases apparently and the total error decreases from 0.0245 to 0.0056.

(2) Figure 7.16 illustrates the comparative result between the collected data and the simulated result. It can be seen that the thrust calculated by the model is very close to the collected data. The maximum of the single error is located at the initial stage of the start-up progress and the single error decreases with the time passing by. However, at the point of $t = 1.7$ s, the error increases and is not according with the developing trend. Referring to the result obtained by BP ANN methods, as shown in Fig. 7.21, it can be seen that both methods share a similar error abrupt at this point. Such phenomenon indicates that this collected value may be incorrect and is not significant for the evaluation of the modeling.

(3) Figures 7.17 and 7.18 illustrate the influence of Num and N on the modeling. It can be seen that the error of the initial stage (0–1.1 s) is very small and corresponding variation is the smallest when Num = 50 and $N = 100$. Figure 7.19 illustrates the total error varying curve obtained at different values

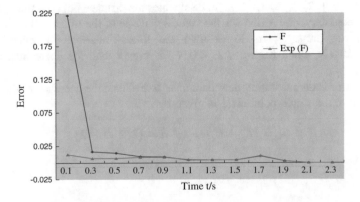

Fig. 7.15 Comparison of the before and after the thrust transform

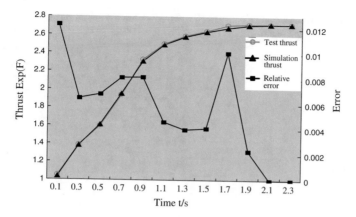

Fig. 7.16 Simulation result and errors when Num = 50, N = 100

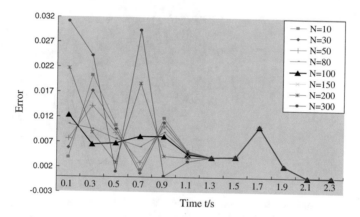

Fig. 7.17 Error variations at different values of Num (N = 50)

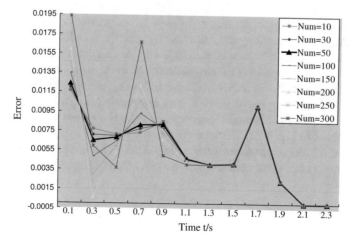

Fig. 7.18 Error variation versus different values of Num (N = 100)

Fig. 7.19 Total error versus
the variation of Num/*N* ration

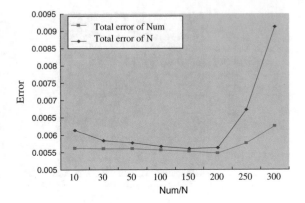

Fig. 7.20 Thrust variation
versus relative error

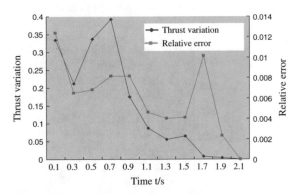

of Num and *N*. It can be seen that the total error is the smallest when
Num = 50 and *N* = 100. It is indicated that Num = 50 and *N* = 100 are the
most suitable for the modeling.

(4) Figure 7.20 illustrates the influence of thrust difference on the error. It can be
seen that the bigger the thrust difference, the greater the error. Therefore,
decreasing the sampling interval may decrease the simulation error.

(5) To verify the performance of the clone selection algorithm, a model based on
BP ANN was used to calculate the thrust of the start-up. With a structure of
4—15—1 net, the BP ANN holds the following settings: the learn efficiency
$\alpha = 0.7$, inertia operator $\eta = 0.9$, and the maximum error is 0.00005. The
training model performed a convergence with 1649 steps and the results are
illustrated in Fig. 7.21. The total error of the BP ANN method is 0.061, while
that of the clone selection algorithm is 0.0056. What is more important is that
the single point error obtained by the clone selection algorithm is smaller than

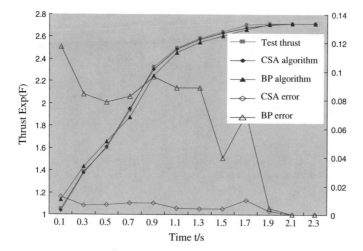

Fig. 7.21 Comparison of the BP algorithm and clone selection algorithm

that obtained by the BP ANN method in the initial stage of the start-up progress. It is indicated that the clone selection algorithm is more powerful than the BP ANN in engineering application.

Chapter 8
Fault Diagnosis Method Based on Fuzzy Theory

Fuzziness is a common phenomenon in the objective world. A large number of facts show that wrong judgment may be caused if the excessive accuracy was demanded, while the fuzzy method may obtain accurate result. Therefore, the fuzzy mathematics came into being.

Fuzzy mathematics has vitality. After decades of development, its practical application is very broad, almost touching the various fields of science and technology, such as artificial intelligence, linguistics, psychology, logic, management, medicine, information science, meteorology, economics science, systems theory, cybernetics, information theory, queuing theory, traffic control, field exploration footprints, fingerprints cluster analysis, and pattern recognition. In recent decades, the fuzzy theory combining with other theoretical methods have produced a series of integrated technology and have good application prospects. For example, the fuzzy pattern recognition, fuzzy wavelet analysis, fuzzy neural network, and fuzzy expert systems have been developed and some of them are hot in today's academic research [127].

8.1 Fuzzy Fault Diagnosis

Different from the nature of randomness, fuzziness is a kind of characteristic of the objective world. The so-called fuzzy refers to the objective things in the form and the properties of the nature is not clear. The root is that a series of transition state exists between the similar things and they penetrate into and communicate with each other, so that there is no clear dividing line between each other. This ambiguity is difficult to describe with the exact scale, the application of the traditional mathematics is also unable to solve, therefore the fuzzy mathematics has emerged as the times require.

© Springer-Verlag Berlin Heidelberg and National Defense Industry Press 2016
W. Zhang, *Failure Characteristics Analysis and Fault Diagnosis
for Liquid Rocket Engines*, DOI 10.1007/978-3-662-49254-3_8

8.1.1   Basic Theory of the Fuzzy

Fuzzy theory is based on the mathematical foundation of fuzzy set theory founded
by Zadeh [127] in 1965 and mainly includes the content of fuzzy set theory, fuzzy
logic, fuzzy reasoning, and fuzzy control. In 1965, Zadeh presented the conception
of membership function to express the fuzzy nature of things [127]. Membership
function, laying the foundation of fuzzy theory, expanded the element membership
degree from the values of 0 and 1 to the random value in [0, 1]. That is to say, we
can quantitatively use the membership degree to describe similarity of elements in
the domain. In this instance, the fuzzy concepts and the continuous transition
characteristics between fuzzy concepts can be described by fuzzy sets and membership function and the absolute membership in the common set has been
expanded.

Definition 8.1 The Fuzzy Sets A in the domain U refers to that any element $x \in U$
corresponds to a value $\mu_A(x) \in [0, 1]$, which was called the membership degree of
x to A. This definition divides a mapping μ_A:

$$\mu_A : U \to [0, 1]$$

$$x \mapsto \mu_A(x),$$

and this mapping is called the function membership of the fuzzy sets A.

It can be seen that the fuzzy set is completely characterized by the membership
function. The value of $\mu_A(x)$ is closer to 1, the membership degree of x to A is
higher; and the value of $\mu_A(x)$ is closer to 0, the membership degree of x to A is
lower. When the range of the membership function is only 0 or 1, the fuzzy set
degenerates to be an ordinary set. Therefore, the general assembly is a special case
of fuzzy sets, and the fuzzy set is a generalization of the ordinary set.

Commonly used membership function includes Gaussian membership function,
bell-shaped membership function, Sigmoid function membership function, and
triangular membership function. The rationality of the membership function usage
directly affects the objectivity of the research object. In the field of fault diagnosis,
the method of expert scoring method, two-element method, fuzzy distribution
membership function, and fuzzy statistical method are commonly used. Expert
scoring method gives a direct scoring based on the experience of experts. The
dualistic contrast compositor method established fuzzy subset membership value
based on the order of priority. The fuzzy distribution membership function is to
specify the distribution function through experience and calculate the corresponding
membership degree value by the given input. The fuzzy statistical method obtains
the membership function by means of fuzzy statistical experiments.

8.1.2 Fault Diagnosis Based on Fuzzy Theory

Fuzzy logic simulates the way of human's logical thinking to deal with the information of fuzzy judgment boundary and is suitable for the qualitative analysis of complex systems. Equipment will be affected in the process of operation by various factors. The relationship between the fault and the symptoms is difficult to be described by the accurate mathematical model, and equipment failure symptoms are vague. In addition, the expression of expert knowledge in fault diagnosis is often fuzzy. Therefore, when compared with the traditional binary logic, the application of fuzzy theory in fault diagnosis is more close to the human language expression and holds greater advantage than the binary one.

Founded on fuzzy pattern matching technology, the fault diagnosis based on fuzzy theory is a high level of intelligence activity. Broadly, there are two methods of fuzzy pattern recognition. Classified according to the "Maximum Degree of Membership principle," the first one is a direct method and mainly used in individual identification. Comparatively, the other one, an indirect method, classifies according to "principle of choosing the nearest is indirect method" and generally used in group pattern recognition.

(1) Maximum degree of membership principle

Set A is a fuzzy subset of the domain U, $u_1, u_2, \ldots, u_n$ are the n objects to be selected, if $\mu_A(u_i) = \max(\mu_A(u_1), \mu_A(u_2), \ldots, \mu_A(u_n))$, then u_i is considered to be a priority of the fuzzy subset A (i.e., the object of the maximum membership degree).

(2) Near rule

Set $A_1, A_2, \ldots, A_n$ are the n fuzzy subset of the domain U, B is also a fuzzy subset of U, if $\sigma(B, A_i) = \max(\sigma(B, A_1), \sigma(B, A_2), \ldots, \sigma(B, A_n))$, then B should belong to the A_i, where σ is a close degree.

8.2 Fault Diagnosis Method Based on Fuzzy Theory

Pattern recognition is a new subject, which belongs to computer science, and has been widely used. Pattern recognition is to judge which category the given object belongs to. The "pattern" refers to the standard sample, style, graphics, symptoms, etc. In daily life, some things hold clear and sure patterns while others do not. For objective things with fuzzy features, people use the machine to simulate human thinking method (i.e., artificial intelligence) to identify and classify the fuzzy objective things [127].

8.2.1   Basic Theory of the Fuzzy Pattern Recognition

Fuzzy pattern recognition is an intelligent activity with higher level based on fuzzy pattern matching technology.

(1) The direct method of fuzzy pattern recognition

Set U is the collection of all the objects to be recognized, and each object in U has p characteristic parameters $u_1, u_2, \ldots, u_p$. A certain feature of the object u is characterized by each characteristic index, so each object u determined by the p characteristic index can be written into $u = (u_1, u_2, \ldots, u_p)$, which is called feature vector. The set U can be divided into n categories, and each category is a fuzzy set of U, written into $A_1, A_2, \ldots A_n$, then they are called fuzzy pattern. The principle of fuzzy pattern recognition is to put the object $u = (u_1, u_2, \ldots, u_p)$ under the control of a similar type of A_i.

When a recognition algorithm acts on the object u, a set of degree of membership is generated. They denote the degree of membership of object u to $A_1, A_2, \ldots A_n$. After the establishment of the fuzzy model of the membership function, the category of object u can be judged according to the principle of membership.

(2) Indirect method of fuzzy pattern recognition

In the direct method of fuzzy pattern recognition, it is clear that the object u to be identified is the single element. But in real life, sometimes the object to be identified is not a single element, but a fuzzy subset of the domain U. At this time, the indirect pattern recognition needs to be used.

Set U is the collection of all the objects to be recognized, of which each object B is a fuzzy subset, and each element in U has p characteristic index $(u_1, u_2, \ldots, u_p)$. The close degree $\sigma(B, A_i)$ needs to be identified when judging which pattern of the known fuzzy pattern $A_1, A_2, \ldots A_n$ the object B belongs to, then judging the object B according to the below principle.

(3) Membership function

In fuzzy mathematics, the membership function, the key to describe the fuzziness, is the foundation of the establishment of fuzzy set theory, and is also the basis of applying fuzzy set theory to practical problems. In the process of pattern recognition, both direct and indirect methods must be carried out in the case of the corresponding degree of membership of each sample.

There are two kinds of different viewpoints and processing methods to obtain the membership function, statistical school and non-statistical school. In practical application, the two methods can be combined. Membership functions are commonly obtained by *fuzzy statistical method, Delphi method, multivariate membership function method, comprehensive weighted method,* and *set method.* In addition, *the model method* is usually used to construct the membership function in practical application. We focus on the construction of two kinds of membership functions:

Table 8.1 Fault simulation data templates

Fault mode	Mode No.	Parameters				
		Pf	Pt	RT	F	$\dot{m}_j$
The main value fault of oxidizer (Mode 1)	1	−0.03167	0.082828	0.022134	0.089583	−0.50626
	2	−0.027767	0.11507	0.02747	0.13003	−0.67914
	3	−0.03287	0.13519	0.035352	0.16569	−0.93825
	4	−0.040916	0.11377	0.038761	0.16887	−1.0445
	5	−0.046473	0.086035	0.038712	0.15695	−1.0471
The main value failure of fuel (Mode 2)	6	−0.55925	−0.24406	0.18893	−0.52203	−0.99625
	7	−0.7962	−0.46441	−0.76329	0.76329	−1.0112
	8	−0.78998	−0.45757	0.55211	−0.76428	−1.0166
	9	−0.78625	−0.44712	0.32316	−0.77006	−1.0247
	10	−0.67641	−0.5335	−0.82756	−0.87323	−1.0123
Fuel injection device blocking fault (Mode 3)	11	0.060397	0.098844	0.0073677	0.16879	−0.2655
	12	0.058449	0.10075	0.0079829	0.16995	−0.2802
	13	0.057042	0.10268	0.008511	0.17147	−0.29333
	14	0.054315	0.10445	0.0092349	0.17181	−0.31055
	15	0.052464	0.10581	0.0097524	0.17224	−0.32284
Combustion chamber throat ablation fault (Mode 4)	16	−0.29493	0.24276	0.0057486	−0.39467	−0.45451
	17	−0.29617	0.24416	0.0058245	−0.39595	−0.45617
	18	−0.29731	0.24546	0.0058952	−0.39714	−0.4577
	19	−0.29837	0.24666	0.0059613	−0.39823	−0.45912
	20	−0.29936	0.24779	0.0060231	−0.39925	−0.46044
Gas generator oxidizer pipeline filter blocking fault (Mode 5)	21	−0.19202	−0.11457	0.021354	−0.22601	−0.25876
	22	−0.22944	−0.13748	0.026884	−0.26737	−0.30609
	23	−0.262	−0.15743	0.032268	−0.30268	−0.34751
	24	−0.29061	−0.17486	0.037536	−0.33307	−0.38429
	25	−0.31594	−0.19007	0.04272	−0.35939	−0.4174
Leak fault at the entrance of gas generator (Mode 6)	26	0.061042	−0.026171	0.0011876	0.10067	0.068499
	27	0.075133	−0.03049	0.0015094	0.12644	0.084483
	28	0.089058	−0.033928	0.0018926	0.15272	0.097904
	29	0.10286	−0.036289	0.0023609	0.17981	0.10777
	30	0.11659	−0.037159	0.0029631	0.20826	0.11214
Oxidizer efficiency loss fault (Mode 7)	31	−0.068385	−0.1202	0.00044623	−0.12291	−0.26855
	32	−0.083019	−0.145	0.00057116	−0.14712	−0.31859
	33	−0.09884	−0.17128	0.00077767	−0.17257	−0.37158
	34	−0.11619	−0.19921	0.001139	−0.19933	−0.42905
	35	−0.13596	−0.22901	0.0018941	−0.22734	−0.49597

model method and *multivariate membership function method*, to identify the simulation data obtained by reference [128]. Fault simulation data are shown in Tables 8.1 and 8.2.

Table 8.2 Simulation data templates of fault measurement

Fault mode	Mode No.	Parameters					Fault mode
		Pf	Pt	RT	F	$\dot{m}_j$	
Measurement templates	1	−0.030355	0.097776	0.024613	0.10767	−0.58057	1
	2	−0.032417	0.12915	0.032209	0.14824	−0.82547	1
	3	−0.03373	0.12744	0.037115	0.17409	−1.0124	1
	4	−0.045381	0.095065	0.038782	0.15986	−1.0435	1
	5	−0.64054	−0.31978	0.26767	−0.61523	−1.0041	2
	6	−0.79303	−0.45674	0.64139	−0.76149	−1.0157	2
	7	−0.76051	−0.46437	−0.14888	−0.79102	−1.031	2
	8	−0.75536	−0.51014	−0.39904	−0.82279	−1.0258	2
	9	0.06018	0.099933	0.0075676	0.17008	−0.27075	3
	10	0.057733	0.10211	0.0083094	0.1712	−0.28858	3
	11	0.056016	0.10385	0.0088496	0.17221	−0.30197	3
	12	0.054049	0.10528	0.009398	0.17267	−0.31496	3
	13	−0.29556	0.24347	0.0057872	−0.39533	−0.45536	4
	14	−0.29675	0.24482	0.0058605	−0.39656	−0.45695	4
	15	−0.29785	0.24607	0.0059288	−0.3977	−0.45842	4
	16	−0.29887	0.24723	0.0059927	−0.39875	−0.45979	4
	17	−0.21141	−0.12643	0.02414	−0.24753	−0.28326	5
	18	−0.24627	−0.1478	0.029592	−0.28571	−0.32745	5
	19	−0.27675	−0.16644	0.034914	−0.31843	−0.36641	5
	20	−0.30365	−0.18272	0.040136	−0.3467	−0.40125	5
	21	0.06811	−0.028433	0.0013418	0.1135	0.076769	6
	22	0.082114	−0.032329	0.0016922	0.1395	0.091563	6
	23	0.09597	−0.035261	0.002114	0.16614	0.10337	6
	24	0.10973	−0.036952	0.0026403	0.19381	0.11084	6
	25	−0.075568	−0.13242	0.00050106	−0.13486	−0.29326	7
	26	−0.090765	−0.15794	0.00066112	−0.15968	−0.34465	7
	27	−0.10729	−0.18503	0.00093117	−0.18578	−0.39959	7
	28	−0.12566	−0.21386	0.0014341	−0.2132	−0.46063	7

8.2.2　Template Method-Based Membership Function Construction and the FPR

Template method-based membership function construction is widely used in engineering. This method, which is easy to understand, easy to calculate, and easy to use, has higher recognition degree of the more concentrated template [129].

1. **Basic process of template method-based membership function construction**

   Select m_i templates from the fuzzy mode A_i.

$$a_{ij} = (a_{ij1}, a_{ij2}, \ldots, a_{ijp},) \quad i = 1, 2, \ldots, n, j = 1, 2, \ldots, m_i$$

where a_{ij} is the characteristic vector of the jth template of the ith fuzzy mode, a_{ijk} is the kth characteristic vector's measurement data of the jth template of the ith fuzzy mode A_i.

Calculate the average value of m_i characteristic vectors $a_{ij}(i = 1, 2, \ldots, n, j = 1, 2, \ldots, m_i)$ of A_i, i.e.,:

$$a_i = (a_{i1}, a_{i2}, \ldots, a_{ip}) \tag{8.2}$$

where $a_{ik} = \frac{1}{m_{i1}} \sum_{j=1}^{m_j} a_{ijk}, k = 1, 2, \ldots, p$, $a_i = (a_{i1}, a_{i2}, \ldots, a_{ip})$ is the mean template of A_i.

Calculate the average distance between object $u = (u_1, u_2, \ldots, u_p)$ and the mean template $a_i = (a_{i1}, a_{i2}, \ldots, a_{ip})$:

$$d_i(u, a_i) = \sqrt{\sum_{j=1}^{p} (u_j - a_{ij})^2}$$
$$= \sqrt{(u_1 - a_{i1})^2 + (u_2 - a_{i2})^2 + \cdots + (u_p - a_{ip})^2} \tag{8.3}$$

Let

$$D_i = \max(d_i(u, a_i)) \quad i = 1, 2, \ldots n, u \in U \tag{8.4}$$

Then the membership function of fuzzy mode A_i is

$$\mu_{Ai}(u) = 1 - \frac{d_i(u, a_i)}{D_i} \tag{8.5}$$

In practical application, the degree of membership of template to each mode is not likely to be 1 or 0, so the formula (8.4) can be written as follows:

$$D_{id} = D_i + \sigma \tag{8.6}$$

where σ is a smaller value. In addition, if $D_{id} = \sum_{i=1}^{n} d_i(u, a_i)$, the similar result can also be achieved.

2. Application examples

As mentioned before, the fuzzy pattern recognition method can be divided into direct method and indirect method. The recognition processes of template method-based membership function construction are as follows:

(1) Direct method of fuzzy pattern recognition

By calculating the degree of membership of the object to be identified directly to determine specific object's belonging. This is called the direct method of fuzzy pattern recognition, of which the effect depends on the accuracy of the fuzzy mode membership function construction.

The mean value template of the simulation data is calculated by the known template. Then, using the membership function and membership principle introduced in Sect. 8.1, the identification of each pattern is obtained.

Fuzzy pattern recognition theory can be used to calculate the degree of membership of a specific fault template to different fault modes (seven kinds of fault modes used), through the principle of membership, we can choose the "most likely" mode as the recognition result. The identification results of direct method are shown in Table 8.3.

From Table 8.3, we know the result of the direct method is effective. The correct results of the 28 identification templates are obtained. But in the verification of known samples, there exists error in the diagnosis of the main valve fault (Mode 1). The degrees of membership of known data in Mode 1 to different modes are shown in Table 8.4.

The result obtained by pattern recognition method shows the most likely template classification. The derivations are evident between all the maximum degrees of membership and the second major subordinate degrees, except for the first set of data, i.e., the occurrence possibility of the second result is very low. Considering

Table 8.3 Results of direct method

Fault mode	Mode 1	Mode 2	Mode 3	Mode 4	Mode 5	Mode 6	Mode 7
Number of known templates	5	5	5	5	5	5	5
Number of known correct recognition templates	4	5	5	5	5	5	5
Number of waited recognition templates	4	4	4	4	4	4	4
Number of correct recognition templates	4	4	4	4	4	4	4

Table 8.4 Degrees of membership of known data in Mode 1 to different modes

Fault mode	Mode 1	Mode 2	Mode 3	Mode 4	Mode 5	Mode 6	Mode 7
1	0.77315	0.13271	0.83776	0.61513	0.64603	0.58409	0.73727
2	0.88951	0.13403	0.73425	0.56726	0.56466	0.4661	0.65002
3	0.93096	0.13507	0.56011	0.46369	0.43329	0.2884	0.50774
4	0.86131	0.13632	0.48404	0.4102	0.37928	0.21205	0.45027
5	0.85735	0.1385	0.4733	0.40275	0.37843	0.19945	0.45023

the reference value of the secondary maximum degree of membership, we know that the membership function constructed by the template method can accomplish the recognition of the first set of data remarkably.

The liquid rocket engine costs much to produce, and requires high accuracy on the fault recognition. The recognition rate (98.41 %) can meet the requirements basically.

In the computer environment of Matlab 6.5, 1.7 GHz Celeron, 256 M RAM, the direct recognition will last 0.0156 s, and the entire training recognition process lasts 0.5313 s.

(2) Indirect method of fuzzy pattern recognition

When the object u is a fuzzy subset, we use the indirect method to deal with the data. The simulated fault data, which are determined, must be fuzzed. There are many fuzzization methods, such as the maximum and minimum values, calibration data, etc. The solution of the membership is actually a data fuzzization process. Here, the indirect method and direct method use different rules to recognize the pattern. The former adopts the principle of selecting near, or the maximum degree of proximity, while the latter uses the maximum membership principle.

Fuzzify the object before using the indirect method, then select near principle to recognize the fault templates. In the recognition process, Euclid close degree is used.

$$N(A, B) \triangleq 1 - \frac{1}{\sqrt{n}} \left(\sum_{i=1}^{n} (A(u_i) - B(u_i))^2 \right)^{1/2} \tag{8.7}$$

Calculate the proximity of the template to the known template. The recognition results are shown in Table 8.5.

Some other close degree (such as Hamming approach degree) is also used for recognizing and calculating, obtaining the similar result. It is obvious that the template-based membership function can be used to identify the fault templates correctly.

In the computer environment of Matlab 6.5, 1.7 GHz Celeron, 256M RAM, the indirect recognition lasted 0.0156 s. This time is the same as that of the direct identification; but the whole program runs for 0.5781 s, and the calculation time is more than that of the direct method. The two methods equally matched in speed.

Table 8.5 Euclid close degree included indirect recognition result

Fault mode	Mode 1	Mode 2	Mode 3	Mode 4	Mode 5	Mode 6	Mode 7
Number of templates to be identified	4	4	4	4	4	4	4
Number of correctly identified templates	4	4	4	4	4	4	4

Both the direct method and the indirect method can recognize the template correctly, showing that the template method-based membership function construction is preferable. The high accuracy and the speediness in the recognition process meet the real-time requirements of the rocket motor fault diagnosis.

8.2.3 Multi-variable Membership Function and the FPR

The multivariate membership function method, also known as the m classification method, is a model of the m fuzzy statistical test. The random sampling is used to construct the method.

1. **The construction theory of multivariate membership function**

Set the mapping of domain U to P_m (fuzzy set is defined as before)

$$
f(u) =
\begin{cases}
1, & \text{when } u \text{ belongs to } A_1 \\
1/2 & \text{when } u \text{ belongs to } A_2 \\
\cdots \\
1/m & \text{when } u \text{ belongs to } A_m
\end{cases}
\tag{8.8}
$$

Let

$$
Y_{n\times1} \triangleq
\begin{pmatrix}
1 \\
\vdots \\
1 \\
1/2 \\
\vdots \\
1/2 \\
\vdots \\
1/m \\
\vdots \\
1/m
\end{pmatrix},
\quad
\beta_{(p+1)\times1} \triangleq
\begin{pmatrix}
\beta_0 \\
\beta_1 \\
\vdots \\
\beta_p
\end{pmatrix}
\tag{8.9}
$$

In $Y_{n\times1}$, the number of each quantity is equal to the number of templates (n_i) in the fuzzy set, $n = \sum_{i=1}^{m} n_i$.

$$
X_{n\times(p+1)} \triangleq
\begin{pmatrix}
1 & x_{11} & \cdots & x_{1p} \\
1 & x_{21} & \cdots & x_{2p} \\
1 & x_{31} & \cdots & x_{3p} \\
1 & x_{41} & \cdots & x_{4p}
\end{pmatrix},
\quad
\varepsilon_{n\times1} \triangleq
\begin{pmatrix}
\varepsilon_1 \\
\varepsilon_2 \\
\vdots \\
\varepsilon_n
\end{pmatrix}
\tag{8.10}
$$

The expression of mapping f can be obtained using the liner model below,

$$y_i = \beta_0 + \beta_1 x_{i1} + b_2 x_{i2} + \cdots + b_p x_{ip} + \varepsilon_i \quad (8.11)$$

or

$$Y = X\beta + \varepsilon \quad (8.12)$$

where ε_i is normal random variable, and its mathematic expectation and variance are $E(\varepsilon_i) = 0, D(\varepsilon_i) = \sigma_2 (i = 1, 2, \ldots, n)$ respectively, σ is a constant.

In the above liner model, when

$$\text{rank}(X'X) = p + 1 \quad (8.13)$$

The least square estimation of β can be obtained

$$\overline{\beta} = (X'X)^{-1} X'Y = (\overline{\beta_0}, \overline{\beta_1}, \ldots, \overline{\beta_P})^T$$

When $\text{rank}(X'X) = p + 1$, $(X'X)^{-1}$ can be instead of $(X'X)^+$,

$$\overline{\beta} = (X'X)^+ X'Y = (\overline{\beta_0}, \overline{\beta_1}, \ldots, \overline{\beta_P})^T \quad (8.14)$$

We can refer to two shapes, standard normal probability distribution function and logistic function, to construct the multivariate membership function.

$$\mu_A(u) = \mu_A(x_1, x_2, \ldots, x_p) = \frac{1}{\sqrt{2\pi}} \int_{-\infty}^{a} \left(\overline{\beta_0} + \sum_{i=1}^{p} \overline{\beta_i} x_i \right) \exp(-t^2/2) dt \quad (8.15)$$

$$\mu_A(u) = \mu_A(x_1, x_2, \ldots, x_p) = \frac{a}{1 + \exp(\overline{\beta_0} + \sum_{i=1}^{p} \overline{\beta_i} x_i)} \quad (8.16)$$

where a is constant, which can be obtained according to specialized knowledge, experience, and physical truth.

Definitively, choose $m - i$ thresholds $\lambda i \in [0, 1]$, $\lambda_i > \lambda_i + 1 (i = 1, 2, \ldots, m - 1)$, cut the fuzzy subset A into m common sets, i.e.

$$A_1^* = \{u | \mu R(u) \geq \lambda_1\}$$
$$A_2^* = \{u | \lambda_2 \leq \mu R(u) \leq \lambda_1\}$$
$$A_m^* = \{u | \mu R(u) \leq \lambda_{m-1}\}$$

Then U can be divided into according to the multivariate membership function value $\mu A(u)$.

(1) Application examples

Recognize the pattern of fault templates in Tables 8.1 and 8.2. The known templates and the templates to be recognized will be put into the model class rather than distinguished from each other.

Here mapping f

$$f(u) = \begin{cases} -1 & \text{when } u \text{ belongs to } A_1(\text{Model}) \\ -0.6 & \text{when } u \text{ belongs to } A_2(\text{Model}) \\ -0.3 & \text{when } u \text{ belongs to } A_3(\text{Model}) \\ 0 & \text{when } u \text{ belongs to } A_4(\text{Model}) \\ 0.3 & \text{when } u \text{ belongs to } A_5(\text{Model}) \\ 0.6 & \text{when } u \text{ belongs to } A_6(\text{Model}) \\ 1 & \text{when } u \text{ belongs to } A_7(\text{Model}) \end{cases} \tag{8.17}$$

$$Y_{63} \times 1 = (-1,-1,-1,-1,-1,\ -1,-1,-1,-1,-0.6,-0.6,-0.6,-0.6,-0.6\ -0.6,-0.6,-0.6,$$
$$-0.6,-0.3,-0.3,-0.3,-0.3,-0.3,\ -0.3,-0.3,-0.3,-0.3,0,0,0,0,0,\ 0,0,0,0,0.3,0.3,$$
$$0.3,0.3,0.3,0.3,0.3,0.3,0.3,0.6,0.6,0.6,0.6,0.6,0.6,0.6,0.6,0.6,1,1,1,1,1,1,1,1,1,]^T$$

$X_{63} \times 6$ can be obtained by substituting templates into $X_n \times (p + 1)$. rank(X' X) = 6 = $p + 1$, then, β = $(0.48122, 1.8735, -2.0825, -1.685, 1.39)^T$ can be obtained by formula (8.17). $a = 5$, the degree of membership can be obtained by the logistic function, of which the result is shown in Table 8.6.

Take $\lambda_1 = 0.88$, $\lambda_2 = 0.8$, $\lambda_3 = 0.5$, $\lambda_4 = 0.4$, $\lambda_5 = 0.3$, $\lambda_6 = 0.1$, then the identification set will be divided into seven parts. The identification results are shown in Table 8.7.

The correct rate of judgment can be calculated

$$\frac{9 \times 6 + 8}{63} = 98.43\,\%$$

From the discrimination results, we know that Mode 2 has a miscarriage of justice. The rest modes can be correctly judged.

Using the standard normal probability distribution function (8.16) could obtain the similar recognition effect.

In the computer environment of Matlab 7, 1.7 GHz Celeron, 256M RAM, the identification lasted 0.5313 s, the whole program run for 0.9844 s, with more time than that of the template method.

The FPR results of multivariate membership function show that the method has better recognition effect, recognition rate of 98.41 %. The recognition accuracy can meet the requirements of liquid rocket engine fault diagnosis. However, the process of constructing the membership function is complicated, which determines that it is time-consuming and can not meet the requirements of the real-time performance in the liquid missile engine fault diagnosis process.

Table 8.6 FPR results on multivariate membership function

Category	Templates serial number	Degree of membership $\mu A(u)$	Category	Templates serial number	Degree of membership $\mu A(u)$	Category	Templates serial number	Degree of membership $\mu A(u)$
Mode 1	1	0.88828	Mode 4	22	0.77005	Mode 6	43	0.16677
	2	0.95968		23	0.7711		44	0.30388
	3	0.98988		24	0.767		45	0.30889
	4	0.99288		25	0.7683	Mode 5	46	0.18255
	5	0.9914		26	0.76949		47	0.15505
	6	0.92718		27	0.77058		48	0.13388
	7	0.98149	Mode 7	28	0.41935		49	0.11745
	8	0.99271		29	0.42543		50	0.10465
	9	0.99174		30	0.43075		51	0.16788
Mode 3	10	0.84155		31	0.43619		52	0.14378
	11	0.85039		32	0.44567		53	0.12515
	12	0.8584		33	0.4225		54	0.11065
	13	0.86722		34	0.42816	Mode 2	55	0.008797
	14	0.87334		35	0.43334		56	0.0094238
	15	0.84548		36	0.43983		57	0.009981
	16	0.85581	Mode 6	37	0.31346		58	0.015678
	17	0.86337		38	0.30683		59	0.065222
	18	0.86998		39	0.30374		60	0.0094251
Mode 4	19	0.76632		40	0.30547		61	0.010619
	20	0.76766		41	0.31478		62	0.0080075
	21	0.7689		42	0.30976		63	0.008797

Table 8.7 Analysis of recognition results

Judgment	Former							
	Mode 1	Mode 3	Mode 4	Mode 7	Mode 6	Mode 5	Mode 2	Total
Mode 1	9	0	0	0	0	0	0	9
Mode 3	0	9	0	0	0	0	0	9
Mode 4	0	0	9	0	0	0	0	9
Mode 7	0	0	0	9	0	0	0	9
Mode 6	0	0	0	0	9	0	0	9
Mode 5	0	0	0	0	0	9	1	9
Mode 2	0	0	0	0	0	0	8	9
Total	9	9	9	9	9	9	8	63

The method is strong skillful, mainly for the expression of mapping f. The expression of mapping f is uncertain defined in range of values and marshaling sequence. In practical application, it can be selected according to need. So, how to select the appropriate value and the corresponding order to achieve a better degree of differentiation is of certainly technical.

8.3 Fault Diagnosis Method Based on Fuzzy Clustering

Clustering is a newly developed mathematical classification method, based on the multivariate analysis method of mathematical statistics. It is a mathematical method to classify things according to the characteristics, the degree of closeness and the similarity relation. Fuzzy clustering methods can be broadly divided into two categories: one is fuzzy equivalence matrix dynamic clustering method; the other is fuzzy ISODATA (Iterative Self-Organizing Data Analysis Technique A) clustering method (also known as fuzzy c-means clustering).

The most commonly used fuzzy clustering method is the transitive closure (i.e., fuzzy equivalence matrix) method, by synthesizing the fuzzy similar matrix *max-min* ($\vee$-$\wedge$) times to obtain the transitive closure and using λ-cut matrix to get clustering graph and clustering results. This method is widely applied, with the mature theory. But the operator $\wedge$ is so strong that fuzzy relation value changed too much when max-min synthesized and lots of information may lose. The clustering results are not accurate enough, even wrong sometimes. Bellman and Ciertz extended the $\wedge$ operation, and promised several t operators such as ∞, $\odot$, ε and $\circledR$, $\infty \leq \odot \leq_{\varepsilon} \leq \circledR \leq \wedge$. So, in order to get a better clustering effect, some other operators were also used. $\wedge$ operator is most commonly used [130].

Respectively, two kinds of fuzzy clustering technique were used to judge the liquid rocket engine fault diagnosis templates, and their value of application on liquid rocket engine fault diagnosis was analyzed. Considering that the max-min ($\vee$-$\wedge$) operator was easy to lose information, the max-$\odot$ based clustering was also used to cluster the template, in order to achieve higher diagnostic accuracy.

8.3.1 Dynamic Clustering Method Based on the Fuzzy Equivalence Matrix

Dynamic clustering method based on the fuzzy equivalence matrix can be divided into the transitive closure of fuzzy clustering, fuzzy clustering method, fuzzy clustering netting method, and fuzzy clustering maximal tree method, etc., according to the specific classification method.

1. **The process of dynamic clustering method-based on the fuzzy equivalence matrix**

Set the object to be classified, X, has n templates, $X = \{x_1, x_2, \ldots, x_n\}$. Each template x_i has m characteristics, i.e., the template x_i can be expressed as the characteristic index vector

$$x_i = (x_{i1}, x_{i2}, \ldots, x_{im})$$

where x_{ij} is the jth index of the ith template.

First, solve the fuzzy similarity matrix R by the data X. If the orders of magnitudes and dimensions between the m characteristic indexes of X are different, the big order of magnitude characteristic parameters will be highlighted on the classification, while the small one will be reduced or even eliminated, resulting in the lack of a unified standard to classify the characteristic index. Therefore, the data involved in the calculation should be unified, i.e., normalization. After normalization, the data has a uniform scale, which can be further compared with the operation.

(1) Data normalization

There are many methods of data normalization, such as data standardization, maximum value, mean value, center specification, and logarithmic specification. The purpose of data normalization is to unify the data with different dimensions and different order of magnitudes to a close range.

(2) Fuzzy similarity relation matrix R construction

For the normalized set X, the similarity relations between different templates can be constructed by multivariate analysis. Calculate the degree of similarity, r_{ij}, between objects to be classified

$$x_i = (x_{i1}, x_{i2}, \ldots, x_{im})$$
$$x_j = (x_{j1}, x_{j2}, \ldots, x_{jm})$$

Make that

$$0 \le r_{ij} \le 1, \quad i,j = 1, 2, \ldots, n$$

So a fuzzy similarity matrix between different templates is obtained

$$R = \begin{pmatrix} r_{11} & r_{12} & \cdots & r_{1n} \\ r_{21} & r_{22} & \cdots & r_{2n} \\ \cdots & \cdots & \cdots & \cdots \\ r_{n1} & r_{n2} & \cdots & r_{nn} \end{pmatrix}$$

It is called calibration to determine the r_{ij}. According to the actual clustering problem, we can choose the similarity coefficient method, the distance method, and the method of close degree method, and so on.

(3) The equivalence transitive closure Rf solution

The equivalent of transitive closure Rf is the smallest transitive relation including R, and it is reflexivity, symmetry, and transitivity.

Definition If $R2m = Rm$, then the $Rf = R2m$, which is called the transitive closure of fuzzy matrix **R**.

The maximum and minimum algorithm of the fuzzy sets rather than the operations of the ordinary set are used to obtain the values of $R2m$ and Rm.

Definition Set $Q \in F(U \times V), R \in F(V \times W)$. The so-called synthesis of the Q to R is a f relation from U to W, denoted by $Q \cdot R$, its membership function is

$$(Q \cdot R)(u, w) = \bigvee_{v \in V} (Q(u, v) \wedge R(v, w))$$

(4) Clustering

Ordinary clustering is taking the λ-cut set of the transitive closure of **R** to obtain the common $\{0, 1\}$ matrix form of Rf, where

$$Rr(i,j) = \begin{cases} 1 & Rf(i,j) \ge \lambda \\ 0 & Rf(i,j) < \lambda \end{cases}$$

The clustering method is taking the appropriate value of λ to obtain Rr. Gather the same corresponding elements in the same column (or row) vector to the same class.

2. **Application**

The no dimensional fault data in Tables 8.1 and 8.2 can be directly calibrated. Angle cosine method was used

Table 8.8 Clustering results at $\lambda = 0.99$

Serial number	Mode 1	Mode 2				Mode 3	Mode 4	Mode 5	Mode 6	Mode 7
Known templates	1–5	6, 7, 9	8		10	11–15	16–20	21–25	26–30	31–35
Templates to be recognized	1–4	5, 6, 7		8		9–12	13–16	17–20	21–24	25–28

$$r_{ij} = \frac{\left| \sum_{k=1}^{m} x_{ik} x_{jk} \right|}{\sqrt{\left(\sum_{k=1}^{m} x_{ik}^2 \right) \left(\sum_{k=1}^{m} x_{jk}^2 \right)}} \tag{8.18}$$

Then the fuzzy similarity matrix R can be obtained, and the equivalent transitive closure of fuzzy matrix R is solved by the fuzzy relation Rf.

The classification results are given in Table 8.8 when $\lambda = 0.99$. Table 8.8 indicates that, in addition to the second class (combustion valve fault), the known templates of the rest fault and template identification can be correctly clustered. The correct clustering ratio is 95.24 %, which is quite high. The dynamic clustering algorithm of fuzzy equivalent matrix has good classification effect on the template.

In the computer environment of Matlab 6.5, 1.7 GHz Celeron, 256M RAM, the whole clustering process lasted 4.172 s. This time was relatively long, which did not meet the real-time requirements of the rocket engine.

8.3.2   Clustering Method of Fuzzy ISODATA

Based on common clustering theory, the fuzzy operators are introduced to solve the classification matrix R and V in the classification space Mc. This algorithm is called fuzzy ISODATA algorithm [131, 132].

1. Basic theory of fuzzy ISODATA

Set the objects to be classified as a collection of

$$X = \{x_1, x_2, \ldots, x_n\}$$

Each of these templates has m characteristics

$$x_i = (x_{i1}, x_{i2}, \ldots, x_{im})$$

The template set X is divided into c kinds ($2 \leq c \leq n$), set the c clustering center vectors

$$
V = \begin{pmatrix} V_1 \\ V_2 \\ \vdots \\ V_c \end{pmatrix} = \begin{pmatrix} v_{11} & v_{12} & \cdots & v_{1m} \\ v_{21} & v_{22} & \cdots & v_{2m} \\ \cdots & \cdots & \cdots & \cdots \\ v_{c1} & v_{c2} & \cdots & v_{cm} \end{pmatrix} \tag{8.19}
$$

In order to obtain an optimal fuzzy classification, we must choose the best fuzzy classification from the fuzzy classification space Mfc according to clustering criteria.

In order to get the optimal classification, the clustering criteria is making the objective function

$$
J(\underset{\sim}{R}, V) = \sum_{k=1}^{n} \sum_{i=1}^{c} (r_{ik})^q \|x_k - V_i\|^2 \tag{8.20}
$$

reach minimum. Where V_i denotes the ith clustering center; and $\|x_k - V_i\|$ stands for the distance between x_k and V_i. q is a parameter greater than zero. In order to change the relative degree of membership, the value of q can be taken as a certain value (generally $q = 2$).

The clustering criterion is getting the R and V to achieve the minimum value of the objective function in the Eq. (8.20).

Bezdek has proved that when $q \geq 1$, $x_k \neq V_i$, the iterative operation can be performed according to the following approach, and operation process is convergent. This is called fuzzy ISODATA method. Basic steps are

(1) Make sure c, ($2 \leq c \leq n$), make the initial classification matrix $R(0) \in$ Mfc, stepwise iterated, $l = 0, 1, 2, \ldots$
(2) For the $R(l)$, calculate the clustering center vector

$$
V^{(l)} = (V_1^{(l)}, V_2^{(l)}, \ldots, V_c^{(l)})^T \tag{8.21}
$$

$$
V_i^{(l)} = \sum_{k=1}^{n} (r_{ik}^{(l)})^q x_k / \sum_{k=1}^{n} (r_{ik}^{(l)})^q \tag{8.22}
$$

(3) Modify the classification matrix $R(l)$

$$
r_{ik}^{(l+1)} = \frac{1}{\sum_{j=1}^{c} \left(\dfrac{\|x_k - V_i^{(l)}\|}{\|x_k - V_j^{(l)}\|} \right)^{\frac{2}{q-1}}}, \quad (\forall i, \forall k) \tag{8.23}
$$

(4) Compare $R(l)$ and $R(l + 1)$, if $\varepsilon > 0$

$$\max\left\{\left|r_{ik}^{(l+1)} - r_{ik}^{(l)}\right|\right\} \leq \varepsilon \tag{8.24}$$

The iteration stops until $R(l+1)$ and $V(l)$ are obtained. Otherwise, $l = l + 1$, back to step 2.

Applying the above algorithm to get the fuzzy classification matrix $R(l+1)$ and the clustering center $V(l)$, which is the optimal solution relative to the number of categories c, an initial fuzzy classification matrix $R(0)$, ε, and q.

Because of the algorithm requirement, $x_k \neq V_i$, and the iterative Eqs. (8.22) and (8.23), the initial fuzzy classification matrix $R(0)$ selection must satisfy two conditions of the three of the fuzzy classification matrix.

(1) $r_{ik} \in [0, 1], \forall i, k$;

(2) $\sum_{k=1}^{n} r_{ik} > 0, \forall i$;

   In addition, it is necessary to restrict the selection of $R(0)$ of the initial classification matrix

(3) The initial classification matrix $R(0)$ cannot be a constant matrix for each element;

(4) The initial matrix $R(0)$ cannot be a matrix of a certain row element;

(5) If there exists one template classification in the initial classification matrix $R(0)$, remove it before clustering, and put it in after clustering.

Only the initial fuzzy classification matrix $R(0)$ satisfies with the above five conditions at the same time, it will not cause the distortion in the process of the fuzzy ISODATA calculation, otherwise it will make the clustering fail.

The new template is identified by the following principles:

Decision principle 1 sets the final clustering center vector as

$$V = (V_1^*, V_2^*, \ldots, V_c^*)^T \tag{8.25}$$

$\forall x_k \in X$, if

$$\left\|x_k - V_i^*\right\| = \min_{1 \leq j \leq c}\left(\left\|x_k - V_j^*\right\|\right) \tag{8.26}$$

The template x_k ranks to class i.

The significance of the decision principle is that the new template will be classified into the category of the closest clustering center.

Decision principle 2 sets the fuzzy classification matrix obtained at last as

$$R^* = \begin{pmatrix} r_{11}^* & r_{12}^* & \cdots & r_{1n}^* \\ r_{21}^* & r_{22}^* & \cdots & r_{2n}^* \\ \cdots & \cdots & \cdots & \cdots \\ r_{c1}^* & r_{c2}^* & \cdots & r_{cn}^* \end{pmatrix} \tag{8.27}$$

$\forall x_k \in X$, in the **k**th row of R^*, if

$$r_{ik}^* = \max_{1 \leq j \leq c} \left(r_{jk}^* \right) \tag{8.28}$$

Then the template x_k will be classified into class i.

The significance of the decision principle is that the new template x_k will be classified into the category of the maximum degree of membership.

2. **Verification of clustering effect**

The fuzzy clustering obtained by fuzzy ISODATA is the optimal solution relative to the number of categories c, an initial fuzzy classification matrix $R(0)$, ε, and q. If change c, $R(0)$, ε and q, we can get many local optimal solution. In order to obtain the best clustering, some indexes are needed to identify the effect of fuzzy ISODATA clustering.

The following indicators can be used to identify the clustering effect,

(1) Classification coefficient,

$$F_c R = \frac{1}{n} \sum_{k=1}^{n} \sum_{i=1}^{c} r_{ik}^2 \tag{8.29}$$

When $R \in M_c$, $F_c(R) \to 1$. Therefore, the closer $F_c(R)$ to 1, the smaller the fuzziness of the final classification is, and the better the effect is.

(2) Average fuzzy entropy,

$$H_c R = -\frac{1}{n} \sum_{k=1}^{n} \sum_{i=1}^{c} r_{ik} \ln(R_{ik}) \tag{8.30}$$

The closer the average fuzzy entropy to zero, the better the clustering effect is.

3. **Application**

Using fuzzy ISODATA method to perform fault diagnosis according to the rocket engine fault data in Tables 8.1 and 8.2.

For the known template X, clustering center V and fuzzy clustering matrix R need to be solved. Template to be identified is clustered according to the identification criteria. Change c, $R(0)$, ε, and q to obtain different results. The best clustering result is shown in Table 8.9.

At this time, $c = 7$, $R(0)$ is the matrix of 7×63, $\varepsilon = 0.00001$, $q = 2$. Respectively, the identification indexes are $F_c(R) = 0.79296600685592$ and $H_c(R) = 0.47036$ 582465065. In terms of the classification results, 60 templates can be correctly clustered in 63 fault templates of 7 fault patterns. The correct clustering ratio is 95.24 %. This recognition rate represents that the fuzzy ISODATA clustering method is effective to these data, but both the indexes are not ideal to cluster.

Table 8.9 Clustering by fuzzy ISODATA

Judged	Former							
	Mode 1	Mode 2	Mode 3	Mode 4	Mode 5	Mode 6	Mode 7	Total
Mode 1	7	0	0	0	0	0	0	7
Mode 2	0	9	0	0	0	0	0	9
Mode 3	2	0	9	0	0	0	0	11
Mode 4	0	0	0	9	0	0	0	9
Mode 5	0	0	0	0	8	0	0	8
Mode 6	0	0	0	0	0	9	0	9
Mode 7	0	0	0	0	1	0	9	10
Total	9	9	9	9	9	9	9	63

In the computer environment of Matlab 6.5, 1.7 GHz Celeron, 256M RAM, the clustering process is 2.546 s. This also lasted too long, which did not meet the real-time requirements of rocket engine.

8.3.3 Fuzzy Clustering Based on Max—⊙ Transitivity and Its Application

The operator $\wedge$ is so strong that the max-$\wedge$ synthesis misses too much information. Sometimes the result of clustering is not ideal or even wrong. Because the max—⊙ synthesis is softer, it loses less information. The max—⊙ clustering is more effective.

1. Basic concepts

There are several definitions of clustering algorithm based on max—⊙.

(1) Max—⊙ synthesis method

An initial fuzzy matrix, $\tilde{R}^{(0)} = (r_{ij}^{(0)})$, is given. Let $R^{(n)} = (r_{ij}^n)$, where

$$r_{ij}^{(n)} = \bigvee_k \left\{ \max(0, r_{ij}^{(n-1)} + r_{ij}^{(n)} - 1) \right\}, \quad n = 1, 2, \ldots \qquad (8.31)$$

This process is called the max—⊙ synthesis.

(2) Transitive closure of fuzzy similarity relation $\tilde{R}$

$\tilde{R}^{(0)}$ is a fuzzy similarity matrix, synthesized by max—⊙,

$$I < \tilde{R}^{(0)} < \tilde{R}^{(1)} < \cdots < \tilde{R}^{(n)} = \tilde{R}^{(n+1)} = \cdots$$

where $\tilde{R}^{(n)}$ is the max—⊙ fuzzy equivalence matrix. If n is infinite, $\lim_{n \to \infty} \tilde{R}^{(n)} = \tilde{R}^{(\infty)}$, where $\tilde{R}^{(\infty)}$ is max—⊙ fuzzy equivalence matrix, i.e..,

$I < \tilde{R}^{(0)} < \tilde{R}^{(1)} < \cdots < \tilde{R}^{(n)} < \tilde{R}^{(n+1)} < \cdots < \tilde{R}^{(\infty)}$. The $\tilde{R}^{(n)}$ or $\tilde{R}^{(\infty)}$ is called max—$\odot$ transitive closure of $\tilde{R}$.

(3) λ-cut matrix

$\tilde{R} = (r_{ij})_{n \times n}$ is a fuzzy matrix in $X \times X$, $\lambda \in [0, 1]$, let

$$r_{ij,\lambda} = \begin{cases} 1 & r_{ij} \geq \lambda \\ 0 & r_{ij} < \lambda \end{cases} \quad (i,j = 1, 2, \ldots, n) \tag{8.32}$$

$\tilde{R} = (r_{ij})_{n \times n}$ is called the λ-cut matrix of $\tilde{R}$. This is not necessarily a common equivalent matrix.

2. Fuzzy clustering based on max—$\odot$ transitivity and its application

Different objects which are more similar to each other should be classified in the same group, on the other hand, the sum of the similarity coefficient of objects in the same class should be the maximum, which of objects in different classes are not the maximum. A max—$\odot$ transitivity-based fuzzy clustering based on the above idea is shown as

(1) The initial fuzzy similarity matrix is $\tilde{R}^{(0)} = (r_{ij}^{(0)})_{n \times n}$. The set $S = \{1, 2, \ldots, n\}$ stands for number set of n elements to be classified, where λ is the given threshold, and $0 < \lambda \leq 1$;

(2) Using max—$\odot$ synthesis method to obtain the transitive closure $\tilde{R}^{(n)}$ of $\tilde{R}^{(0)}$;

(3) In $\tilde{R}^{(n)}$, change all the elements in diagonal to 0, and change all the elements less than λ in other positions to 0;

(4) Find l and k in S to make $r_{lk} = \max\{r_{ij} | i < j, i, j \in S\}$. If the maximum value is not the only one, choose one randomly. If $r_{lk} \neq 0$, then classify l and k to a same class $C = \{l, k\}$; If $r_{lk} = 0$, regard each element in S as a class, and export them respectively. End the operation.

(5) Select the u in the S/C to meet $\sum_{i \in C} r_{iu} = \max\left\{\sum_{i \in C} r_{ij} | j \in S \backslash C, \text{ to all } i \in C, r_{ij} \neq 0\right\}$. If more than one j satisfies the condition, choose one as u randomly. Join u to the collection C, i.e., $C = \{l, k, u\}$, then turn to the fifth step. If there is no such u, export the elements in the set;

(6) Make $S = S/C$, then turn to the fourth step.

3. Application examples

Clustering the rocket motor fault data in Tables 8.1 and 8.2 to diagnose the fault of the template. The results are shown in Table 8.10. From the results, we know the fuzzy clustering based on max—$\odot$ transitivity has a high accuracy, and the correct recognition rate is 98.41 %.

In the computer environment of Matlab 6.5, 1.7 GHz Celeron, 256M RAM, the clustering process lasts 3.518 s, which is more than that of fuzzy ISODATA method and not suitable for a relatively high real-time situation such as the rocket motor.

Table 8.10 Result of the fuzzy clustering based on max—⊙ transitivity

Judge	Former							
	Mode 1	Mode 2	Mode 3	Mode 4	Mode 5	Mode 6	Mode 7	Total
Mode 1	8	0	0	0	0	0	0	8
Mode 2	0	9	0	0	0	0	0	9
Mode 3	1	0	9	0	0	0	0	10
Mode 4	0	0	0	9	0	0	0	9
Mode 5	0	0	0	0	9	0	0	8
Mode 6	0	0	0	0	0	9	0	9
Mode 7	0	0	0	0	0	0	9	10
Total	9	9	9	9	9	9	9	63

8.4 Fault Diagnosis Method Based on FNN

When the mechanical equipment operates, the condition from faultless to fault is a gradual process. During this period, it is neither a complete "normal" nor a complete "failure", but in an intermediate state. The traditional fault diagnosis methods, such as fault tree, fault dictionary, are relatively easy to achieve for simple diagnosis objects, but are difficult and less effective for complex diagnosis objects. With the rapid development of the technology of artificial intelligence, especially further application of knowledge engineering, expert system, and artificial neural network in fault diagnosis field, people make more in-depth and systematic research on intelligent diagnosis. Now, the application of fuzzy diagnosis method, the diagnosis method based on neural network, expert diagnosis method or hybrid intelligent diagnosis method for complex object of fault diagnosis has become a trend and research hotspot.

8.4.1 Fuzzy Neural Network

Fuzzy neural network (FNN) is capable of dealing with fuzzy information. It is a network system usually composed of a large number of linked fuzzy or non-fuzzy neurons. It is a product of the combination of fuzzy theory and neural network. It is of the advantages of both neural network and fuzzy theory. It integrates learning, association, recognition, adaptive and fuzzy information processing.

1. Fusion of fuzzy system and artificial neural network

Nowadays, with the in-depth research of fuzzy information processing technology and neural network technology, combing fuzzy technology with neural network technology to construct an automotive neural network or an adaptive fuzzy system to deal with fuzzy information, is attracting the attention of more and more scientists, and also has become a hot research in current.

Table 8.11 Connections and differences between fuzzy system and neural network

	Items	Fuzzy system	Neural network
Differences	Mapping set	Mapping between regional blocks and blocks	Mapping between a couple
	Knowledge storage way	Storing knowledge by principle way	Storing in the middle of the right of learning knowledge distribution
	Link way	Uncertain link structure	Certain link structure
	Structure physic meaning	Clear physical meaning	Unclear mapping meaning between neurons
	Knowledge acquire way	Export offering rules	Weight coefficient obtained by learning
Connections	Not need accurate mathematic model	Both calculate the nonlinear mapping relation between input and output by numerical method	
	Both can realize by hardware	Fuzzy chip and neurons already developed	
	Both can imitate human intelligence behavior	Both solve information according to human language thinking and brain physical	
	Data parallel processing	Both use parallel processing	
	Strong fault-tolerant ability	A deviation of the rules in fuzzy system affects little on result; a little fault between neurons and link in neural network affect little on network function	

Through the analysis of the characteristics of the fuzzy theory and neural network theory, it can be found that from the point of view of the information processing and control, there are many similarities and complementarity between neural network and fuzzy system, and the complementarity between them can be infiltrated and promoted by neural network and the fuzzy inference technique to make new processing architectures and algorithms.

From the difference and connection between the fuzzy system and the neural network shown in Table 8.11, we can see that the couple has common grounds as well as differences with respective advantages. The former, characterized by human thought, has advantage of description of the high-level knowledge, which can deal with fuzzy information processing problems more easily than the conventional mathematical methods through imitating human integrated inference; the latter is based on biological neural network, trying to step forward in simulated perception, cognition, automatic learning and so on, making that the artificial intelligence is more close to the human brain self-organization and parallel processing functions.

According to the form and function of the fuzzy system and neural network connection, the fusion of the two types can be divided into five categories,

Fig. 8.1 Structure equivalent
FNN model

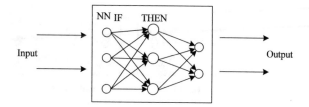

(1) Loose type: In the same system, using fuzzy system to describe the part which can be expressed by "IF-THEN" rule, and using neural network to describe the part which is difficult to be expressed by "IF-THEN" rule. There is no direct link between them.

(2) Parallel type: Fuzzy systems and neural networks are connected in parallel in the system. According to the importance of the two systems, they also can be divided into equal type and subsidy type.

(3) Series connection: The fuzzy system and neural network are connected in series in the system, i.e., the output of one party is the input of the other. This situation can be considered as retreatment part of the series whose former is the input signal of the latter.

(4) Web-based learning: The system is represented by a fuzzy system, but the membership function and fuzzy inference of the fuzzy system are generated and adjusted by neural network.

(5) Structure equivalent combination: The fuzzy system is represented by a neural network equivalent structure. All the nodes and parameters of the neural network have certain significance that corresponds to a fuzzy system of membership function and reasoning process. The combination can best represent the fusion of neural network theory and fuzzy rule, and it is the most widely used fuzzy neural network model, whose structure is shown in Fig. 8.1.

2. Basic theory of fuzzy neural network

Fuzzy neural network is a traditional neural network theory importing fuzzy operator, so that the application of neural network algorithm is extended, and the accuracy of solving problem has been greatly increased. In 1987, B. Kosko took the lead in combining fuzzy theory with neural network, and proposed the concept of fuzzy neural network. In 1993, Jang proposed the concept of fuzzy reasoning based on the structure of the network, and designed the network structure model, which is the embryonic form of the fuzzy neural network. After this, fuzzy theory and application of neural network developed rapidly. A variety of new fuzzy neural network model and learning algorithm accelerated the improvement of fuzzy neural network theory and has been very widely used in practice.

Fuzzy neural network has the ability of learning and optimization, and retains the nature and structure of the neural network. It makes the element fuzzy so that the precise neurons become fuzzy. In domain knowledge, when the fuzzy set is formed, it can enhance the learning ability of the neural network and expand its explanatory

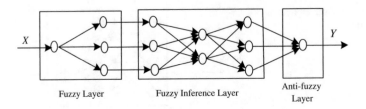

Fig. 8.2 Structure chart of FNN

power. In this case, the network structure does not change while the numerical control parameters connecting the lower level with the top changes. So it has stronger function of expression, faster training speed and is more robust than the conventional neural network.

There are many types of fuzzy neural network, which are similar to the neural network, and are usually divided into two categories, forward type fuzzy neural network and feedback type fuzzy neural network,

(1) Forward type fuzzy neural network: It is a kind of fuzzy neural network, which can obtain the fuzzy mapping relationship. It has a clear hierarchical structure, and the fuzzy inference is easy to be realized by neural network.
(2) Feedback type fuzzy neural network: It can realize the fuzzy associative memory and mapping, also known as fuzzy associative memory. Different from common feedback neural network, feedback type fuzzy neural network uses fuzzy neuron to process information rather than ordinary threshold neurons. It is fuzzy associative and storage has a larger domain of attraction and the ability of fault tolerance than the general association and mapping.

Network topology description, different fuzzy neural network models have different network topology structures, whose structure construction method is also different. By comparing various network structures, we can find that the fuzzy neural network is basically composed of three layers, which can be expressed in the form of unified structure, as shown in Fig. 8.2.

The first layer is the fuzzy layer. It realizes the ambiguity of the input variables, completing a membership function, and calculates the degree of membership of the variable to each fuzzy subspace. It is a necessary part of each kind of fuzzy neural network.

The second layer is the fuzzy inference layer. It is the core part of the network structure, linking fuzzy reasoning premise with conclusion. It can execute network fuzzy relation mappings in simulation, in order to realize the fuzzy pattern recognition, fuzzy inference, and fuzzy associative. Its structure is various, including BP network, RBF networks, or other forms of network; different structures correspond to different algorithms, which is just the reflection of different models. Among them, there are two ways to obtain the fuzzy IF-THEN rule, expert consultation method and data measurement-based study method. Fuzzy IF-THEN rules are acquired based on data measurement.

The third layer is the anti-fuzzy layer. It transforms the basic fuzzy state of the variable distribution type into certain state, which can make sure the output convenient for system excitation. This is necessary in the field of identification and control, because the system sometimes requires deterministic result for the execution of the system. However, in some specific networks, we need not construct the anti-fuzzy layer.

In general, in order to enhance the adaptability of the fuzzy neural network, the fuzzy layer, the fuzzy inference layer, and the anti-fuzzy layer are all made of the multi-layer network. Thus, the membership and fuzzy rule in fuzzy inference model can automatically adjust themselves through the network study.

Fuzzy neural network learning algorithm is usually conventional neural network learning algorithm or its extended form. Common learning algorithms include back propagation learning algorithm, fuzzy back propagation learning algorithm, and the stochastic search algorithms. In recent years, people make deep research on the learning algorithm and obtain a variety of models, including the fuzzy min-max neural network classifier (FMM), fuzzy adaptive resonance theory (F-ART), fuzzy learning vector quantization classifier (FLVQ), fuzzy neural network (FNN), and fuzzy multilayer perceptron (FMLP).

3. **Fault diagnosis based on Fuzzy Neural Network**

In recent years, arbitrary nonlinear function approximation ability of fuzzy neural network has been extensively studied and confirmed. Compared with artificial neural network, fuzzy neural network parameters have certain physical meanings; the choice of initial parameters is of certain basis, which is better than others.

Fuzzy neural network fault diagnosis model is a general diagnosis model based on knowledge. It belongs to the category of artificial intelligence, reflecting the combining diagnosis strategy of problem-solving and inverse problems-solving. It provides the basis for the research of varying initial knowledge and studying method. The irreversible problems in large rotating machinery fault diagnosis is considered in fuzzy neural network diagnosis model, and the fuzzy approach is used for organizational learning templates, so that the fuzzy diagnosis model can give the results by blur estimation, and has fuzzy diagnosis ability.

The fault diagnosis of liquid rocket engine was studied using the fuzzy neural network in the reference [133] of Huang Minchao. Based on the fuzzy hypersphere neural network, this paper puts forward a real-time detection system of a liquid rocket engine fault. Engine test data analysis showed that the fuzzy hypersphere neural network was very sensitive to the input templates. Based on fuzzy min-max neural network, the vibration fault of rocket engine was detected. By aggregating many smaller hypercube fuzzy set, the nonlinear boundary of each kind of engine operating modes was formed to realize the classification, achieving satisfied simulation results. So, the fuzzy neural network is of great application value in the diagnosis of the complex system of liquid rocket engine.

8.4.2 Fuzzy RBF ANN and Its Application in the Fault Diagnosis of LRE

Poor working conditions, easily affected working process, and strong coupling between subcomponents make the failure mechanism of liquid rocket engine unclear and the fault mode hard to express by clear characteristic, i.e., there is no clear separation boundary between failure modes. The engine changes from normal work state to abnormal working state until the final failure. Not only the fault accumulative process exist fuzziness, but also the fault modes, all of which cause great difficulties in fault diagnosis. In addition, the limitation of people's knowing levels makes the observation of the fault symptom unclear, with too much subjection. Therefore, based on the fuzziness of engine failure mode and fault isolation results, the fuzzy RBF neural network-based fault diagnosis method is used to study the fault mode of liquid rocket engine.

1. **Fuzzy RBF neural network structure**

Fuzzy RBF is the structure equivalence type combination of fuzzy inference system and RBF neural network. The RBF neural network nodes and parameters correspond to membership function and reasoning process of the fuzzy inference system. It fully reflects the fusion of the RBF neural network principle and the fuzzy rules.

Radial basis function (RBF) network is a kind of neural network with high quality. It has been proved that RBF network can approximate any nonlinear function with arbitrary precision, and there is no local minimum problem.

In 1993, Jang and Sun proposed fuzzy inference system and RBF network function equivalence theorem in the reference [134]. In 1996, Hunt extends Jang and Sun's equivalence theorem in the reference [135], making equivalence theorem more widely used. The extension theorem is as follows:

(1) The number of the receiving area units is equal to the number of fuzzy "IF-THEN" rules;
(2) The membership function of each rule is chosen to have the same form as the Gauss function;
(3) T-norm operator is the product of the starting strength of each rule;
(4) RBF network and fuzzy inference system are derived by the same method.

Functional equivalence based on fuzzy inference system and RBF neural network introduces fuzzy reasoning theory to RBF network, applying RBF network mode to construct fuzzy neural network, i.e., fuzzy RBF neural network.

The fuzzy RBF neural network is a three-layer network structure, which is composed of input layer, rule layer, and output layer, as shown in Fig. 8.3. The basic feature is the premise and conclusion, and each part contains the adjustable parameters set. Because in reality, the fuzzy rules' consequent are fuzzy variables, if the consequents are different constants, they can be considered as special cases of fuzzy variables, so the fuzzy inference system is used whose consequent are fuzzy variables.

Fig. 8.3 Structure chart of fuzzy RBF neural network

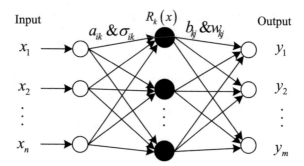

Define the fuzzy domain of input vector $X = \{x_1, x_2, \ldots, x_n\}$ is $U = U_1 \times U_2 \times \cdots \times U_n$; the fuzzy domain of output vector $Y = \{y_1, y_2, \ldots, y_n\}$ is $V = V_1 \times V_2 \times \cdots \times V_n$. The **k**th fuzzy rule owing **S** IF-THEN rules can be described as follows:

$$\text{Rule } k : \text{IF } (x_1 \text{ is } A_{1k}) \text{ and } (x_2 \text{ is } A_{2k}) \text{ and} \ldots \text{and } (x_n \text{ is } A_{nk})$$
$$\text{THEN } (y_1 \text{ is } B_{k1}) \text{ and } (y_2 \text{ is } B_{k2}) \text{ and} \ldots \text{and } (y_m \text{ is } B_{km})$$

where A_{ik} is a fuzzy set on the domain U_i, B_{kj} is a fuzzy set on the domain V_j. The membership of the language variable x_i to fuzzy set A_{ik} is

$$u_{Ai}(x_i) = \exp[-(x_i - a_{ik})^2/\sigma_{ik}^2] \tag{8.33}$$

The membership of the language variable y_i to fuzzy set B_{ik} is

$$u_{Bi}(y_i) = \exp[-(y_i - b_{kj})^2/w_{kj}^2] \tag{8.34}$$

The following operations are performed on each layer of the forward propagation.

(1) The rule layer nodes use the product operation on the known input and the storage condition to obtain the starting strength. The output of the **k**th rule node is

$$R_k(x) = \prod_i u_{A_{ik}}(x_i) \tag{8.35}$$

(2) Perform the gravity anti-fuzziness operation on output layer node. The output of the **j**th rule node is

$$\tilde{y}_j = \frac{\sum_k R_k(x) b_{kj} w_{kj}}{\sum_k R_k w_{kj}} \tag{8.36}$$

2. **Learning algorithm of fuzzy RBF neural network**

At present, the learning strategy of fuzzy RBF neural network can be divided into three categories

(1) Simply considering the accuracy of the approximation of the function. It presets a sufficient number of hidden nodes, initializing each hidden layer nodes' Gaussian function mean by equal intervals way in the input domain and using learning algorithm to adjust the connection weights and node of hidden layer parameters. This method is easy to use. But to achieve certain accuracy, more hidden layer nodes is needed, which is because of the unknown prior exact number. So, more space is needed when initializing the network structure. This method is suiTable for many cases, and has no special requirement for the number of hidden layer nodes.

(2) The hidden layer node number and parameters of the Gauss function are adaptively determined by other methods. If the hidden layer node number is appropriate and initialization parameters are accurate, the parameters of hidden layer nodes need not adjust much, and mainly to learn weights. !!!Because the weight learning is a linear problem, training speed is very fast, but it is difficult to make sure the hidden layer nodes and parameters. Until now, there is no effective method.

(3) Variable structure learning strategy: The idea is to gradually add hidden layer nodes in the training process until meeting the accuracy index of error. The initialization number of hidden layer nodes is 1, then the mean square error of the sample set is calculated. Theoretically, this method can guarantee to find the minimum number of hidden layer nodes in the case of satisfying certain error.

The learning strategy of variable structure is based on the increase of the number of hidden layer nodes, thus the network structure is determined. Its learning algorithm is divided into two stages, learning each template respectively and sequentially at the first stage to determine the network hidden layer nodes and the parameters of network at this stage; the second stage is using the full template to enhance the study of network to eliminate the effect of template order. In the two stages, the steepest gradient descent method of error back propagation is used.

(1) Parameter correction formula of the error back propagation

N training templates is $\{Xp, Yp\}$, the template serial number $p = 1, 2, \ldots,$ N. $X^p = \{x_1^p, \ldots, x_m^p\}$, $Y^p = \{y_1^p, \ldots, y_n^p\}$. If the actual output of the pth training template is $\tilde{y}_j^p$ ($j = 1, 2, \ldots, n$), then the error objective function of the pth training template is

$$E^p = \frac{1}{2} \sum_{j=1}^{n} (y_j^p - \tilde{y}_j^p)^2 \qquad (8.37)$$

Error objective function of all training templates is

$$E = \sum_{p=1}^{N} E^p \tag{8.38}$$

The network training should refer to the steepest descent algorithm. Learning step size should be adjusted dynamically according to the gradient absolute value to improve the convergence rate. The calculation formula of the learning step size and the formula of the parameters are as follows:

① The iterative formula of the parameters of each training template in turn

When the **p**th template is trained. The corresponding dynamic parameters learning step vector can be calculated according to the gradient decline method

$$\eta(t) = \{\eta_a, \eta_\sigma, \eta_b, \eta_w\}^T = \alpha(E^p(t))^{1/r}\left\{e^{-\left|\frac{\partial E^p}{\partial a_{kj}}\right|}, e^{-\left|\frac{\partial E^p}{\partial \sigma_{kj}}\right|}, e^{-\left|\frac{\partial E^p}{\partial b_{kj}}\right|}, e^{-\left|\frac{\partial E^p}{\partial w_{kj}}\right|}\right\}^T \tag{8.39}$$

where the constants $\alpha, r > 0$.

Parameters correction formula is as follows:

$$a_{ki}(t+1) = a_{ki}(t) - \eta_a(t)\frac{\partial E^p}{\partial a_{ki}} \tag{8.40}$$

$$\sigma_{ki}(t+1) = \sigma_{ki}(t) - \eta_\sigma(t)\frac{\partial E^p}{\partial \sigma_{ki}} \tag{8.41}$$

$$b_{kj}(t+1) = b_{kj}(t) - \eta_b(t)\frac{\partial E^p}{\partial b_{kj}} \tag{8.42}$$

$$w_{kj}(t+1) = w_{kj}(t) - \eta_w(t)\frac{\partial E^p}{\partial w_{kj}} \tag{8.43}$$

The gradient calculation formula of the above is as follows:

$$\frac{\partial E^p}{\partial a_{ki}} = \sum_{j=1}^{n} \frac{2(y_j^p - \tilde{y}_j^p)(b_{kj} - \tilde{y}_j^p)(x_i^p - a_{ki})w_{kj}R_k}{\sigma_{ki}^2 \sum_{k=1}^{K} w_{kj}R_k} \tag{8.44}$$

$$\frac{\partial E^p}{\partial \sigma_{ki}} = \sum_{j=1}^{n} \frac{2(y_j^p - \tilde{y}_j^p)(b_{kj} - \tilde{y}_j^p)(x_i^p - a_{ki})w_{kj}R_k}{\sigma_{ki}^3 \sum_{k=1}^{K} w_{kj}R_k} \tag{8.45}$$

$$\frac{\partial E^p}{\partial b_{kj}} = \sum_{j=1}^{n} \frac{(\tilde{y}_j^p - y_j^p) w_{kj} R_k}{\sum_{k=1}^{K} w_{kj} R_k} \tag{8.46}$$

$$\frac{\partial E^p}{\partial w_{kj}} = \sum_{j=1}^{n} \frac{(\tilde{y}_j^p - y_j^p)(b_{kj} - \tilde{y}_j^p) R_k}{\sum_{k=1}^{K} w_{kj} R_k} \tag{8.47}$$

② An iterative formula for the parameters of network training for all templates

According to the gradient descent method, parameters vector corresponding to a dynamic learning step can be calculated as follows:

$$\eta(t) = \{\eta_a, \eta_\sigma, \eta_b, \eta_w\}^T = \alpha (E^p(t))^{1/r} \left\{ e^{-\left|\frac{\partial E^p}{\partial a_{kj}}\right|}, e^{-\left|\frac{\partial E^p}{\partial \sigma_{kj}}\right|}, e^{-\left|\frac{\partial E^p}{\partial b_{kj}}\right|}, e^{-\left|\frac{\partial E^p}{\partial w_{kj}}\right|} \right\}^T \tag{8.48}$$

where the constants α, $r > 0$.

Parameters correction formula is as follows:

$$a_{ki}(t+1) = a_{ki}(t) - \eta_a(t) \sum_{p=1}^{N} \frac{\partial E^p}{\partial a_{ki}} \tag{8.49}$$

$$\sigma_{ki}(t+1) = \sigma_{ki}(t) - \eta_\sigma(t) \sum_{p=1}^{N} \frac{\partial E^p}{\partial \sigma_{ki}} \tag{8.50}$$

$$b_{kj}(t+1) = \sigma_{kj}(t) - \eta_b(t) \sum_{p=1}^{N} \frac{\partial E^p}{\partial b_{kj}} \tag{8.51}$$

$$w_{kj}(t+1) = w_{kj}(t) - \eta_w(t) \sum_{p=1}^{N} \frac{\partial E^p}{\partial w_{kj}} \tag{8.52}$$

(2) Improved hierarchical self-organizing learning algorithm

One of the important factors that restrict the development of neural network is the lack of theoretical guidance. In general, the network structure determination has a great relationship with the distribution of the template set, and it is closely related to the training speed and accuracy. The distributed template is not conducive to the network learning; too much hidden layer neurons will reduce the iteration speed, and too little will not reach the error accuracy. Therefore, a new hierarchical self-organizing learning algorithm is adopted to determine the network structure according to the distribution characteristics of the template.

Fuzzy system-based radial basis function neural network learning process is an algorithm composing of determined minimum knowledge rules (the number of nodes in hidden layer) and adjusting hidden layer parameter vector, making the network estimate the unknown rules from the template data. Whether the network will produce new radial nodes is determined by the size of the effective radius. The effective radius of the kth hidden layer node can be expressed by a super ball domain $R(r_k)$

$$R(r_k) = \left\{ x \middle| d(x, a, \sigma) = \sum_{i=1}^{m} \frac{(x_i - a_{ik})^2}{\sigma_{ik}^2} \leq r_k^2 \right\} \tag{8.53}$$

In order to take into account all the learning samples and accelerate the convergence speed, the standard deviation initial value σ_{init}, w_{init} of the addition hidden layer nodes are the standard deviation of various components in overall learning templates. The effective radius r_k of the super ball domain should be changed according to the characteristics of the system. Too small or too big are not conducive to network learning, and cannot reflect the distribution of the template set.

The steps of the improved hierarchical self-organizing learning algorithm are as follows:

Step 1   Set $k = 1$, $p = 1$ (k stands for the number of hidden layer nodes, p stands for the pth training template). The mean vector a_{ki} is the first template input, b_{kj} is the first template output, the standard vector deviation $\sigma_{ki} = \sigma_{init}$, $w_{ki} = w_{init}$;

Step 2   Calculate the output error of the pth training template;

Step 3   If $Ep > $ ERROR1, (ERROR1 is the error bound), then turn to Step 4, otherwise turn to Step 6;

Step 4   If the pth template falls into the super sphere R of a hidden layer node, then turn to the Step 5, otherwise, produce the new hidden layer nodes, $k = k + 1$, the parameters of the new hidden layer nodes are determined by the following method,

Set $k = k + 1$; the mean vector a_{ki} is the training template input, b_{kj} is the training template output; standard deviation vector $\sigma_{ki} = \sigma_{init}$, $w_{ki} = w_{init}$, turn to Step 2.

Step 5   Applying correction method of Eqs. (8.7)–(8.11) to adjust the correction parameter vector, until $Ep \leq $ ERROR1;

Step 6   Waiting for the new input template, if the template has been input, then turn to Step 7. Otherwise, $p = p + 1$, and turn to Step 2;

Step 7   Reenter all the templates;

Step 8   Calculate error $E = \sum_{p=1}^{N} E^p$;

Step 9   If $E > $ ERROR2 (ERROR2 is the error bound), turn to Step 10, otherwise turn to Step 11;

Step 10   Applying the correction method of Eqs. (8.16)–(8.20) to adjust the vector, until $E \leq \text{ERROR2}$;
Step 11   The end.

When the above algorithm respectively learns for each single sample, the number of the hidden rule node must keep certain to determine the network size. But the memory knowledge of network prefers recent template knowledge to the prior learning template. So if the obtained network parameters are the initial values, we can use all samples to enhance learning, which can improve the generalization ability of the network.

3. Liquid rocket engine fault diagnosis based on fuzzy RBF neural network

In the process of monitoring and diagnosis of equipment running status, monitoring and diagnosis are complementary to each other. The purpose of monitoring is to detect the fault early, and the purpose of diagnosis is to find out the cause of the failure. From this point of view, there is no essential difference between the classification and monitoring, diagnosis. According to classification results, we can determine whether the running state of the equipment changes through current running status of equipment operation categories; in addition, we can find the cause of the fault diagnosis. Therefore, the importance of correct classification is self-evident.

Currently in monitoring and diagnosis process, people are exploring appropriate methods of fault classification. From the traditional pattern recognition methods such as k-means, clustering analysis to fuzzy classification expert system classification were carried out with a lot of research, but these classification techniques are run according to the prior to the design of the case, it is difficult to achieve dynamic modification, with great fragility. And in the process of fault diagnosis, the classification system in order to meet the complexity of the field operation should be easy to modify, and should have greater adaptation range. You can add new content and modify inappropriate characteristics. Therefore, the neural network for fault classification is attracting more and more attention.

In addition, fault diagnosis of equipment condition can be determined by the relationship research between faults and symptoms. As there is a fuzzy relation between fault and symptom, the fuzzy logic should be used to study fuzzy diagnosis. Therefore, fuzzy RBF neural network is used for fault diagnosis, which can combine the advantages of fuzzy logic and neural network, and give full play to the advantages of the two in fault diagnosis.

In the application of neural network to identify the failure mode of a liquid rocket engine, an important problem is the acquisition of fault data. There are mainly two ways to solve this problem, one is a ground test and flight test in the recorded fault data; the other is constructing the fault model of the engine and reaching closely to the fault data of actual process through the simulation of the model. Due to the lack of the number of tests about the failure and fault reported not usually containing all possible failure modes, so it is the most economical method to obtain the failure data by establishing accurate and reliable fault model, and is also an important link of fault analysis and the main means to analyze engine fault mode and effect. Therefore, using the fault template data obtained from the engine fault simulation model as the network training learning template, we can study the application of Fuzzy RBF neural network in the fault diagnosis of liquid rocket engine.

A fuzzy RBF neural network model is established based on the fault sample data of reference [128]. The network input layer contains five neurons corresponding to the selected five monitoring parameters, respectively, fuel pump power Pf, turbine power Pt, thrust F, the calorific value of the gas turbine RT, and the sonic nozzle flow $\dot{m}_j$. The number of hidden layer nodes is network rule number, which is initialized to 1. Output layer contains five neurons corresponding to the five kinds of degree of the membership of failure mode, respectively, the fuel injector blocking fault (Mode 1), combustion chamber throat erosion fault (Mode 2), gas generator oxidizing agent pipeline filter blocking fault (Mode 3), gas generator fuel inlet leak fault (Mode 4), and oxidant efficiency loss fault (Mode 5). If a failure occurs, the target output corresponding to the output unit is 1, otherwise it is 0.

The simulation data in Table 8.10 is used as the fault training template data (20 sets). Because the influence of interference factors exist in accessing the different failure patterns, which have direct relationship with the sensor performance, test environment, immature software, and so on, so adding 5 % random noise to the simulation template data in Table 8.11 to simulate the situation, and as to be a fault test template. The fuzzy RBF neural network model was trained by the learning algorithm. After the completion of the training, save the center, width and network rule number of the input and output membership functions (the number of nodes in hidden layer). The number of the network rules obtained is 5. The training error curve is shown in Fig. 8.4. The fuzzy neural network training speed is very fast. When the training steps are 137, the global error accuracy of the fuzzy RBF neural network is 0.005. Then recognize fault pattern of the 20 groups of test templates

Fig. 8.4 Network training error curve

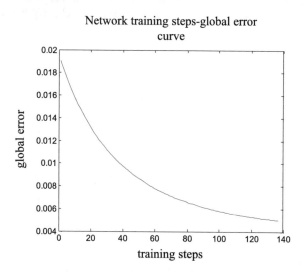

Network training steps-global error curve

Table 8.12 Diagnosis results of test templates

Template serial number	Fault mode	Output results of fuzzy RBF neural network	Diagnosis results
1	Mode 1	(0.9911, 0.0000, −0.0000, 0.0087, 0.0002)	Mode 1
2	Mode 1	(0.9941, 0.0000, 0.0000, 0.0049, 0.0009)	Mode 1
3	Mode 1	(0.9683, 0.0000, 0.0000, 0.0311, 0.0006)	Mode 1
4	Mode 1	(0.9899, 0.0005, 0.0012, 0.0026, 0.0061)	Mode 1
5	Mode 2	(0.0000, 1.0000, 0.0000, 0.0000, 0.0000)	Mode 2
6	Mode 2	(0.0000, 0.9999, 0.0000, 0.0000, 0.0001)	Mode 2
7	Mode 2	(0.0000, 0.9997, 0.0002, 0.0000, 0.0001)	Mode 2
8	Mode 2	(0.0000, 0.9998, 0.0001, 0.0000, 0.0001)	Mode 2
9	Mode 3	(−0.0000, 0.0002, 0.9952, 0.0001, 0.0044)	Mode 3
10	Mode 3	(−0.0001, 0.0001, 0.9990, 0.0001, 0.0009)	Mode 3
11	Mode 3	(−0.0001, 0.0000, 0.9991, 0.0001, 0.0008)	Mode 3
12	Mode 3	(−0.0001, 0.0000, 0.9992, 0.0001, 0.0008)	Mode 3
13	Mode 4	(0.0137, 0.0000, −0.0000, 0.9850, 0.0013)	Mode 4
14	Mode 4	(0.0317, 0.0000, −0.0000, 0.9673, 0.0011)	Mode 4
15	Mode 4	(0.0293, 0.0000, −0.0000, 0.9705, 0.0003)	Mode 4
16	Mode 4	(0.0019, 0.0000, −0.0000, 0.9980, 0.0001)	Mode 4
17	Mode 5	(0.0089, 0.0002, 0.0263, 0.0164, 0.9481)	Mode 5
18	Mode 5	(0.0101, 0.0002, 0.0584, 0.0028, 0.9271)	Mode 5
19	Mode 5	(−0.0015, 0.0001, −0.0005, 0.0020, 0.9997)	Mode 5
20	Mode 5	(−0.0020, 0.0000, 0.0004, 0.0017, 0.9997)	Mode 5

according to the trained network model. The pattern was recognized by maximum degree of membership principle. The results of the fault diagnosis are shown in Table 8.12. From Table 8.12, we can see that the 20 sets of test templates can be correctly identified, which indicates that the network model is reasonable and effective, and has a very strong ability to classify the fault patterns.

Chapter 9
Fault Analysis and Diagnosis Method Based on Statistical Learning Theory

Nowadays, the development of intelligent fault diagnostics has been restricted by two aspects. The first one is the serious deficiency of typical fault data samples and the other one is the difficulty of fault feature discovery. The traditional intelligent diagnosis methods, such as fuzzy diagnosis technology, expert system, and ANN, require a great number of fault samples or rely on the statistic character of samples. The generalization performance of the classifier learned from the finite samples may be not good enough. In this instance, it is very important to develop an intelligent diagnosis method based on the machine learning method suitable for the small sample and holding good generalization performance.

Fortunately, the statistical learning theory, born in 1970s, has investigated the machine learning problem systematically and provides an efficient way for the machine learning with finite samples. Compared with the traditional statistic learning methods, the statistic learning theory has devoted to the studies of statistic principles and learning methods with finite samples.

9.1 Statistical Learning Theory and Support Vector Machine [136–140]

Involving only small samples, the program of fault diagnosis is essentially equivalent to the program of fault mode recognition. Generally, many faults may occur abruptly and are difficult to be repeated. Signals of these faults are very difficult to be collected and valuable for the fault diagnosis. On the other hand, many valuable machines cannot be allowed to run with faults in long time because they are related to the production efficiency of the company. Third, the relationship between the fault and the signal may be not accurate and many faults may share a similar signal. When a fault occurred in one element of the system, conditions of the relative parts may be changed and a new fault may happen, for example, several faults may exist

© Springer-Verlag Berlin Heidelberg and National Defense Industry Press 2016
W. Zhang, *Failure Characteristics Analysis and Fault Diagnosis for Liquid Rocket Engines*, DOI 10.1007/978-3-662-49254-3_9

at the same time. Therefore, fault samples are very few for the fault diagnosis in most cases and the support vector machine (SVM) developed based on the small sample theory may be efficient for the pattern recognition of the fault. Furthermore, compared with the recognition of the speech voice and image identification with very high feature dimension, the fault mode recognition has less fault features and samples. All these elements make a big challenge for the fault diagnosis.

With the increasing complexity of the LRE, the analytical model-based method is greatly restricted and methods based on the signal processing and intelligence are more suitable for the cases with much complex characters. In nature, both methods are parts of the data-driven-based method and develop based on the statistic learning theory. Theoretically, the statistical principle can be accurately expressed only when the number of sample is close to infinite. However, the number of collected sample is finite in application and the fault diagnosis method developed based on the theory with infinite sample number cannot provide good classification performance enough.

Devoting to the learning principle of the small group samples, the statistic learning theory has constructed a new theoretical structure which considered both convergence and optimal results obtained with finite information. In contrast, the traditional learning methods using the empirical risk minimization (ERM) criterion cannot minimize the generalization error of the classifier and the expected error may be much bigger than the ERM error; for example, the excessive learning of the ANN is one of the typical cases. In this instance, a classifier with good generalization performance is required for the usage of finite samples and the structural risk minimization (SRM) criterion was proposed by Vapnik. By minimizing the superior boundary of the confidence error, the SRM criterion can maximize the generalization capability of the classifier [136–138].

As the expression of the SRM criterion, SVM also has excellent performance on the pattern recognition problems involving nonlinear, high dimension, and finite number of samples which attracted more attention in the pass decades [139–140].

Compared with the other machine learning method, SVM holds the following advantages.

The structure of the classifier is very simple and particularly suitable for the problem with small number of samples.
The generalization performance has been increased.
Suitable for the problem involving high dimension
There is no direct relation between the computation complexity and the dimension of the input vector and the dimension disaster.
Only one minimum point exists.
Machine learning can be completed within a very short time.
Different classifying plane can be obtained by the usage of different kernel function.

From the mid-1990s, the SVM has attracted great many attentions owing to its simple structure and good generation capability. In general, the SVM is proposed on the basis of a systematical and normative theory which reduces the optional design of the algorithm. Actually since the SVM involves, small group of samples with low dimension is also suitable for the pattern recognition with multiple classes and has great application potentiality in the fault mode identification and condition monitoring.

9.1.1 Machine Learning

As one of the data-driven methods, machine learning can predict the unknown data based on a model developed by the known data. Generally, the basic model of the machine learning can be illustrated as follows: (Fig. 9.1).

For l independent identical samples,

$$(x_1, y_1), (x_2, y_2), \ldots, (x_l, y_l) \tag{9.1}$$

variables $x \in Rn$ and $y \in R$ follow an unknown unite probability density of $F(x, y)$. Search an optimal function of $f(x, w0)$ in the group of $\{f(x, w) : w \in \Lambda\}$ to estimate the relationship between variables x and y based on the expectation risk minimum criterion, namely,

$$R(w) = \int L(y, f(x, w))dF(x, y) \tag{9.2}$$

where $\{f(x, w) : w \in \Lambda\}$ denotes the set of predicted functions (learning function, learning model, or learning machine), w is the general parameter of $f(x, w)$, and $L(y, f(x, w))$ denotes the loss produced by the prediction of x using $f(x, w)$. It can be seen that the expectation risk describes the averaged risk of the learning machine on the given samples and indicates the actual generation capability of the learning machine.

Generally, different types of learning problems, such as pattern recognition, regression and probability density estimation, involve different loss functions. For the pattern recognition, the output y is the label of class. When there is only two modes, the output y can be denoted by $\{0, 1\}$ or $\{-1, 1\}$. Then the loss function is given by

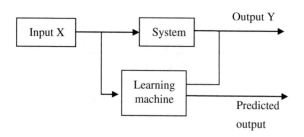

Fig. 9.1 Basic model of the machine learning

$$L(y, f(x, w)) = \begin{cases} 0 & y = f(x, w) \\ 1 & y \neq f(x, w) \end{cases} \tag{9.3}$$

When the output y is continuous, the loss function can be determined by the least square error criterion and denoted as

$$L(y, f(x, w)) = (y - f(x, w))^2 \tag{9.4}$$

For the probability density estimation problem, the probability density $p(x, w)$ can be estimated by the given samples and the loss function is denoted as

$$L(p(x, w)) = -\log p(x, w) \tag{9.5}$$

9.1.2 Statistical Learning Theory

Focused on the machine learning of small group of samples, the statistical learning theory (SLT) is very suitable for the statistic estimation and prediction. From the mid-1990s, SLT has attracted more attentions and a novel generalized machine method, SVM, has been developed [138].

As the kernel concept of SLT, the Vapnik–Chervonenkis dimension provides the best description, so far, for the learning performance of function set. For a group of indication function, if all functions can be scattered by h samples with the $2\,h$ ways, then the VC dimension equals the maximum samples which can scatter the indication function group. It can be seen that the VC dimension provides a description for the performance of the learning machine and is the kernel conception of the SLT. Generally, the bigger the VC dimension, the more powerful and complex the learning machine is.

For different types of function sets, the relationship between the empirical risk and actual risk has been discussed in the SLT and been named as the boundary of the generalization. Based on the results of SLT, the empirical risk Remp (a) and the actual risk $R(a)$ satisfy the following equation with the probability of $1-\eta$.

$$R(\alpha) \leq R_{\text{emp}}(\alpha) + \frac{\varepsilon}{2} \left[1 + \sqrt{1 + \frac{4R_{\text{emp}}(\alpha)}{\varepsilon}} \right] \tag{9.6}$$

where $\varepsilon = 4 \frac{h\left(\ln \frac{2l}{h} + 1\right) - \ln \eta}{l}$ h denotes the VC dimension and n denotes the sampling amount. It can be seen that the actual risk is composed of two parts, the first part is the empirical risk of the training samples and the other one is the believe range which is determined by the believe level, VC dimension, and the amount of the samples. Generally, Eq. (9.6) can be simplified as

$$R(\alpha) \leq R_{\text{emp}}(\alpha) + \Omega(h/l) \tag{9.7}$$

It can be seen that the higher the VC dimension, the wider the believe range. The difference between the actual risk and empirical is bigger when the training sample is finite and it the reason that a complex learning machine does not hold good performance. It is indicated that both empirical risk and believe rang should be minimized in the design program of the learning machine, then the expectation risk can be minimized and the generalization of the machine can be maximized. In this instance, a generalized criterion, named as structure risk minimization (SRM), has been proposed to solve the above-mentioned problem in the SLT. For a function set $S = \{f(x, \omega), \omega \in \Omega\}$, the whole set can be divided in to a series of subset of

$$S_1 \subset S_2 \subset \cdots \subset S_k \subset \cdots \subset S \tag{9.8}$$

and these set can be sorted with the values of their VC dimensions.

$$h_1 \leq h_2 \leq \cdots h_k \leq \cdots \tag{9.9}$$

With the minimization of the empirical risk of every subset at first, the expectation risk minimization is minimized by considering the believe range of every sub set and the idea is shown in Fig. 9.2 and named as SRM. It can be seen that the SRM

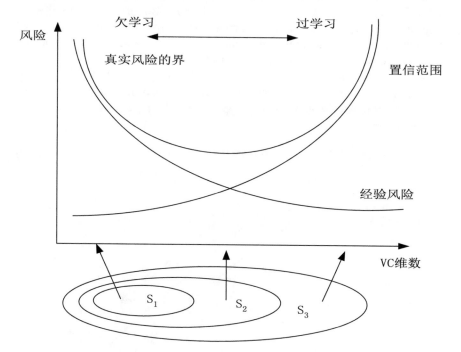

Fig. 9.2 Structure risk minimization criteria

introduced a temporization of the precision and complexity of the model. Apparently, there are two ways to perform the SRM: the first one is choosing a suitable function set and making the believe range minimizing as little as possible. The other one is making the empirical risk of every sub set equalization and minimizing the believe range. The SVM is performed as the latter way and provides a way to solve the problems of the over learning, local optimization, dimension tragedy, and so on, especially for the small sample learning problem.

9.1.3 Support Vector Machine

Introduced from the implement of the SRM, SVM is a novel algorithm of the SLT and widely used in the pattern classification and regression of nonlinear problems. Owing to the powerful learning capability, SVM has attracted much attention in many areas, such as hand writing recognition of the numbers and the face recognition.

At first SVM is rooted from the calculation of the optimization hyper-plane of the linear problems. As Fig. 9.3 [141] illustrated, the square and circle denote two different samples, H denotes the classification line, and $H1$ and $H2$ denote the samples with the nearest distance in the two classes, respectively. The distance between $H1$ and $H2$ is named as margin. The optimization classification line does not only disjoin the two types of samples, but also maximize the margin. Apparently, disjoining the different samples minimizes the empirical risk and maximizing the margin minimizes the believe range of the generalization boundary. Both parts make the expectation risk minimization together.

For the linear separable sample $(x_i, y_i), i = 1, 2, \cdots, n, x \in R^d, y \in \{+1, -1\}$, where n denotes the sample amount, d denotes the dimension of the sample space, and y denotes the expectation output. Suppose that all samples can be separated by a hyper-plane

$$(\omega \cdot x) + b = 0 \tag{9.10}$$

Fig. 9.3 Optimal classification plane

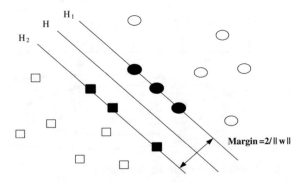

where "•" denotes the inner product. Further the hyper-plane can be normalized and satisfies

$$yi[(\omega \cdot x) + b] - 1 \geq 0, i = 1, 2, \ldots, n \tag{9.11}$$

In this instance, the margin equals $2/\|\omega\|$ and the maximizing of the margin is equivalent to the minimization of $\|\omega\|^2$. The hyper-plane which satisfies Eq. (9.11) and minimizes $\frac{1}{2}\|\omega\|^2$ is named as optimal hyper-plane. The samples located on $H1$ and $H2$ are called support vectors (SVS). In the space of N dimension, supposed that the samples are scattered in a hyper-sphere with the radius of R, the regular hyper-planes satisfying the condition of $\|\omega\| \leq A$ compose the indication function set

$$f(x, \omega, b) = \text{sign}\{(\omega \cdot x) + b\} \tag{9.12}$$

VC dimension satisfies the following equation

$$h \leq \min\left(\left[R^2 A^2\right], N\right) + 1 \tag{9.13}$$

From Eq. (9.13), it can be seen that minimizing $\|\omega\|^2$ is equivalent to minimizing the upper boundary of the VC dimension and the achievement of the SRM is that it performs choosing of the function complexity.Then the classification problem is transformed to solve the optimization function

$$\min \frac{1}{2}\omega^T \omega + C \sum_{i=1}^{n} \xi_i$$
$$s.t. \quad y_i(\omega^T f(x_i) + b) \geq 1 - \xi_i \tag{9.14}$$
$$\xi_i \geq 0, i = 1, 2, \cdots, n$$

where ξ_i denotes the slack variable, $C \geq 0$ denotes the penalty term, and n denotes the amount of the samples. By solving Eq. (9.14), an optimal classification function can be obtained as follows:

$$f(x) = \text{sign}\{(\omega \cdot x) + b\} = \text{sign}\left\{\sum_{i=1}^{n} \alpha_i y_i(x_i \cdot x) + b\right\} \tag{9.15}$$

For the samples which are not support vectors, the coefficient $ai = 0$, therefore, only support vectors that need to be calculated in Eq. (9.15).

For the problems which are not linearly separable, Vapnik introduced the kernel theory which mapped the input data from the low-dimension space to the high-dimension space by a nonlinear kernel function. It is proved that the nonlinear problem can be transformed to be a linear separable problem when a suitable kernel function is used. As Fig. 9.4 illustrated, the input data are transformed to a

Fig. 9.4 Structure of the
SVM

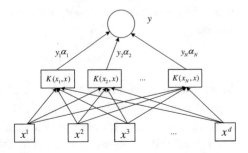

high-dimension space by a kernel function at first and the output data are the linear combination of the intermediate nodes and every node is corresponding to a support vector.

Different kernel functions implement SVM differently. At present, three types of kernel functions are widely used, such as polynominal function, radius-based function, and igmoid function.

9.1.4   Kernel Function and the Parameter Optimization

9.1.4.1   Kernel Function

The usage of kernel function provides an effective way for the nonlinear problem of the pattern classification. However, the performance of the classifier is greatly affected by the selection of the kernel parameters and the optimization of the kernel parameter is the key point of the SVM application. Since with the difference of the object or data, the parameters of SVM classifier may differ from each other and there is no determined way for the parameters selection. In most instances, the parameters are determined by the method of trial and error. Of course, the trial and error method is time consuming for the reason that all points of the combination of the parameters have to be used to estimate the best performance of the classifier. Comparatively, the artificial methods, including genetic algorithm, artificial neural net, test design, etc., are introduced in the determination of the kernel parameters.

It is known that once the function $K(x, y)$ satisfies the Mercer condition, it can be used as the kernel function and the Mercer condition is described as follows:

Mercer theorem: In mathematics, a real-valued function $K(x, y)$ is said to fulfill Mercer's condition if all square integrable functions $\phi(x)$ has

$$\iint K(x, y)\phi(x)\phi(y)\mathrm{dxdy} > 0 \qquad (9.16)$$

namely, $K(x, y) = K(y, x)$ and fulfill the inequality of Cauchy–Schwartz

$$K2(x, y) \leq K(x, x)K(y, y)$$

The kernel method provides a mapping from the input space to the feature space. By considering the data $x_i \in R^{d_L}$ in the input space, an arbitrary function $K(x, y)$ fulfilling the Mercer's condition should satisfy

$$K(x, y) = \sum_{n=1}^{d_F} \gamma_n \Phi_n(x)\Phi_n(x) \tag{9.17}$$

where $\Phi : R^{d_L} \rightarrow H^{d_F}$ is a mapping function in the Hilbert space and dF denotes the dimension of the space H, and $\gamma_n \geq 0$ is used to make sure that the above equation fulfills the inner product form. Apparently, the kernel function of the input space is equivalent to the inner product of the feature space. The advantage of the kernel method is that there is no need to know the detailed form of the mapping $\Phi(x)$ and the input data can be mapped to the feature space by the simple kernel function. At present, the most widely used kernel functions are list as follows:

1. Polynomial kernel function

$$K(x, x_i) = [(x \cdot x_i) + 1]^d \tag{9.18}$$

where d denotes the order of the polynomial. When $d = 1$, the polynomial creates a linear classifier.
2. Radius-based function

$$K(x, x_i) = \exp(-\frac{||x - x_i||^2}{2\sigma^2}) \tag{9.19}$$

3. Multilayer perception kernel function

$$K(x, x_i) = \tanh(k(x \cdot x_i) + v) \tag{9.20}$$

4. B spline kernel function

$$K(x, x_i) = B_{2p+1}(x - x_i) \tag{9.21}$$

where $B2p + 1(x)$ denotes a spline function with an order of $2p + 1$.
5. Exponential radius-based function

$$K(x, x_i) = \exp(-\frac{||x - x_i||}{\sigma^2}) \tag{9.22}$$

6. Fourier series kernel function

$$K(x, x_i) = \frac{\sin(N+0.5)(x - x_i)}{\sin(0.5(x - x_i))} \qquad (9.23)$$

From these kernel functions, it can be seen that the approximation function is the indication function in the pattern recognition. Compared with the approach of the indication function, the approach of the real-value function is more difficult and different real-value function needs different indication function set. Therefore, it is very important to construct the kernel function which can reflect the feature of approach function. Generally, this kind of kernel function can be constructed by the first N term of orthogonal polynomial $Pi(x)$, $i = 1, 2, ..., N$, such as Chebyshev and Hermite polynomial by the equation of Christoffel–Darboux.

$$K_n(x, y) = \sum_{k=1}^{n} P_k(x)P_k(y) = a_n \frac{P_{n+1}(x)P_n(y) - P_n(x)P_{n+1}(y)}{x - y} \qquad (9.24)$$

$$K_n(x, x) = \sum_{k=1}^{n} P_k^2(x) = a_n[P'_{n+1}(x)P_n(x) - P'_n(x)P_{n+1}(x)] \qquad (9.25)$$

where a_n is a constant and relies on the polynomial type and series number of the orthogonal basis. However, with the increase of n, the kernel function $K(x, y)$ trends to Dirac function and the kernel function can be constructed by the modified function.

9.1.4.2 Parameter Optimization

The margin between two classes is usually used to judge the generalization performance of the classifier in the traditional SVM. However, the margin is a rough estimation of the risk upper boundary and its value is not proportion with the performance of the classifier. On the other hand, different kernel functions transform input data into different feature space and the margin in different space is not comparable. In this instance, the margin cannot evaluate the usage of the kernel function effectively and a more precise estimation of the risk should be used. Vapnik introduced the value of $R2/M2$ to the evaluation of the SVM classifier performance, where $R2$ denotes the smallest radius of the training sample in the feature space. It is worth noting that $R2/M2$ does not have dimension and is more suitable than the margin for the evaluation of the kernel function in different feature spaces. Latter, several indexes were proposed to improve the evaluation of the SVM performance. At present, the generalization error estimation theory, proposed by Luntz and Brailovsky, was widely used in the evaluation of the model selection. The equation is shown as follows:

$$EP_{\text{err}}^{n-1} = \frac{1}{n} E(L((x_1, y_1), (x_2, y_2), \ldots, (x_n, y_n))) \tag{9.26}$$

where EP_{err}^{n-1} denotes the expectation value of the error ratio when $n-1$ samples were used for the training of the classifier. $L((x1, y1), (x2, y2), \ldots, (xn, yn))$ is the error number of the classifier applied on all samples non-recurringly. For the reason that $n-1$ samples were used for classifier training and only one sample was used to evaluate the performance, this method is named as leave-one-out (LOO) method. It can be seen that the error ratio obtained by LOO method is the unbiased estimation of the true error ratio of the classifier. However, the LOO method needs lots of computing amounts when there is large amount of samples. Therefore, the k-fold cross-validation method is used as an estimation of the LOO method. In the k-fold cross-validation method, all samples are divided into k unfold subset, $S1$, $S2$, ..., Sk, at first. All subsets share nearly equivalent amount of samples. Both training and testing are performed k times. For example, when the classifier is created by the training of $S2$, ..., Sk, then the subset $S1$ is used for test. Similarly, when the classifier is obtained by the training $S1$, $S3$, ..., Sk, then the subset $S2$ is used for test. Finally, when all subsets are used for the performance test, the error ratio is the mean value of the k time tests. Apparently, the LOO method is a special case when $k = n$ in the k-fold cross-validation.

Although the k-fold cross-validation method saves much computing amount than the LOO method and is used in many SVM softwares, such as LibSVM etc., it cannot satisfy the demand and some more effective methods were proposed to estimate the error ratio. Many methods were devoted to the estimation of the upper boundary of $L((x1, y1), (x2, y2), \ldots, (xn, yn))$ as little as possible.

Suppose that $\psi(\cdot)$ is a ladder function with one order, $f0$ denotes the judge function obtained by using all training samples, and fi denotes the judge function obtained by using all samples except the ith sample, then

$$L((x_1, y_1), (x_2, y_2), \ldots, (x_n, y_n)) = \sum_{i=1}^{n} \psi(-y_i f^i(x_i))$$

$$= \sum_{i=1}^{n} \psi(-y_i f^0(x_i) + y_i(f^0(x_i) - f^i(x_i))) \tag{9.27}$$

Suppose that Ui is an upper bound of $y_i(f^0(x_i) - f^i(x_i))$ and $y_i f^0(x_i) \geq 1$ when the samples are linearly separable, then the estimation upper bound of $L((x1, y1), (x2, y2), \ldots, (xn, yn))$ is

$$L((x_1, y_1), (x_2, y_2), \ldots, (x_n, y_n)) \leq \sum_{i=1}^{n} \psi(-1 + U_i) \tag{9.28}$$

It is worth noting that most risk estimation can be calculated using Eq. (9.28). The reasons are listed as follows:

For the samples which are not support vectors, $y_i(f^0(x_i) - f^i(x_i)) = 0$ and $L((x_1, y_1), (x_2, y_2), \ldots, (x_n, y_n)) \leq N_{SV}$, where NSV denotes the amount of the support vector, then $T = NSV/n$ is one of the risk indices of the SVM; namely, the less support vector may be corresponding to lower error rate.

Jaakkola and Haussler had proved the following inequality in Ref. [142]

$$y_i(f^0(x_i) - f^i(x_i)) = \alpha_i^0 K(x_i, x_i) = U_i \tag{9.29}$$

And the estimation of the error is calculated by

$$T = \frac{1}{n} \sum_{i=1}^{n} \psi(\alpha_i^0 K(x_i, x_i) - 1) \tag{9.30}$$

where α_i^0 denotes the solution of the SVM obtained by all training samplings.

In Ref. [143], Opper and Winther had proved that if the support vectors do not change when one of the samples is removed, then

$$y_i(f^0(x_i) - f^i(x_i)) = \alpha_i^0 / (K_{SV}^{-1})_{ii} \tag{9.31}$$

where KSV denotes the matrix constructed by the inner product of the support vectors and corresponding error estimation is calculated by

$$T = \frac{1}{n} \sum_{i=1}^{n} \psi(\alpha_i^0 / (K_{SV}^{-1})_{ii} - 1) \tag{9.32}$$

Meanwhile, Vapnik and Chapelle had introduced the concept of span of support vectors, Ref. [144], for the error estimation. At present, the error can be estimated most accurately by the span of support vectors. Apparently, once the measure index has been determined, then the calculation of the kernel parameter was transformed in to an optimization problem.

9.2   Application of the SVM in the Fault Diagnosis of LRE

9.2.1   Fault Character Analysis and Diagnosis of LRE in Steady State Based on SVM

Based on the analysis in Chap. 2, it is known that the parameter variation can be obtained by solving the stable model of the LRE by replacing the normal parameter by a fault value. For a LRE, there are many fault modes and parts of the fault modes are produced by the same parameter variation. In this instance, the more parameters have to be used for the classification. In this section, seven fault modes were involved in the case study.

1. Primary valve failure simulation of oxidant or fuel

   The primary valve may be not fully open and cause increasing of the local pressure loss. The pressure loss is related to the angle of the valve and the pipe flow. Referring to the Appendix A (A.8), (A.9) and changing the values of $a9$ and $a10$, we can obtain the primary valve fault. Suppose that when the primary valve is fully open it is corresponding to the angle of zero, then in the fault condition

$$a'_k = a_k \left(\frac{1}{\cos^4 \theta} \right), \quad k = 9, 10 \tag{9.33}$$

   where ak denotes the normal value and θ denotes the angle of the valve.

2. Simulation of the thrust chamber fault

   Apparently, the fault effect of oxidizer injector choke can be simulated by changing the value of $a13$ in the Appendix A (A.11) and the ablation of chamber throat can be obtained by changing $D15$ in Eqs. I (1–5).

3. Simulation of the gas generator tubing fault

   By adjusting the value of $a17$ in the Appendix Eq. (A.14), we can simulate the fault effect caused by the oxidizer pipeline block of the gas generator. Similarly, the leak at the fuel entrance can be simulated by adjusting the values of $a1$ and $a19$ in the Appendix Eqs. (A.2) and (A.15).

4. Simulation of the turbo pump fault

   By adjusting the factor of $D1$, we can simulate the efficiency loss of oxidizer pump.

   In this instance, seven different fault modes can be obtained by the calculation of the stable model of the LRE and the variations of difference variables are listed in Table 9.1. For the sake of later fault diagnosis, nine samples are obtained in every mode and all data are normalized by

$$\text{Sampling value} = (\text{fault value} - \text{normal value})/\text{normal value}$$

In the test, the first five groups are used for the classifier training and the later four groups are used for the test. The detailed test data are listed in Table 9.2.

In this case, the SVM was used for the fault classification and the exhaustive method of the training data was used to construct the classifier. In the SVM classifier construction program, the gauss kernel with the kernel parameter σ limited in the range of 1–8 was used and the step was set to be 1. The penalty factor C was limited in the range of 10–100 and the step is set to be 10. Test results obtained at different combinations of the σ and C were listed in Table 9.3. It can be seen that the classification accuracy is not greatly influenced when C is suitably selected.

For the reason that the collected signals are influenced by many factors, the training data are added with white noise to consider the effect of the unknown

Table 9.1 Training data obtained by the simulation (35 groups)

Fault modes	Sample number	Variable				
		Pf	Pt	RT	F	$\dot{m}_j$
Primary valve fault of the oxidizer (Mode 1)	1	−0.03167	0.082828	0.022134	0.089583	−0.50626
	2	−0.027767	0.11507	0.02747	0.13003	−0.67914
	3	−0.03287	0.13519	0.035352	0.16569	−0.93825
	4	−0.040916	0.11377	0.038761	0.16887	−1.0445
	5	−0.046473	0.086035	0.038712	0.15695	−1.0471
Primary valve fault of the fuel (Mode 2)	6	−0.55925	−0.24406	0.18893	−0.52203	−0.99625
	7	−0.7962	−0.46441	−0.76329	0.76329	−1.0112
	8	−0.78998	−0.45757	0.55211	−0.76428	−1.0166
	9	−0.78625	−0.44712	0.32316	−0.77006	−1.0247
	10	−0.67641	−0.5335	−0.82756	−0.87323	−1.0123
Block fault of the fuel injector (Mode 3)	11	0.060397	0.098844	0.0073677	0.16879	−0.2655
	12	0.058449	0.10075	0.0079829	0.16995	−0.2802
	13	0.057042	0.10268	0.008511	0.17147	−0.29333
	14	0.054315	0.10445	0.0092349	0.17181	−0.31055
	15	0.052464	0.10581	0.0097524	0.17224	−0.32284
Ablation fault of the combustor throat (Mode 4)	16	−0.29493	0.24276	0.0057486	−0.39467	−0.45451
	17	−0.29617	0.24416	0.0058245	−0.39595	−0.45617
	18	−0.29731	0.24546	0.0058952	−0.39714	−0.4577
	19	−0.29837	0.24666	0.0059613	−0.39823	−0.45912
	20	−0.29936	0.24779	0.0060231	−0.39925	−0.46044
Oxidant pipeline block fault of gas generator (Mode 5)	21	−0.19202	−0.11457	0.021354	−0.22601	−0.25876
	22	−0.22944	−0.13748	0.026884	−0.26737	−0.30609
	23	−0.262	−0.15743	0.032268	−0.30268	−0.34751
	24	−0.29061	−0.17486	0.037536	−0.33307	−0.38429
	25	−0.31594	−0.19007	0.04272	−0.35939	−0.4174
Fuel entrance leakage fault of gas generator (Mode 6)	26	0.061042	−0.026171	0.0011876	0.10067	0.068499
	27	0.075133	−0.03049	0.0015094	0.12644	0.084483
	28	0.089058	−0.033928	0.0018926	0.15272	0.097904
	29	0.10286	−0.036289	0.0023609	0.17981	0.10777
	30	0.11659	−0.037159	0.0029631	0.20826	0.11214
Efficiency loss fault of oxidizer pump (Mode 7)	31	−0.068385	−0.1202	0.00044623	−0.12291	−0.26855
	32	−0.083019	−0.145	0.00057116	−0.14712	−0.31859
	33	−0.09884	−0.17128	0.00077767	−0.17257	−0.37158
	34	−0.11619	−0.19921	0.001139	−0.19933	−0.42905
	35	−0.13596	−0.22901	0.0018941	−0.22734	−0.49597

disturbs and make the results of SVM more real. In this test, the strength of the noise is set in the range of 5–35 % and the classification accuracies corresponding to different ranges of kernel parameter are listed in Table 9.4. Figure 9.5 illustrates

Table 9.2 Test data obtained by the simulation

	Sample number	Variable					Fault mode
		Pf	Pt	RT	F	$\dot{m}_j$	
Test samples	1	−0.030355	0.097776	0.024613	0.10767	−0.58057	1
	2	−0.032417	0.12915	0.032209	0.14824	−0.82547	1
	3	−0.03373	0.12744	0.037115	0.17409	−1.0124	1
	4	−0.045381	0.095065	0.038782	0.15986	−1.0435	1
	5	−0.64054	−0.31978	0.26767	−0.61523	−1.0041	2
	6	−0.79303	−0.45674	0.64139	−0.76149	−1.0157	2
	7	−0.76051	−0.46437	−0.14888	−0.79102	−1.031	2
	8	−0.75536	−0.51014	−0.39904	−0.82279	−1.0258	2
	9	0.06018	0.099933	0.0075676	0.17008	−0.27075	3
	10	0.057733	0.10211	0.0083094	0.1712	−0.28858	3
	11	0.056016	0.10385	0.0088496	0.17221	−0.30197	3
Test samples	12	0.054049	0.10528	0.009398	0.17267	−0.31496	3
	13	−0.29556	0.24347	0.0057872	−0.39533	−0.45536	4
	14	−0.29675	0.24482	0.0058605	−0.39656	−0.45695	4
	15	−0.29785	0.24607	0.0059288	−0.3977	−0.45842	4
	16	−0.29887	0.24723	0.0059927	−0.39875	−0.45979	4
	17	−0.21141	−0.12643	0.02414	−0.24753	−0.28326	5
	18	−0.24627	−0.1478	0.029592	−0.28571	−0.32745	5
	19	−0.27675	−0.16644	0.034914	−0.31843	−0.36641	5
	20	−0.30365	−0.18272	0.040136	−0.3467	−0.40125	5
	21	0.06811	−0.028433	0.0013418	0.1135	0.076769	6
	22	0.082114	−0.032329	0.0016922	0.1395	0.091563	6
	23	0.09597	−0.035261	0.002114	0.16614	0.10337	6
	24	0.10973	−0.036952	0.0026403	0.19381	0.11084	6
	25	−0.075568	−0.13242	0.00050106	−0.13486	−0.29326	7
	26	−0.090765	−0.15794	0.00066112	−0.15968	−0.34465	7
	27	−0.10729	−0.18503	0.00093117	−0.18578	−0.39959	7
	28	−0.12566	−0.21386	0.0014341	−0.2132	−0.46063	7

the variation of the kernel parameter range along with the strength of the noise. It can be seen that the more powerful the added noise, the lower the tolerance of the kernel parameter; namely, the high accuracy of the SVM classification needs an accurate selection of the value of kernel parameter.

Figure 9.6 illustrates the variation trend of the kernel parameter along with the penalty factor when the classification accuracy is acceptable. It can be seen that the range of kernel parameter increases with the increase of the penalty factor. However, the calculation amounts increase with the increase of the penalty factor. In this case, the system needs to be updated to increase the calculation speed.

Table 9.3 Influence of the kernel parameter and penalty factor on the accuracy

C	σ (Parameter of the gauss kernel)							
	1.0	2.0	3.0	4.0	5.0	6.0	7.0	8.0
10	0.96429	0.96429	0.92857	0.82143	0.78571	0.78571	0.78571	0.78571
20	1.000	1.000	0.96429	0.89286	0.78571	0.78571	0.78571	0.78571
30	1.000	1.000	1.000	0.89286	0.85714	0.78571	0.78571	0.78571
40	1.000	1.000	1.000	0.92857	0.85714	0.82143	0.78571	0.78571
50	1.000	1.000	1.000	1.000	0.89286	0.85714	0.82143	0.78571
60	1.000	1.000	1.000	1.000	0.92857	0.85714	0.82143	0.78571
70	1.000	1.000	1.000	1.000	0.92857	0.85714	0.85714	0.82143
80	1.000	1.000	1.000	1.000	0.96429	0.92857	0.85714	0.82143
90	1.000	1.000	1.000	1.000	1.000	0.92857	0.85714	0.82143
100	1.000	1.000	1.000	1.000	1.000	0.92857	0.85714	0.82143

Table 9.4 Range variation of the kernel parameter at different noise strength

Noise strength	5 %	10 %	15 %	20 %	25 %	30 %	35 %
Range of kernel parameter	0.8–2.0	1.0–1.6	1.2–1.6	1.3–1.4	1.3–1.4	–	–
Classification accuracy	1.000	1.000	1.000	1.000	1.000	0.933	0.917

Fig. 9.5 Variation of the kernel parameter range along with the noise

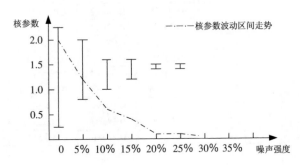

Fig. 9.6 Variation of the kernel parameter along with the penalty factor

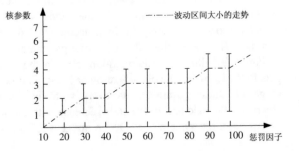

It can be seen that the determination of both kernel parameter and penalty factor is important for the accuracy of the SVM classifier. In this case, the GA algorithm may provide a way for the optimization of the determining of both parameters.

9.2.2 Fault Diagnosis of LRE Based on GA—SVM

In order to increase the efficiency of the classification, the GA was used and the method is listed step by step as follows:

1. Code the parameters by gray code and construct the initial population randomly. By constructing the strength of noise and the kernel parameter with the length of 10, the individual is with the length of 20. The amount of the initial population is 50 and the penalty factor $C = 20$.
2. Calculate the fitness of every individual. In this test, the fitness is defined by the classification accuracy of the test data.
3. Perform the selection, crossover, and mutation operations on every individual based on the fitness. Terminate the iteration when the fitness fulfills the convergence condition or the iteration reaches the maximum iteration value. In this test, the maximum iteration is set to be 40.

Figure 9.7 illustrates the classification accuracy variation curve in the iteration calculation. It can be seen that the classification accuracy weakens to the value of 98 % when the kernel parameter of SVM is 1.56 and the anti-noise strength is 31 %. Apparently, such result is near to that obtained by the exhaustive method. Although the accuracy varies with the iteration, the calculation speed is much faster than that of the exhaustive method.

Fig. 9.7 Classification accuracy variation curve along with the iteration

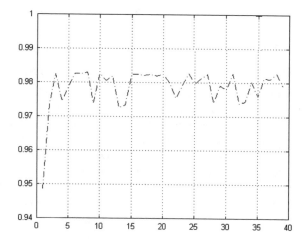

9.2.3   Fault Modeling and Analysis of LRE Based on SVM

The modeling and analysis of the LRE fault is greatly significant for the selection of the monitoring parameter, character analysis of the fault and diagnosis, etc. However, an accurate model of LRE is difficult to be obtained since the power system has become more and more complicated. Although some artificial intelligence methods, such as ANN, do not involve an accurate model, the inherent shortages and high demands on the data restrict their applications in the modeling of LRE. In this section, based on the comparison between the SVM and ANN mechanisms, the difference of both methods applied in the nonlinear system modeling is deeply analyzed. Finally, the dynamic fault modeling and analysis of LRE are performed by the SVM to get good performance.

9.2.3.1   SVM and ANN

As the SLT illustrated, the SRM is to minimize the risk boundary by minimizing both empirical risk and confidence range together. From the view of SRM, the function set $\{f(x, w) : w \in \Omega\}$ is constructed by the nested function subsets [138].

$$S_n := \{f(x, w) : w \in \Omega_n\}, \ S_1 \subset S_2 \subset \ldots \subset S_n \subset \ldots \tag{9.34}$$

The VC dimension of each subset fulfills

$$h_1 \leq h_2 \leq \ldots \leq h_n \leq \ldots \tag{9.35}$$

Then select an element S_k and a function $f(x, w) \in S_k$ to minimize a boundary as the following.

$$R(w) \leq R_{\text{emp}}(w) + P(\frac{l}{h_k}) \tag{9.36}$$

It can be seen that there are two terms used in the minimization of the boundary. The first one is to minimize the confidence range $P\left(\frac{l}{h_k}\right)$ of the learning machine by the determination of the VC dimension of the admissible function set. However, a complex learning machine is corresponding to a large range of the confidence range. In this instance, even if the empirical risk, the first term at the right of Eq. (9.36), is set to be zero, the error ratio of the machine on the training set may be high. Such phenomenon is called over learning and is one of the shortages of the ANN. To avoid the over learning phenomenon, the learning machine should be constructed with small VC dimension. However, the training set may not be fitted by the machine with too small VC dimension. In this instance, both empirical risk and confidence range should be considered together while constructing the learning machine and the prior knowledge of the training set can be fully used. The SVM is performed by the minimization of both empirical risk and confidence range.

In summary, the ANN fixes the confidence range and minimizes the empirical risk, while the SVM fixes the empirical risk and minimizes the confidence range. In the ANN method, the weight value of the neurons is determined by the minimization of the empirical risk. Generally, the neuron is described by the sigmoid function of $f(x, w) = S\{(w \cdot x)\}$, where w denotes the weight.

Sigmoid is a smooth monotonic function and fulfills the conditions of $S(-\infty) = -1$ and $S(+\infty) = +1$. For example

$$S(u) = \tanh u = \frac{\exp(u) - \exp(-u)}{\exp(u) + \exp(-u)}$$

For the Sigmoid function set, the empirical risk functional is determined by

$$R_{emp}(w) = \frac{1}{l} \sum_{j=1}^{l} (y_i - S\{(w \cdot x_j)\})^2 \tag{9.37}$$

and is smooth for the weight w, whose gradient is

$$\text{grad}_w R_{emp}(w) = -\frac{2}{l} \sum_{j=1}^{l} [y_j - S(w \cdot x_j)] S'\{(w \cdot x_j)\} x_j^T \tag{9.38}$$

It can be seen that the minimization can be performed by the standard gradient-based method, for example the gradient descent method.

$$w_{new} = w_{old} - \gamma(\cdot) \text{grad} R_{emp}(w_{old}) \tag{9.39}$$

where $\gamma(\cdot) = \gamma(n) \geq 0$ is relied on the iteration number n.

The sufficient condition of the gradient descent method weakening to the local minimum is the gradient value bounded and the coefficient $\gamma(n)$ fulfills the following condition.

$$\sum_{n=1}^{\infty} \gamma(n) = \infty, \sum_{n=1}^{\infty} \gamma^2(n) < \infty \tag{9.40}$$

Finally, the problem can be solved by the function approach with the sigmoid function in the parameter estimation stage and using a threshold function for the last neuron at the identification stage.

The empirical risk gradient calculated by the weakening of sigmoid function in ANN is named as backward propagation (BP) algorithm and can be used for the correction of the weights w which is iteratively based on the standard gradient-based method.

By the BP ANN, one local minimum point of the empirical risk functional can be calculated, although lots of local minimum points exist. The actual calculation is influenced by many factors, especially the weight matrix $w(k)$ and the initial parameters.

9.2.3.2   Application on the Nonlinear System Modeling of SVM and ANN

With the properties of nonlinear, global, unsteady nonconvex, etc., ANN is suitable for the description of the LRE with nonlinear and time varying characters. So far, lots of ANN algorithms, such as the BP ANN and RBF ANN, have been proposed and widely used in the engineering application. Figure 9.8 illustrates the mechanism of the SVM. By comparing SVM with the ANN, it can be seen that the structure of SVM is similar to that of the ANN with three layers and the amount of the support vector equals the node number of the hidden layer. Although the BP ANN algorithm is very slow and may weaken the local minimum, it is still widely used in the applications owing to the simplicity and easy to be realized.

However, the development of ANN is based on the minimization of the empirical risk, which may restrict the generalization capability of the method and produce the over learning phenomenon. Generally, the structure of ANN classifier is sensitive to the training samples and the pre-processing of the data is very strict. Actually, the ANN has been applied widely in engineer application, since the empirical knowledge of the user has been well used in the design of the ANN structure.

Comparatively, the SVM, based on the SRM, has strictly mathematical support and do not extremely rely on the empirical knowledge. Aiming to the problem with small amount of samples, the optimal solution of SVM is calculated based on the finite amount of samples and is not on the infinite samples. Such character is

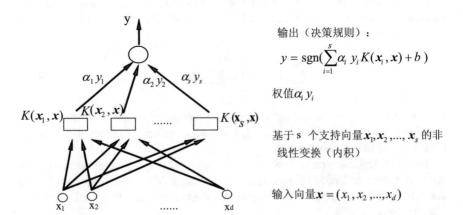

Fig. 9.8 Mechanism of the SVM

suitable for the engineering application. Additionally, there are several advantages listed as follows:

1. The relationship between the algorithm complexity and the input vectors has been fixed in the SVM.
2. By introducing the kernel function, the nonlinear problem is mapped to be a linear problem in the higher dimension feature space.
3. By transforming the problem to be a convex optimization problem, the global optimization of the algorithm can be ensured and avoid the local minimum phenomenon in the ANN.

In summary, the SVM is more suitable than the ANN for the simulation of LRE in the view of training speed, accuracy, and the algorithm complexity when only small amount of samples used.

9.2.3.3 Case Study

The dynamic process of LRE involves the process of starting up, shutting down, over loading, and the transforming caused by the faults and is the main stage of the occurrence and development of the fault. In this instance, the dynamic process simulation of the LRE is performed by both SVM and ANN in order to obtain the difference between these two methods.

Since the condition of LRE can be reflected by the thrust variation, calculating the thrust at arbitrary moment by the model is very necessary for the design, detection, and monitoring of the LRE.

When the environment is determined, the thrust of LRE is the function of oxidant flow mo, fuel flow mf, and chamber pressure pc. In the modeling, both training and test data are collected by the ground test and normalized first. As listed in Tables 9.5 and 9.6, there are 25 training and test data together with the collected interval of 0.1 s and corresponding thrust variation in the start-up program is illustrated in Fig. 9.9.

In order to investigate the performance of SVM in the modeling, the exhaustive method is used for the parameter analysis. The Gauss kernel function with $C = 100$ and $\varepsilon = 0.01$ was used. As shown in Fig. 9.10, the test error reaches the minimum when the kernel parameter equals to 3.5. The reflection of the ε variation on the error is listed in Table 9.7. It can be seen that the error is controlled by the width ε. The smaller the width ε, the smaller the error is and the longer the averaged training time. Only 47 microseconds have been consumed when the width ε equals 0.01 in the modeling of the thrust as shown in Fig. 9.9. Figure 9.11 illustrates the reflection of C on the results. It can be seen that the error is developed to be stable when C reaches a determined value. In this instance, a suitable value of C should be used for the balance of the calculation efficiency and hardware demand.

Table 9.5 Training samples

Time t (s)	Fuel flow ($\dot{m}_f$)	Oxidant flow ($\dot{m}_o$)	Combustion chamber pressure (p_c)	Thrust (F)
0	0.3179	0.4011	0.0098	0.0000
0.2	0.5780	0.5339	0.2238	0.1418
0.4	0.6012	0.5658	0.6170	0.4113
0.6	0.6763	0.6374	0.7832	0.5674
0.8	0.8555	0.8238	0.8951	0.7730
1.0	0.9306	0.9149	0.9580	0.8865
1.2	0.9665	0.9593	0.9874	0.9362
1.4	0.9827	0.9810	0.9944	0.9574
1.6	0.9442	0.9919	0.9972	0.9787
1.8	0.9977	0.9973	1.0000	0.9929
2.0	1.0000	1.0000	1.0000	1.0000
2.2	1.0000	1.0000	1.0000	1.0000
2.4	1.0000	1.0000	1.0000	1.0000

Table 9.6 Test samples

Time t (s)	Fuel flow ($\dot{m}_f$)	Oxidant flow ($\dot{m}_o$)	Combustion chamber pressure (p_c)	Thrust (F)
0.1	0.3931	0.4661	0.0559	0.0496
0.3	0.6532	0.5935	0.4643	0.3262
0.5	0.6012	0.5572	0.7049	0.4681
0.7	0.7746	0.6721	0.8447	0.6596
0.9	0.9017	0.8753	0.9301	0.8440
1.1	0.9526	0.9420	0.9790	0.9163
1.3	0.9769	0.9702	0.9930	0.9504
1.5	0.9884	0.9864	0.9951	0.9716
1.7	0.9965	0.9957	0.9986	0.9858
1.9	0.9988	0.9995	1.0000	0.9986
2.1	1.0000	1.0000	1.0000	1.0000
2.3	1.0000	1.0000	1.0000	1.0000

Fig. 9.9 Thrust variation in the start-up process

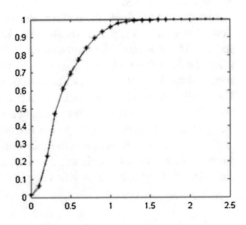

Fig. 9.10 Influence of the
kernel parameter on the error

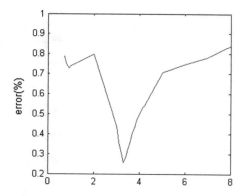

Table 9.7 Influence of ε on
the error

ε	0.1	0.01	0.001	0.0001
Error (%)	7.78	0.28	0.07	0.023
Training time (ms)	46	47	62	94

Fig. 9.11 Influence of the
penalty factor on the error

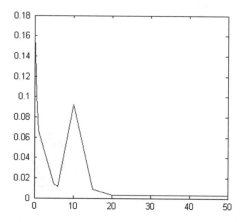

Table 9.8 Simulation error used different kernel function

Kernel	Linear kernel	Polynomial kernel	Gauss kernel	RBF
Error (%)	0.2292	0.2793	0.2162	0.1386

Table 9.8 illustrates the influence of different kernel functions on the test error
when both C and ε are determined. It can be seen that all kernel functions are useful
for modeling the nonlinear description of the LRE and the Gauss and RBF kernel
have higher accuracy than the polynomial kernel function in the simulation.

Comparatively, the BP ANN with the structure of 4-15-1 net was also used for
modeling the LRE. The learning ratio $\alpha = 0.7$, inertia factor $\eta = 0.9$, and the
maximum error is limited to be 0.00005. Consuming iteration of 1649, the BP ANN

Fig. 9.12 Errors of the BP
algorithms

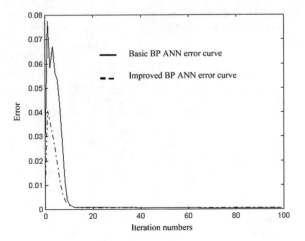

Fig. 9.13 Results of the BP
algorithms

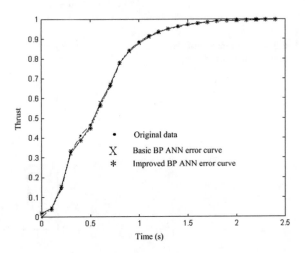

weakens and the test error is 0.0028. Although the BP ANN has reached a very high
accuracy, the SVM is more powerful owing to the higher accuracy, faster speed
(Figs. 9.12 and 9.13).

Chapter 10
Fault Diagnosis Method Based on Hidden Markov Model

As a statistical model of time series, the hidden Markov model (HMM) is suitable for the dynamic time series analysis, especially for the signal with a large amount of information, and nonstationary and low reproducibility. It is a dynamic pattern recognition tool, which could gather statistic models, classify the information of a time span, and expand the fault diagnosis method which is only based on the static observation. Studying the method of HMM fault diagnosis is to observe and evaluate the equipment status in a dynamic environment. It can detect the fault development signs at an early stage and eliminate the failure. Therefore, it is meaningful for the prediction and prevention of the potential failure of the rocket weapon system. In addition, HMM is based on the statistical pattern recognition theory and the theory is a branch of machine learning. Therefore, it is also very important to study the HMM fault diagnosis method to monitor and estimate the running state of the equipment.

10.1 Fault Diagnosis Method Based on HMM

HMM is a kind of intelligent algorithms with a solid statistical basis and effective training method and is widely used to describe the sequence data or processes [145]. The basic theory of HMM was first proposed by Baum in 1970s and was soon applied on the speech recognition [146]. After being spread in the middle of 1980s, HMM had been almost widely used in all aspects of speech processing and is successfully applied in genetics, molecular biology, and biological information science and other disciplines [147]. In view of the successful application of HMM in other fields, the HMM theory can be applied to turbopump fault diagnosis considering the similar characteristics of certain dynamic processes of turbopump and voice signal. To study the application of HMM in the fault diagnosis of turbopump, the basic theory, algorithm, and problems of HMM are studied first.

© Springer-Verlag Berlin Heidelberg and National Defense Industry Press 2016
W. Zhang, *Failure Characteristics Analysis and Fault Diagnosis
for Liquid Rocket Engines*, DOI 10.1007/978-3-662-49254-3_10

10.1.1 Basic Ideology of HMM

1. Markov chain

As a special case of the Markov stochastic process, the Markov chain is a process in which both the state and the time parameters are discrete. From the mathematical point of view, the definition of a Markov chain can be defined as follows [148,149]:

A random sequence X_n may be in any state of θ_1, ..., θ_N at the time of n. The state at the $m + k$ moment equals the probability of q_{m+k} which is related to the state q_m at the moment of m and has nothing to do with the status before the moment of m. Namely

$$
\begin{aligned}
P(X_{m+k} = q_{m+k}|X_m = q_m, X_{m-1} = q_{m-1}, \ldots, X_1 = q_1) \\
= P(X_{m+k} = q_{m+k}|X_m = q_m)
\end{aligned} \tag{10.1}
$$

where $q_1, q_2, \ldots, q_m, q_{m+k} \in (\theta_1, \theta_2, \ldots, \theta_N)$, X_n is called the Markov chain.

$$
P_{ij}(m, m+k) = P(q_{m+k} = \theta_j|q_m = \theta_i), \quad 1 \le i, j \le N, \quad m, k \in Z_+ \tag{10.2}
$$

is called k-step transfer probability. When $P_{ij}(m, m + k)$ is independent from m, the Markov chain is homogeneous. In this instance,

$$
P_{ij}(m, m+k) = P_{ij}(k) \tag{10.3}
$$

When $k = 1$, $P_{ij}(1)$ is called a one-step transfer probability or the transfer probability and denoted by a_{ij}. All the transfer probabilities a_{ij}, $1 \le i$, and $j \le N$ can form a transition probability matrix, i.e.,

$$
A = \begin{bmatrix} a_{11} & \cdots & a_{1N} \\ \cdots & & \cdots \\ a_{N1} & \cdots & a_{NN} \end{bmatrix} \tag{10.4}
$$

and

$$
0 \le a_{ij} \le 1, \quad \sum_{j=1}^{N} a_{ij} = 1 \tag{10.5}
$$

It can be seen that a Markov model is described by a Markov chain and a transfer probability matrix.

Since the k-step transfer probability $P_{ij}(k)$ can be obtained by the transfer probability a_{ij}, the most important parameter of the Markov chain is the transfer probability matrix A. However, A was unable to determine the initial distribution, namely the probability of $q_1 = \theta_i$ cannot be obtained by A. Therefore, in order to fully describe the Markov chain, we also must introduce the initial probability vector $\pi = (\pi_1, \pi_2, \ldots, \pi_n)$, in addition to the matrix A, which

$$\pi_i = P(q_1 = \theta_i), \quad 1 \le i \le N \tag{10.6}$$

It is obvious that

$$0 \le \pi_i \le 1, \quad \sum_i \pi_i = 1 \tag{10.7}$$

Each state of the Markov chain can correspond to an observable physical event, such as rain, sunshine, snow, and so on in the weather forecast. Then it can be called the Markov chain model of weather forecast. Based on this model, the probability of the weather (state) at any one certain moment can be calculated.

2. Basic concepts of HMM

HMM is developed on the basis of the Markov chain. As the actual problem is more complex than described by the Markov chain model, the events and the observed states are not corresponding, but associated through a set of probability distributions. Such a model is called HMM. It is a double random process. One of them is the Markov chain, which is the basic random process. It describes the state of the transfer. Another random procedure describes the statistical correspondence between the states and the observed values. Thus, in the view of observer, we can only see the observations, unlike the observations and states correspondence of the Markov model. Therefore, we could not see the states directly, but sense the presence and characteristics of the state through a random process, which is called "hidden" Markov model, HMM for short. Here is a famous example of the HMM— jar model as an example to illustrate the basic concepts of HMM.

There are N jars. Each jar has a lot of colorful balls. As shown in Fig. 10.1, the color of the balls is described by a group of probability distribution vectors. Experiments were conducted as follows: According to an initial probability distribution, select one of the N jars randomly. For example, the i-th jar is selected. Then according to the probability distribution of the color ball in the jar, select a ball, denote the color of the ball as o_1. Then put the ball back in the jar, and

Fig. 10.1 Jar model

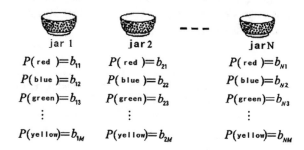

according to the transfer probability distribution of the jar, randomly choose the next jar. For example, the j-th jar is selected. We choose a ball from the jar again and denote the color of the ball as o_2. Proceeding in this way, we can get a color sequence $o_1, o_2, \ldots, o_N$. There are a number of observed events, and hence it is called the observation sequence of values. But the transfer between the jars and each of the jars chosen were hidden and could not be observed directly. And the color of the ball is not corresponding to the jar one by one. The probability distribution of the ball in the jar was determined randomly. In addition, each chosen jar was determined by a group of transfer probability.

3. Definition of HMM

The jar model above had given the basic idea of HMM. Now we will give the definition of HMM.

A HMM can be described by the following parameters:

(1) N: The number of states of the Markov chain in the model. Denote N states as $\theta_1, \theta_2, \ldots, \theta_N$. Denote the status of the Markov chain at the t moment as q_t. It is obvious that $q_t \in (\theta_1, \theta_2, \ldots, \theta_N)$. In the jar model, the jar is corresponded to the state of HMM.

(2) M: The number of the possible observations per state. Denote M observation values as $v_1, v_2, \ldots, v_M$ and the observation values at the moment as t for o_t. The color selected in the experiment of the jar and ball is the observation value in the HMM model.

(3) The initial probability distribution vector, $\pi = (\pi_1, \pi_2, \ldots, \pi_N)$, where $\pi_i = P(q_1 = \theta_i), 1 \leq i \leq N$. The initial probability distribution vector is the probability of a certain chosen jar at the beginning of the experiment of the jar and ball.

(4) A: The state transition probability matrix is $A = (a_{ij})_{N \times N}$, where $a_{ij} = P(q_{t+1} = \theta_j | q_t = \theta_i), 1 \leq i, j \leq N$. It refers to the probability of selecting the next jar in the experiment of the jar and ball.

(5) B: The observation probability matrix is $B = (b_{jk})_{M \times N}$. In the experiment of the jar and ball, b_{jk} is the probability of the color k of the ball in the j pot.

In summary, denote HMM as $\lambda = (N, M, \pi, A, B)$, or $\lambda = (\pi, A, B)$ for short.

To be more vivid, HMM can be divided into two parts. One is the Markov chain, which is described by π and A. And the output is the state sequence. The other random process is described by B. And the output is the sequence of observations. The composition of a HMM is shown in Fig. 10.2, where T is the time length of the observation sequence.

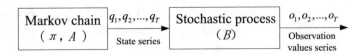

Fig. 10.2 Sketch map of HMM

The difference between HMM and Markov model are that Markov model can be observed directly, and HMM must be describe by two stochastic processes. One is the state sequence and the other is observation value sequence. The model and the algorithm are more complex than the Markov model.

10.1.2   Basic Algorithm of HMM

To build a HMM, three basic questions must be solved. Based on these three basic questions, people studied three basic algorithms. The three questions are

Problem 1: HMM probability calculation

When the observation sequence $O = \{o_1, o_2, ..., o_T\}$ and model λ are given, how to calculate the probability $P(O|\lambda)$ of the variable sequence O in the given model.

Problem 2: HMM optimal state sequence

When the observation sequence $O = \{o_1, o_2, ..., o_T\}$ and model λ are given, how to choose a corresponding state sequence $q = \{q_1, q_2, ..., q_T\}$ to make it optimal in a sense (for example, better to explain the observation variable).

Problem 3: HMM training (parameter estimation)

The problem is solved by Baum–Welch algorithm. The algorithm can identify a model $\lambda = (\pi, A, B)$, and make $P(O|\lambda)$ maximum for the given observation sequence $O = \{o_1, o_2, ..., o_T\}$. It is a functional extremum problem, and so there is no optimal scheme to estimate the λ.

1. Forward–backward algorithm

The forward–backward algorithm is the solution to HMM probability computation problem. The algorithm is used to calculate the probability $P(O|\lambda)$ generated from λ, when an observation value sequence $O = \{o_1, o_2, ..., o_T\}$ and a model $\lambda = (\pi, A, B)$ are given. According to the composition of HMM in Fig. 10.2, the most direct method to calculate $P(O|\lambda)$ is as follows: For a fixed states sequence, $S = q_1, q_2, ..., q_T$, there

$$P(O|S, \lambda) = \prod_{t=1}^{T} P(o_t|q_t, \lambda) = b_{q_1}(o_1)b_{q_2}(o_2)...b_{q_T}(o_T) \qquad (10.8)$$

where

$$b_{q_t}(o_t) = b_{jk}|q_t = \theta_j, o_t = v_k, \quad 1 \leq t \leq T \qquad (10.9)$$

And for a given λ, the probability of generating the state sequence S is

$$P(S|\lambda) = \pi_{q_1} a_{q_1 q_2}...a_{q_{T-1} q_T} \qquad (10.10)$$

Therefore, the probability is

$$P(O|\lambda) = \sum \pi_{q_1} b_{q_1}(o_1) a_{q_1 q_2} b_{q_2}(o_2) \ldots a_{q_{T-1} q_T} b_{q_T}(o_T) \qquad (10.11)$$

The computational quantities of the above formula are enormous, about 2TNT order of magnitude. When $N = 5$, and $T = 100$, the amount of computation reached 1072, which is totally unacceptable. In this case, it is required to find a more efficient algorithm to calculate the $P(O|\lambda)$, which is the forward–backward algorithm proposed by Baum et al.

(1) Forward algorithm

The forward variable in the forward algorithm is first defined as:

$$\alpha_t(i) = P(o_1, o_2, \ldots, o_t, q_t = \theta_i|\lambda), \quad 1 \le t \le T \qquad (10.12)$$

On this basis, the algorithm is divided into three steps:

① Initialization:

$$\alpha_1(i) = \pi_i b_i(o_1) \qquad (10.13)$$

② Recursive:

$$\alpha_{t+1}(j) = \left[\sum_{i=1}^{N} \alpha_t(i) a_{ij} \right] b_j(o_{t+1}), \quad 1 \le t \le T, \ 1 \le j \le N \qquad (10.14)$$

③ Finality:

$$P(O|\lambda) = \sum_{i=1}^{N} \alpha_T(i) \qquad (10.15)$$

In forward algorithm, recursion is the kernel step, and the arithmetic diagram is shown in Fig. 10.3. This figure illustrates how the state $i(1 \le i \le N)$ at the moment t changes to state j at the moment $t + 1$. The forward algorithm is shown in Fig. 10.4.

Through the forward algorithm, it can greatly reduce the amount of computation. Still $N = 5$, $T = 100$, for example, the amount of computation is about 3000 times, much less than 1072, making it possible to calculate the $P(O|\lambda)$.

Fig. 10.3 Recurrence
relations of the moments t

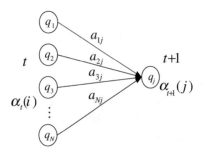

Fig. 10.4 Lattice structure of
forward algorithm

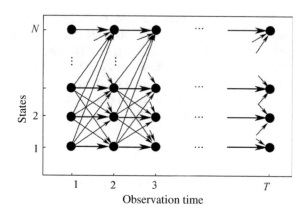

(2) Backward algorithm

Like the forward algorithm, the definition of the backward variable is:

$$\beta_t(i) = P(o_{t+1}, o_{t+2}, \ldots, o_T | q_t = \theta_i, \lambda), \quad 1 \leq t \leq T - 1 \tag{10.16}$$

Among them, $\beta_T(i) = 1$, the computational procedure of the backward algorithm is as follows:

① Initialization:

$$\beta_T(i) = 1, \quad 1 \leq i \leq N \tag{10.17}$$

② Recursive:

$$\beta_t(i) = \sum_{j=1}^{N} a_{ij} b_j(o_{t+1}) \beta_{t+1}(j), \quad t = T - 1, T - 2, \ldots, 1, \ 1 \leq i \leq N \tag{10.18}$$

③ Finality:

$$P(O|\lambda) = \sum_{i=1}^{N} \beta_1(i) \tag{10.19}$$

When initialized in the backward algorithm, $\beta_T(i)$ is equal to 1 for all the states of i. Similar to the forward algorithm, the backward algorithm is also a lattice structure algorithm.

2. Viterbi algorithm

The Viterbi algorithm is used to determine the problem of an optimal state sequence $Q^* = q_1^*, q_2^*, \ldots, q_T^*$, when an observation sequence $O = \{o_1, o_2, \ldots, o_T\}$ and a model of $\lambda = (\pi, A, B)$ are given. The optimal Q^* here refers to the sequence of the state when the maximum of $P(Q|O, \lambda)$ is defined. It can be realized by Viterbi algorithm. The description is as follows:

$\delta_t(i)$ is defined as the maximum probability of $o_1, o_2, \ldots, o_t$ produced by moments t along a path $q_1, q_2, \ldots, q_t$, and $q_t = \theta_i$, i.e.:

$$\delta_t(i) = \max_{q_1,\ldots,q_{t-1}} P(q_1, q_2, \ldots, q_t, q_t = \theta_i, o_1, o_2, \ldots, o_t|\lambda) \tag{10.20}$$

The process to calculate the optimal state sequence Q^* is divided into four steps:

(1) Initialization:

$$\delta_t(i) = \pi_i b_i(o_1), \quad 1 \le i \le N \tag{10.21}$$

$$\varphi_t(i) = 0, \quad 1 \le i \le N \tag{10.22}$$

(2) Recursive:

$$\delta_t(j) = \max_{1 \le j \le N_{t-1}} \left(\left[\delta_{t-1}(i) a_{ij} \right] \right) b_j(o_t), \quad 2 \le t \le T, \ 1 \le j \le N \tag{10.23}$$

$$\varphi_t(j) = \arg \max_{1 \le j \le N} \left[\delta_{t-1}(i) a_{ij} \right], \quad 1 \le t \le T, \ 1 \le j \le N \tag{10.24}$$

(3) Finality:

$$P^* = \max_{1 \le i \le N} \left[\delta_T(i) \right] \tag{10.25}$$

$$q_T^* = \arg \max_{1 \le i \le N} \left[\delta_T(i) \right] \tag{10.26}$$

(4) Determine the optimal sequence of States

$$q_t^* = \varphi_{t+1}(q_{t+1}^*), \quad t = T - 1, T - 2, \ldots, 1 \tag{10.27}$$

In the Eq. (10.24), the symbol arg max is defined as follows: if $i = I$, $f(i)$ reaches the maximum, then $I = \arg \max_{1 \le i \le N} [f(i)]$.

It should be showed that the actual Viterbi algorithm (except for the last step) is similar to the forward computation in the Eqs. (10.13)–(10.15). The main difference is the maximum of the former state in the Eq. (10.23). While the Eq. (10.14) uses summation.

3. Baum–Welch algorithm

This algorithm is used to solve the problem of HMM training, that is, the estimation of HMM parameters. The Baum–Welch algorithm can be described as follows: the algorithm can determine a model $\lambda = (\pi, A, B)$ to make the $P(O|\lambda)$ maximum when a sequence of observed values of $O = \{o_1, o_2, \ldots, o_T\}$ was given. This is a functional extremum problem. Hence there is no optimal scheme to estimate the λ. In this case, the Baum–Welch algorithm makes use of the recursion idea to make the $P(O|\lambda)$ reach its local maximum, and finally get the parameters of the model.

$\xi_t(i, j)$ is defined as the probability when the Markov chain is in the θ_i state at the time t and the θ_j state at the time $t + 1$ in a given training sequence O and model λ, i.e.,

$$\xi_t(i,j) = P(O, q_t = \theta_i, q_{t+1} = \theta_j | \lambda) \tag{10.28}$$

Based on the definition of the forward variable and the backward variable, we can derive:

$$\xi_t(i,j) = \left[\alpha_t(i) a_{ij} b_j(o_{t+1}) \beta_{t+1}(j) \right] / P(O|\lambda) \tag{10.29}$$

Then at the t moment, the probability of Markov chain at the θ_i state is:

$$\xi_t(i) = P(O, q_t = \theta_i | \lambda) = \sum_{j=1}^{N} \xi_t(i,j) = \alpha_t(i) \beta_t(i) / P(O|\lambda) \tag{10.30}$$

Therefore, $\sum_{t=1}^{T-1} \xi_t(i)$ represents the expectation of the state θ_i moving away. $\sum_{t=1}^{T-1} \xi_t(i,j)$ represents the expectation of the state transferring from θ_i to θ_j. So, the famous revaluation formula of Baum–Welch algorithm is derived:

$$\bar{\pi}_i = \xi_t(i) \tag{10.31}$$

$$\bar{a}_{ij} = \sum_{t=1}^{T-1} \xi_t(i,j) \bigg/ \sum_{t=1}^{T-1} \xi_t(i) \tag{10.32}$$

$$\bar{b}_j(k) = \sum_{t=1}^{T} \xi_t(j) \bigg/ \sum_{t=1}^{T} \xi_t(j), \quad \text{and} \quad o_t = v_k \tag{10.33}$$

So the process of calculating HMM parameters $\lambda = (\pi, A, B)$ is: According to the observation sequence O and the selection of initial model $\lambda_0 = (\pi, A, B)$, we can obtain a new set of parameters $\bar{\pi}_i$, $\bar{a}_{ij}$ and $\bar{b}_j(k)$ from the reestimation Eqs. (10.31), (10.32) and (10.33). At the same time, we can get a new model $\bar{\lambda} = (\bar{\pi}, \bar{A}, \bar{B})$. It can be proved that $P(O|\bar{\lambda}) > P(O|\lambda)$, that is, $\bar{\lambda}$ obtained by the reestimation formulas is better than λ representing the observation sequence O. Repeat this process and improve the model parameters gradually until $P(O|\bar{\lambda})$ can satisfy certain convergence condition, that is $P(O|\bar{\lambda})$ no longer enlarges obviously. At this time λ is the model we want.

10.1.3 The Type of HMM

HMM has some different models like continuous hidden Markov model (CHMM), discrete hidden Markov model (DHMM), semi-continuous hidden Markov model (semi-continuous HMM), linear prediction Markov model (Linear Predictive HMM), autoregressive hidden Markov model (autoregressive HMM), and so on. There are some schematic diagrams of common HMM structure in Figs. 10.5, 10.6, 10.7, and 10.8 as follows.

In the left and right type of HMM models above, the state transfer can only be from left to right and cannot be reversed. This model is well suitable for time-varying signals like speech signals etc. So it is common in practice application.

Fig. 10.5 Left and right type of HMM with five states

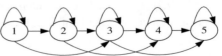

Fig. 10.6 First order across of the left and right HMM with five states

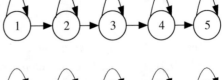

Fig. 10.7 Ergodic HMM
with four states

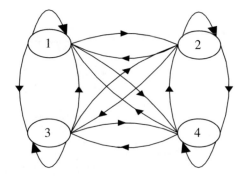

Fig. 10.8 Parallel path of left
and right HMM with six
states

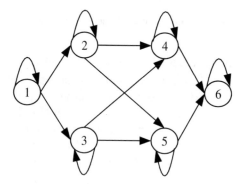

10.1.4 Improvement Measures of HMM in Practical Application

1. Selection of initial model

According to the Baum–Welch algorithm, when obtaining the parameters of the HMM from the training data set, an important problem is the selection of the initial models. Different training results will be obtained from different initial models. Because the parameters of the model are obtained by the algorithm, it makes $P(O|\lambda)$ reach its local maximum. Therefore, it is meaningful to select the good initial model, which makes the local maximum and the global maximum close at last. However, there is still no perfect answer to this question. Some experiences are often used in practice. Generally, the parameters of π and A values had little influence on the initial selection and can be selected randomly or uniformly. However, the initial value of B has a greater influence in the training of HMM. And a more complicated initial value selection method would be adopted generally. Based on this consideration, a typical HMM parameter estimation procedure is shown in Fig. 10.9. The initial model can be arbitrary selected here. However, due to $P(O|\hat{\lambda}) > P(O|\lambda), \hat{\lambda}$ is the improved model of the model λ. Then make $\hat{\lambda}$ as the initial value and use revaluation formula and $\bar{\lambda}$ will be obtained, avoiding the error

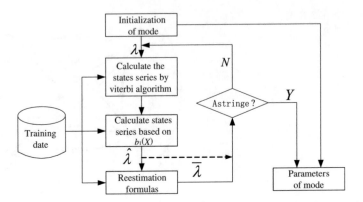

Fig. 10.9 Improved HMM parameter estimation method sketch map

initial value of the selection and changing the classic training process $\lambda \to \bar{\lambda}$ to $\lambda \to \hat{\lambda} \to \bar{\lambda}$. The dotted line in the graph indicates that the $\hat{\lambda}$ can be approximated as a model parameter without revaluation.

Of course, the Markov chain in HMM has different shapes. Therefore, different effective selection methods of initial values can be used for different forms of HMM. In the next content, the HMM initial conditions are set according to the constraints of the left and right type of HMM.

2. The processing of underflow problem

In the forward-backward algorithm and Baum–Welch algorithm, they both have the recursive computations of $\alpha_t(i)$ and $\beta_t(i)$. Since all quantities are less than 1, the $\alpha_t(i)$ (increasing with t) and the $\beta_t(i)$ (decreasing with the t) tend to be zero rapidly. In order to solve the underflow problems, the method of increasing the proportion of factor must be used to revise the algorithm. The process is:

(1) The processing of α

$$\alpha_1(i) = \pi_i b_i(o_1) \tag{10.34}$$

$$\alpha_1^* = \frac{\alpha_1(i)}{\sum_{i=1}^{N} \alpha_1(i)} \approx \frac{\alpha_1(i)}{\Phi_1}, \quad 1 \leq i \leq N \tag{10.35}$$

$$\tilde{\alpha}_{t+1}(j) = \left[\sum_{i=1}^{N} \alpha_t^*(i) a_{ij} \right] b_j(o_{t+1}), \quad 1 \leq j \leq N, \ t = 1, 2, \ldots, T-1 \tag{10.36}$$

$$\alpha_{t+1}^*(j) = \frac{\alpha_{t+1}(j)}{\sum_{j=1}^{N} \tilde{\alpha}_{t+1}(j)} \approx \frac{\alpha_{t+1}(j)}{\Phi_{t+1}}, \quad 1 \leq j \leq N, \ t = 1, 2, \ldots, T-1 \tag{10.37}$$

(2) The processing of β

$$\beta_T(i) = 1, \quad 1 \leq i \leq N \tag{10.38}$$

$$\beta_T^*(i) = 1, \quad 1 \leq i \leq N \tag{10.39}$$

$$\tilde{\beta}_t(i) = \sum_{j=1}^{N} a_{ij} b_j(o_{t+1}) \beta_{t+1}^*(j), \quad 1 \leq i \leq N, \ t = 1, 2, \ldots, T - 1 \tag{10.40}$$

$$\beta_t^*(i) = \frac{\tilde{\beta}_t(i)}{\Phi_{t+1}}, \quad 1 \leq i \leq N, \ t = 1, 2, \ldots, T - 1 \tag{10.41}$$

(3) The processing of the calculation formula of the probability $P(O|\lambda)$

After the above processing, in order to keep the calculation results of the original formula, the corresponding processing of the formula of probability $P(O|\lambda)$ is necessary to eliminate the influence of scale factor.

It is easy to launch by a process of α:

$$\alpha_t^*(i) = \alpha_t(i)/\Phi_1\Phi_2\ldots\Phi_t \tag{10.42}$$

$$\Phi_t = \sum_{j=1}^{N} \tilde{\alpha}_t(j) = \sum_{j=1}^{N} \left[\sum_{i=1}^{N} \alpha_{t-1}^*(i) a_{ij} \right] b_j(o_t) = \sum_{j=1}^{N} \alpha_t(j)/\Phi_1\Phi_2\ldots\Phi_{t-1} \tag{10.43}$$

Therefore

$$\sum_{j=1}^{N} \alpha_t(j) = \Phi_1\Phi_2\ldots\Phi_t \tag{10.44}$$

That is

$$P(O|\lambda) = \sum_{j=1}^{N} \alpha_T(j) = \Phi_1\Phi_2\ldots\Phi_T \tag{10.45}$$

Or

$$\log P(O|\lambda) = \sum_{t=1}^{T} \log \Phi_t \tag{10.46}$$

(4) The processing of Viterbi algorithm

Add a logarithmic processing in the original Viterbi algorithm. That is defined

$$\delta_t(i) = \max_{q_1,\dots,q_{t-1}} \log P(q_1,\dots,q_t, q_t = \theta_i, o_1, o_2,\dots,o_t|\lambda) \tag{10.47}$$

Then the initialized formula becomes:

$$\delta_1(i) = \log \pi_i + \log b_i(o_1), \quad 1 \le i \le N \tag{10.48}$$

Recursive arithmetic formula becomes:

$$\delta_i(i) = \max_{1 \le i \le N_1} \left[\delta_{t-1}(i) + \log a_{ij} \right] + \log \left[b_j(o_t) \right] \tag{10.49}$$

The ending formula becomes:

$$\log P^* = \max_{1 \le i \le N_1} \left[\delta_t(i) \right] \tag{10.50}$$

The value is the logarithm value of P^*, rather than the value of P^*. It should be pointed out that in order to avoid the probability values calculated is too small, log $P(O|\lambda)$ is always used. In fact, the value of a single $P(O|\lambda)$ is somewhat different from that of the normal probability. The single concrete $P(O|\lambda)$ is always so small that it is not too much meaningful. In general, comparing the different values of $P(O|\lambda)$ is meaningful.

3. Improved method of classical training algorithm

The HMM training method (Baum–Welch) is a maximum likelihood estimation (ML algorithm) algorithm essentially. It may be possible to reach the local minimum in the training process. So people put forward a lot of improvement measures. The most representative of them is the maximum mutual information (MMI) criteria [146]. However, the effective valuation method for MMI similar to the forward-backward algorithm for the ML has not been found at present. Therefore, the algorithm to make the mutual information maximum is the classical maximum gradient algorithm. People have done a lot of work. There is an example that an effective method of MMI training is the gradient method to speed up the convergence and so on.

There are also many improvements to the classical Baum–Welch algorithm such as the optimization of the Vector Quantization (VQ) together in the discrete HMM training, segmental K-means training algorithm, the competition training algorithm similar to neural network, the linear interpolation method by decompounding HMM into several more accurate models; the method not only assesses the HMM parameters, but also optimizes the HMM structure; the method statistical matrix quantization (MQ) method replaces the general VQ to improve the discrete HMM method; the method using the most likely state sequence instead of the all state sequence in classic ML method. However, the work is not over, and it is still an active research direction to search for more effective parameters.

10.1.5 A Pump Fault Diagnosis Turbo Based on HMM

There are many multiple faults in the operation of turbopump system such as friction, loosening, cracks, breakage, rotating stall, oil film vortex motion, and oil film oscillation. All the faults will lead to the appearance of dynamic signal non-stationary. And the nonlinearity of the driving force, the damping force, and the elastic force and the structure of the device are also nonlinear in the dynamic signal. Even for a steady operation turbine pump, when there are some failures caused by some faults such as rub, impact, etc., rotor damping, stiffness, and elastic force will change and exhibit a nonlinear, and nonstationary vibration signals. The dynamic signals obtained from the engineering are relative and local. And the nonstationary nonlinear is absolute and wide. Because it is difficult to obtain the mechanical pump test data, the vibration signal of the turbopump is considered of low repeatability. The HMM is a self learning probability classification statistical model for dynamic time series. And it is suitable for the analysis of some nonlinear, nonstationary, and reproducible bad signals. The HMM method was introduced to the field of state monitoring [151] and fault diagnosis [152] circa 2000, and it has achieved good effects. In HMM, if the observed variables are continuous, the model is continuous HMM and is called CHMM in short. If the observation variable is a discrete symbol, the model is discrete HMM, DHMM in short. Because the DHMM algorithm is stable, simple, and intuitive, according to the idea of DHMM, a HMM fault diagnosis method for a turbopump system is proposed based on the normalized eigenvalue of fault frequency.

1. HMM method for fault diagnosis of turbo pump system

Many fault phenomena can be described by the plurality of dynamic variables; so, these fault diagnosis problems can be seen as a dynamic pattern recognition problem. Dynamic pattern recognition problem can be defined as a classification problem to seek fault type ω_j pattern with given input mode X_t. At the given time t, in math, X_t is defined as

$$X_t = (x_1, x_2, \ldots, x_d) \tag{10.51}$$

Enter all possible modes collection of the model space. It is $X_t \subset R_d$, R_d is a d-dimensional real vector space. The observed sequence k at the time t of arrival is defined as:

$$\Phi_{t-k} = \{X_{t-k+1}, \ldots, X_{t-1}, X_t\} \tag{10.52}$$

The possible fault types of the ω_j form the space Ω of the classification is

$$\Omega(t) = \{\omega_1, \omega_2, \ldots, \omega_c\} \tag{10.53}$$

c is the number of classification. The task of fault classification is to find a mapping from input pattern space Φ_{t-k} to space classification Ω:

$$f: \Phi_{t-k} \rightarrow \Omega(t) \tag{10.54}$$

If the probability of $P(\omega)$ is known, the probability is called priori probability. According to the Bayes theory of probability, posterior probability $P(\Phi_{t-k}|\omega)$ can provide a reliable basis for the classification of events. According to Bayes rule, the posteriori probability is

$$P(\hat{\omega}|\Phi_{t-k}) \propto \max\{P(\Phi_{t-k}|\omega)P(\omega)\} \tag{10.55}$$

In fact, the prior probability $P(\omega)$ is hard to obtain. It is assumed that the event of occurrence is equal probability in general. Then the size of the conditional probability $P(\Phi_{t-k}|\omega)$ determine the input space pattern classification completely.

HMM is a double random process, which can realize the classification of the above patterns under the framework of probability. The parameters of HMM can be estimated from the observation data by Baum–Welch algorithm. In the sense of maximum likelihood, the algorithm can guarantee that the parameters of the model can be improved in every step of the iterative process. The classification of the input pattern space can be realized by computing the highest probability of the input mode.

The general process of the DHMM fault diagnosis method for the turbopump is shown in Fig. 10.10.

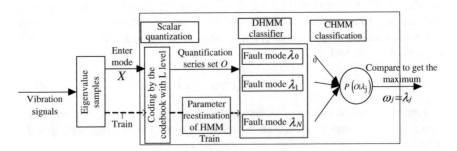

Fig. 10.10 DHMM fault diagnosis design block diagram

2. Selection of fault feature variables

The vibration signal of the turbopump reflects the running state of the equipment. The results of time domain, frequency domain, and amplitude analysis can be used as the fault signs. Fault diagnosis is a problem of state recognition and classification in essence. Because the type of fault is closely related to the vibration frequency of the unit and vibration, the signal of the turbopump has obvious characteristics in the frequency domain. The frequency domain characteristic of vibration signal is the main fault symptom [153].

Since the occurrence of the fault is often characterized by a certain spectrum distribution, the frequency range is divided into several characteristic frequency bands. Take the rotational frequency f of the multiple into consideration, as shown in Table 2.2. The table represents the frequency features and some main phenomena of common faults of turbine pump which has been normalized. The first 10 frequency features are extracted from the Table 2.2 and constitute the 10 dimensional feature vectors. Note that the first three failures of the table have the same frequency characteristics. So when you distinguish the three faults, you need the following nine features. The first three kinds of faults are referred to the rotor unbalance fault.

3. Scalar quantization of feature vector

When establishing the DHMM, it is required that the observation values are limited discrete values, while the vibration signal is continuous signal. So the observed feature vector is normalized and scalar quantization is used to form the discrete coding set. This process is called the scalar quantization of vector. The quantized sequence can be trained and classified as DHMM training code.

Scalar quantization is based on the amplitude of the signal. The signal value is divided into $N - 1$ regions. And the regions are ranged in ascending order. The N regions nearby of the signal are mapped to N discrete values. The index value $\mathrm{indx}(x)$ of the signal x in each region is defined as:

$$\mathrm{indx}(x) = \begin{cases} 1 & x \leq \mathrm{partition}(1) \\ i & \mathrm{partition}(i) < x \leq \mathrm{partition}(i+1) \\ N & \mathrm{partition}(N-1) < x \end{cases} \quad (10.56)$$

i is the nature number in the formula. In order to achieve scalar quantization, a partition vector (partition) with a length of $N - 1$ and a codebook (codebook) with a length of N are need be defined. The signal is divided into N regions according to the $N - 1$ partition vector ascending order. The value of each region is decided by the codebook vector based on the endpoint value or the average of the region. Since each region has an output value, the actual number of partitions is equal to the length of the code.

The quantized distortion is defined as the average of the D-value square of the original input signal sig and the quantized signal quan, i.e.,:

$$\text{distortion} = \frac{1}{N} \sum_{i=1}^{M} (\text{sig}(i) - \text{quan}(i))^2 \tag{10.57}$$

where M is the sampling number of the source signal.

The important step of scalar quantization is to generate the partition and codebook parameters. This process can be implemented using Lloyds algorithm [154]. The basic idea is to get the best partition and codebook parameters by reducing the quantization distortion of the input set. The processes of scalar quantization coding and decoding are shown in Fig. 10.11. Because the topological features and the size of the original signal are retained, the quantization index indx(x) can be used as the integer quantization codebook of the signal x.

According to the actual effect of data discrimination and DHMM fault classification, the maximum code word is set to 20 and the minimum code word is 1, that is, the code level is 20. Figure 10.12 reflects the code's length and distortion in speech recognition. In the speech recognition, the length of the code is generally selected between 32 and 256. Because of the similarity of the vibration signal and the speech signal, we set the length of the code 32.

4. Improvement of HMM algorithm for multi observation sample sequences

When establishing the HMM, the single observation sample sequence is not conducive to the revaluation of the model parameters. Because it can get more reliable model parameters by using enough data set, it is necessary to use the HMM with multiobservation sample sequence train. However, the training algorithm of

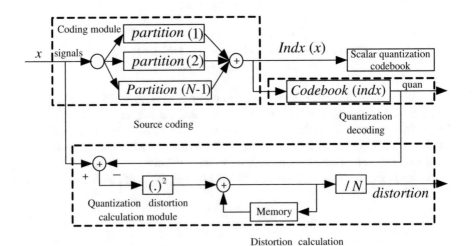

Fig. 10.11 The process of the vector scalar quantization coding and decoding calculation

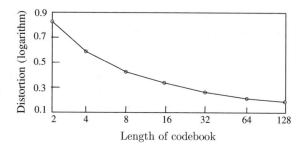

Fig. 10.12 The relationship between the length and the distortion of the code

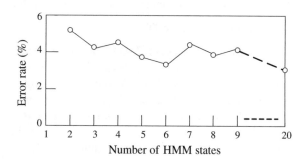

Fig. 10.13 Relationship between HMM state number and error rate

multisample model could not be realized by using the single-sample calibration technology. In practice, the algorithm should be improved to adapt to the situation of multiobservation sample sequence and solve the problem of overflow.

5. The establishment of HMM training and fault diagnosis model

Before the fault diagnosis, it is necessary to train the sample of the fault mode of the turbine pump and establish the corresponding DHMM model to constitute a fault diagnosis model library.

(1)   Selection of DHMM initial model

Figure 10.13 is the relationship between the number of states and the error rate in HMM [155]. It can be seen from the graph that when the state number is 6, the error rate reaches the local minimum. Then the error rate will grow with the increasing of the states' number. When the number of states is 20, a local minimum is reached again.

By referring to the relevant references and the actual situation, the model with five hidden states is adopted in the modeling of DHMM.

The initial probability distribution vector is $\pi = [1, 0, 0, 0, 0]$. In the model, the initial observation probability matrix is determined according to the uniform selection method. That is, it assumes that there are 20 codes appeared with the form of equal probability. Since the initial model chose the ergodic model with 5 states, the initial value of the probability matrix of the state transition is:

$$A = \begin{bmatrix} 0.2 & 0.2 & 0.2 & 0.2 & 0.2 \\ 0.2 & 0.2 & 0.2 & 0.2 & 0.2 \\ 0.2 & 0.2 & 0.2 & 0.2 & 0.2 \\ 0.2 & 0.2 & 0.2 & 0.2 & 0.2 \\ 0.2 & 0.2 & 0.2 & 0.2 & 0.2 \end{bmatrix}$$

After establishing the initial conditions of the model, the Baum–Welch revaluation process uses the modified multiobservation sample revaluation formula.

(2)   Training process of DHMM

In the training process of HMM, the maximum likelihood value of the maximum likelihood estimates also increases with the increasing iterative times until the convergence error is reached. Choose the convergence error with 10–4. The main parameters of the model after training are the probability matrix of the state transfer and the probability matrix of the observation value. The HMM training curves of the fault models are shown in Fig. 10.14. And from the iterative curve, we could see that the HMM has a strong learning ability [154].

6. HMM fault diagnosis

Based on the fault diagnosis process of HMM, the quantification series of the observation value is used as the input vector. Then, the forward–backward algorithm or Viterbi algorithm is used to calculate the probability $P(O|\lambda)$. We make a decision by comparing the maximum output. In Table 2.2, the partial mode sample is added with 5 % of the random noise, which form the diagnostic test sample, as shown in Table 10.1. There are 10 test samples [154].

The content in Table 10.2 is the similar probabilities and diagnostic results of the HMM models. As shown in the black part of the table, the test sample can be classified

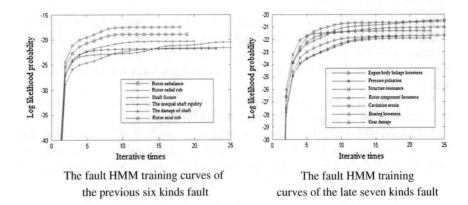

The fault HMM training curves of   The fault HMM training
the previous six kinds fault   curves of the late seven kinds fault

Fig. 10.14 The HMM training curves of each fault model

Table 10.1 The test sample with 5 % random noise added

Sample	0.00–0.39fr	0.40–0.49fr	0.50fr	0.51–0.99fr	1.0fr	2.0fr	3.0–5.0fr	0ddfr	>5.0fr	Meshing frequency
1	0	0	0	0.04728	0.90183	0.05555	0.02408	0	0.03481	0
2	0	0	0.02612	0	0.94369	0.09275	0.04464	0	0.01664	0.02
3	0.07578	0.05649	0.0077	0.07757	0.31392	0.05501	0.12734	0.11668	0.08608	0
4	0	0	0	0.02947	0.00794	0.77422	0.16392	0.01569	0.02130	0
5	0.13678	0.09218	0	0	0.4444	0.15617	0.19867	0.00386	0	0
6	0.1888	0.20913	0.04517	0.03403	0.28852	0.13413	0.14987	0.07275	0.0316	0
7	0.22652	0.19279	0	0.02907	0.06523	0.06421	0.34185	0.00980	0.14377	0
8	0.45333	0.33221	0	0	0	0.03768	0.05548	0.05455	0.04122	0
9	0.74529	0.02933	0	0.04332	0.02926	0.24077	0	0.0413	0.05932	0
10	0.04064	0	0	0.03686	0.24332	0.00709	0.02382	0	0	0.76

Table 10.2 HMM output values and diagnostic results

Sample	Fault 1	Fault 4	Fault 5	Fault 6	Fault 7	Fault 8	Fault 9	Fault 10	Fault 11	Fault 12	Fault 13	Fault 14	Fault 15	Results
1	**-28.92**	-109.6	-Inf	-118.2	-159.6	-65.39	-65.03	-67.19	-61.89	-69.64	-111.9	-110.5	-129.1	1
2	**-39.44**	-98.85	-381.5	-129.7	-169.2	-277.4	-242.1	-259.4	-217.4	-245.4	-111.9	-111.5	-151.6	1
3	-Inf	**-21.42**	-119.3	-147.1	-172.4	-105.4	-161.6	-90.45	-225.9	-120.6	-120.2	-121.2	-189.0	4
4	-Inf	-99.58	-Inf	-149.3	**-29.28**	-69.49	-69.95	-67.41	-62.81	-65.62	-119.7	-112.1	-29.45	7
5	-Inf	-89.61	-427.9	-135.2	-190.7	**-32.58**	-159	-88.11	-239.5	-119.2	-119.1	-119.7	-208.9	8
6	-Inf	-110.6	-388.9	-95.91	-109.2	-287.1	**-33.08**	-35.07	-39.84	-248.5	-117.4	-119.5	-59.42	9
7	-Inf	-78.65	-428.3	-160.7	-139.3	-109.5	-128.4	**-35.02**	-195.5	-119.5	-117.4	-119.5	-151.8	10
8	-Inf	-169.8	-Inf	-300.1	-309.5	-282.1	-259.4	-265.2	-111.2	-249.9	**-33.27**	-119.4	-279.2	13
9	-Inf	-209.8	-Inf	-267.2	-208.6	-278.8	-218.9	-228.6	-209.3	-249.3	-119.2	**-32.22**	-188.6	14
10	-Inf	-99.59	-Inf	-85.88	-27.28	-69.48	-39.74	-39.79	-59.01	-65.62	-119.6	-112.1	**-29.05**	15

Table 10.3 HMM classification results of 5 % noise intensity

Sample	Fault 1	Fault 4	Fault 5	Fault 6	Fault 7	Fault 8	Fault 9	Fault 10	Fault 11	Fault 12	Fault 13	Fault 14	Fault 15
1	50												
4		46	3			1							
5		2	47			1							
6				50									
7					50								
8		1	2			46	1						
9						1	49						
10								50					
11									50				
12										49	1		
13										2	48		
14												50	
15													50

Table 10.4 HMM classification results of 10 % noise intensity

Sample	Fault 1	Fault 4	Fault 5	Fault 6	Fault 7	Fault 8	Fault 9	Fault 10	Fault 11	Fault 12	Fault 13	Fault 14	Fault 15
1	50												
4		40	7			1	1	1					
5		4	42	1		1	1		1				
6		2		46		1	1						
7					50								
8		1	1			46	1		1				
9		2	1			2	45						
10			1					48		1			
11		1		1		2	2		44				
12										47	3		
13										4	46		
14												50	
15													50

correctly. Meanwhile, the diagnostic results show that the HMM method has a high diagnostic accuracy. Taking sample 3 as an example, the logarithmic likelihood probability of the radial rub impact of the rotor is -21.4168. But the maximal probability of the other 12 faults may be the pressure fluctuations, whose log likelihood is -90.4489. Therefore, it can be judged that the possibility of radial rub is about 1069 times than that of the radial rub, thus proving the strong classification ability of HMM. Of course, like the similar samples 10, the probability of the test sample for the gear damage is -26.05. And the unequal probability of the shaft stiffness is -27.28. The difference is 101.23 times. That is the probability of damage of the gear in the sample 10 is 5.56 %, and the unequal possibility of the shaft rigidity is 99.4 %.

7. Diagnostic tests of HMM method

Add each of the fault mode samples in the Table 2.2 (the first three merging models for a class) with 5 and 10 % random noise, respectively, forming the testing samples. Then proceed with the classification test. Each pattern was tested 50 times. The results are shown in Tables 10.3 and 10.4, respectively. In Tables 10.3 and 10.4, the number represents the times of testing sample diagnosis as each fault mode and the vacancy is 0.

It shows that the average recognition rate of the fault pattern classification with the 5 % noise intensity can reach 97.69 % by using the HMM method from Table 10.3. From Table 10.4, the average recognition rate of the fault pattern with the 10 % noise intensity is 92.92 % by using the method of HMM. The experimental results show that the fault diagnosis method based on HMM has the advantages of stability, high classification accuracy, and so on.

10.2   HMM-SVM Hybrid Fault Diagnosis Model and Its Application

The fault diagnosis based on HMM is determined by the maximum output probability in all HMM models. However, there may be not much output probability difference between a contaminated pending recognition signal and a number of possible HMM models. There exists potential risk of a miscarriage of justice, when only based on the maximum probability. But the fault diagnosis based on support vector machine (SVM) is based on the signal features of the current signal, neglecting the relationship of the moment. The results which will win in the competition would not happen actually, resulting in miscarriage of justice [157].

SVM reflects the differences of the categories in greater degree and HMM reflects similarity within classes more. So the shortage caused only by the maximum probability of HMM can be overcome by the advantages of the classification by SVM. SVM has a great advantage on the statistical estimation and learning forecasting of small sample data. Because the difficulty of the fault acquisition is relatively large in the practical application, the SVM can play a good effect in the case of limited sample. And the scalar quantization can cause distortion to certain

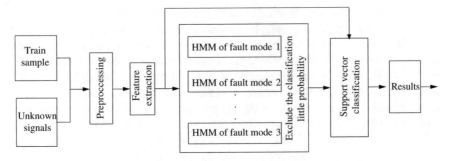

Fig. 10.15 The implementation of the HMM-SVM hybrid model block diagram

degree in the process of HMM fault diagnosis. But the SVM can complete the classification and judgment without distortion. Based on the above analysis, the HMM-SVM mixed fault diagnosis model is available. The implementation of the hybrid model block diagram is showed in Fig. 10.15.

10.2.1   SVM Training

Add 5 % noise in the standard sample of the fault mode 4, 5, and 8, and take each 10 group of data for SVM training, respectively. In Fig. 10.16, the left part is the process of parameters continuous adjustment in the training process. The results of training are showed in the right part. When the optimum C and γ values were 2 and 0.0078125, the accuracy of the classification can reach 100 %.

10.2.2   HMM-SVM Fault Diagnosis Application   Examples

Six fault modes of fault classification in Table 10.3 are as the study objects. That is, fault 4, fault 5, fault 8, fault 9, fault 12, and fault 13. According to the practical

Fig. 10.16 The process of SVM training and its results

Table 10.5   HMM-SVM classification results of 5 % noise intensity

	Fault 4	Fault 5	Fault 8	Fault 9	Fault 12	Fault 13
Sample 4	50					
Sample 5		50				
Sample 8			50			
Sample 9				50		
Sample 12					50	
Sample 13						50

Table 10.6   HMM-SVM classification results of 10 % noise intensity

	Fault 4	Fault 5	Fault 8	Fault 9	Fault 12	Fault 13
Sample 4	46	4				
Sample 5	3	47				
Sample 8	1		48	1		
Sample 9				50		
Sample 12					48	2
Sample 13					3	47

diagnosis effect, when HMM excluded half of the minimum probability HMM-SVM has the best effect on classification. 5 % noise and 10 % noise, are respectively, added to the above six kinds of fault mode samples. A new sample is formed. Each pattern is tested 50 times. And the results are shown in Tables 10.5 and 10.6.

The results showed that the HMM-SVM model can improve the classification accuracy. The correct recognition rates of two kinds of noise intensity recognized only by HMM are raised. One is raised from 95 % (The six kinds of error classification rates in Table 10.3) to 100 % and another one is raised from 88.67 % (as above) to 95.33 %, respectively. When the classification errors of HMM and SVM appear or HMM excludes the correct classification, wrong judgment of MM-SVM will appear.

Chapter 11
Fault Prediction Methods of Liquid Rocket Engine (LRE)

As the core of the rocket power system, LRE has direct influence on the reliability of the whole power system, the reliability of the whole rocket system, and the launch of the rocket. With the increase of payload, the working condition of the LRE becomes more and more complex and execrable. For example, it has to face the ablation of airflow with the high temperature, high pressure and high speed, the erosion of the propellant, and the vibration excitation with large amplitude and wide spectrum, etc. The reliability and security of LRE has faced more and more serious challenge. Therefore, corresponding condition monitoring system, especially the rapid, real-time, intelligent fault detection, and prediction technology, has attracted wide attentions in the field of aerospace and missile technology, and the prediction technology investigation has important scientific significance and practical value for the improvement of LRE reliability and security.

As an important guarantee for the reliability and security of LRE, the fault prediction technology provides an important way for the real-time fault detection and prediction and is the main technical basis of the state maintenance. Therefore, it is to realize that the fault prevention, the important method to improve real-time fault diagnosis system, is the main technical basis for state maintenance. In-depth developments through the study of liquid missile engine fault prediction technology improve the intelligent level of missile weapon system fault forecast technology, promote the application of artificial intelligence technology in the field of space, improve the reliability and security of the missile weapon system, and improve the operation and maintenance support capability of forces, which is of great significance.

This chapter denotes the investigation of the nonlinear time series prediction technology by the usage of artificial intelligence method. Combining the technology of ANN, gray system theory, and SVM with the prediction method, we developed the prediction model of the LRE to improve the discover capability of the initial fault.

© Springer-Verlag Berlin Heidelberg and National Defense Industry Press 2016
W. Zhang, *Failure Characteristics Analysis and Fault Diagnosis
for Liquid Rocket Engines*, DOI 10.1007/978-3-662-49254-3_11

The prediction model is used to estimate the future state based on the historical statistics data. In the view of control, the prediction is to model the dynamic process of the system. Generally, the measured value of dynamic process or sequence is named as a time series. Various prediction methods have been developed in the past several decades. The first one, named time extrapolation model method, constructs the model by the past statistic data. The second one, named causal extrapolation model method, constructs the model by the causal relationship, obtained by the regression analysis, of the data. The third method is to analyze the specific causal relationship and constructs corresponding causal model at first. Then the prediction model is obtained by the determination of the model parameters. No matter which kind of prediction method is used, the core of prediction is to construct the mapping relationship between the input and output data. In the engineering application, many mapping relationships are nonlinear and dynamic [158].

11.1 Fault Prediction Method Based on Time Series Analysis

The time series is a sequence of random variable sorted by the time order and contains the change information of the object. Influenced many accidental factors, the time series is commonly random and has only statistic relationship between each other. Apparently, the investigation of the time series analysis is to find out the embedded static characters and development rules. The basic idea is to establish a reasonable statistical model as much as possible according to the characteristics of the observed data and to explain the statistical rules in order to achieve the purpose of control or forecast.

11.1.1 Time Series Analysis

Generally, the time series analysis is mainly composed of random time series analysis model, model identification, parameter estimation, model test, etc.

(1) Random time series analysis model

The random time series analysis model is composed of the autoregressive (AR) model, moving average (MA) model, and autoregressive moving average (ARMA) model [158].

1. AR model

Suppose that the time sequence $\{x_t\}$ can be linearly expressed by the early terms, namely

$$X_t = \varphi_1 x_{t-1} + \varphi_2 x_{t-2} + \cdots + \varphi_p x_{t-p} + \mu_t \tag{11.1}$$

then the time series $\{x_t\}$ is named as autoregressive series and the model is called AR model of p order, denoted by AR(p). The parameters φ_1, φ_2, ..., φ_p are the weight coefficients. The random term μ_t denotes the white noise and obeys the normal distribution with the mean of 0 and variance of σ_μ^2. The random term μ_t is not relevant to the terms of x_{t-1}, x_{t-2}, ..., x_{t-p}.

Denoting B^k for the lag operator of k steps, namely $B^k x_t = x_{t-k}$, then Eq. (10.1) can be expressed as

$$x_t = \varphi_1 B x_t + \varphi_2 B^2 x_t + \cdots + \varphi_p B^p x_t + \mu_t \tag{11.2}$$

It can be transformed to be

$$(1 - \varphi_1 B - \varphi_2 B^2 - \cdots - \varphi_p B^p) x_t = \mu_t \tag{11.3}$$

Let $\varphi(B) = (1 - \varphi_1 B - \varphi_2 B^2 - \cdots - \varphi_p B^p)$, then Eq. (11.3) can be rewritten as

$$\varphi(B) x_t = \mu_t \tag{11.4}$$

2. MA model

If the time sequence $\{x_t\}$ can be linearly expressed by the early random term, namely

$$x_t = \mu_t - \theta_1 \mu_{t-1} - \theta_2 \mu_{t-2} - \cdots - \theta_q \mu_{t-q} \tag{11.5}$$

then the $\{x_t\}$ is called moving average series; the model is named as MA model of q order and denoted by MA(q). Parameters of θ_1, θ_2, ..., θ_q are weight coefficients.

Similarly, Eq. (11.5) can be rewritten as

$$(1 - \theta_1 B - \theta_2 B^2 - \cdots - \theta_q B^q) \mu_t = x_t \tag{11.6}$$

Let $\theta(B) = (1 - \theta_1 B - \theta_2 B^2 - \cdots - \theta_q B^q)$, then Eq. (11.6) can be

$$x_t = \theta(B) \mu_t \tag{11.7}$$

3. ARMA model

As one of the most common statistical analysis methods, the ARMA model can fit the time series by the usage of finite samples and has been one of the most mature methods in the time series analysis. First, a mathematical model is used to describe the determination performance of the random time series. Then, by analyzing the model, the structure and characteristics of time series can be understood more naturally and achieve the prediction of the time series with the minimum variance. By referring to Eqs. (11.1) and (11.5), the following equation exists:

$$x_t - \varphi_1 x_{t-1} - \varphi_2 x_{t-2} - \cdots - \varphi_p x_{t-p} = \mu_t - \theta_1 \mu_{t-1} - \theta_2 \mu_{t-2} - \cdots - \theta_q \mu_{t-q}$$

$$(11.8)$$

and the time series $\{x_t\}$ is called ARMA series; the model is called ARMA model with the order of (p, q) and denoted by ARMA(p, q). Parameters φ_1, φ_2, ..., φ_p are called regression parameters and θ_1, θ_2, ..., θ_q are called moving average parameters.

Introducing the lag operator B, Eq. (11.8) can be expressed as

$$\varphi(B)x_t = \theta(B)\mu_t \tag{11.9}$$

(2) Model identification

Model identification is used to determine the model structure and order by the usage of the autocorrelation function (ACF), partial correlation function (PCF), AIC, or FPE of the known time series and obtains an equivalent mathematical description for the physical process or dynamic system as possible as we can [158, 159]. Apparently, the basis task is to determine the parameters p and q.

1. ACF

ACF is able to measure the correlation property of the k-adjacent segment of the time series:

$$\rho_k = \frac{r_k}{r_0} \tag{11.10}$$

where $r_k = E(x_t, x_{t+k})$; $r_0 = E(x_t^2)$.

① The ACF of AR(p)

$$\rho_k = \frac{r_k}{r_0} = \varphi_1 \rho_{k-1} + \varphi_2 \rho_{k-2} + \cdots + \varphi_p \rho_{k-p} \tag{11.11}$$

The ACF of AR(p) is a truncated sequence and is called trail sequence.

② The ACF of MA(q)

$$\rho_k = \frac{r_k}{r_0} = \begin{cases} 1 & (k = 0) \\ (-\theta_k + \theta_1 \theta_{k+1} + \cdots + \theta_{q-k}\theta_q)/(1 + \theta_1^2 + \cdots + \theta_q^2) & (1 \leq k \leq q) \\ 0 & (k > q) \end{cases}$$

$$(11.12)$$

When $k > q$, x_t is not related to x_{t+k}, namely the truncated phenomenon. It is indicated that the order of $MA(q)$ model can be estimated by the observation of the zero start point of the ACF.

③ The ACF of ARMA(p, q)

The ACF of ARMA(p, q) is the mixture of the ACFs of $MA(q)$ and AR (p) models of the autocorrelation function of mixture. When $p = 0$, it denotes a truncated sequence, and when $q = 0$, it denotes a trailing sequence. When both p and q are not zero, it denotes a trailing sequence. If $k > q$, $r_k = \varphi_1 r_{k-1} + \varphi_2 r_{k-2} + \cdots + \varphi_p r_{k-p}$.

The ACF of ARMA(p, q) is

$$\rho_k = \frac{r_k}{r_0} = \varphi_1 \rho_{k-1} + \varphi_2 \rho_{k-2} + \cdots + \varphi_p \rho_{k-p} \tag{11.13}$$

When $k > q$, the ACF ρ_k only depends on the model parameters φ_1, φ_2, ..., φ_p and ρ_{k-1}, ρ_{k-2}, ..., ρ_{k-p}.

(2) PCF

For the time series $\{x_t\}$, suppose that the kth value can be linearly determined by x_{t-1}, x_{t-2}, ..., x_{t-k}. The minimum variance estimation of $\{x_t\}$ can be denoted as $Q = E(x_t - \sum_{l=1}^{k} \varphi_{k-l} \cdot x_{t-l})^2$. The determination of the coefficients $\varphi_{k-l}, l = 1, \ldots, k$ can be performed by the calculation of the partial derivation $\frac{\partial Q}{\partial \varphi_{k-l}} \frac{2Q}{2\varphi_{kl}}$. Namely

$$\begin{bmatrix} 1 & \rho_1 & \cdots & \rho_{k-1} \\ \rho_1 & 1 & \cdots & \rho_{k-2} \\ & & \vdots & \\ \rho_{k-1} & & \cdots & 1 \end{bmatrix} \begin{bmatrix} \varphi_{k1} \\ \varphi_{k2} \\ \vdots \\ \varphi_{kk} \end{bmatrix} = \begin{bmatrix} \rho_1 \\ \rho_2 \\ \vdots \\ \rho_k \end{bmatrix} \tag{11.14}$$

Equation (11.14) is the Yule–Walker (Y–K) equation, and the coefficient matrix is Toeplitz matrix.

Using the Cramer law, the following equations exist:

$$\varphi_{11} = \rho_1, \quad \varphi_{22} = \frac{\begin{vmatrix} 1 & \rho_1 \\ \rho_1 & \rho_2 \end{vmatrix}}{\begin{vmatrix} 1 & \rho_1 \\ \rho_1 & 1 \end{vmatrix}}, \quad \varphi_{33} = \frac{\begin{vmatrix} 1 & \rho_1 & \rho_1 \\ \rho_1 & 1 & \rho_2 \\ \rho_2 & \rho_1 & \rho_3 \end{vmatrix}}{\begin{vmatrix} 1 & \rho_1 & \rho_2 \\ \rho_1 & 1 & \rho_1 \\ \rho_2 & \rho_1 & 1 \end{vmatrix}}, \ldots,$$

$$\varphi_{kk} = \frac{\begin{vmatrix} 1 & \rho_1 & \rho_2 & \cdots & \rho_{k-2} & \rho_1 \\ \rho_1 & 1 & \rho_1 & \cdots & \rho_{k-3} & \rho_2 \\ \rho_2 & \rho_1 & 1 & \cdots & \rho_{k-4} & \rho_3 \\ \vdots & \vdots & \vdots & \cdots & \vdots & \vdots \\ \rho_{k-1} & \rho_{k-2} & \rho_{k-3} & \cdots & \rho_1 & \rho_k \end{vmatrix}}{\begin{vmatrix} 1 & \rho_1 & \rho_2 & \cdots & \rho_{k-1} \\ \rho_1 & 1 & \rho_1 & \cdots & \rho_{k-2} \\ \vdots & \vdots & \vdots & \cdots & \vdots \\ \rho_{k-1} & \rho_{k-2} & \rho_{k-3} & \cdots & 1 \end{vmatrix}} \tag{11.15}$$

When $k > p$, φ_{kk} is zero and $\varphi_{kk}(k = 1, 2, 3, \ldots)$ is called the partial correlation functions of time sequence $\{x_t\}$. Obviously, the partial correlation function φ_{kk} is constructed by autocorrelation function ρ_k. Partial correlation function is another statistical characteristic of the ARMA(p, q) model and is the measurement of the relationship between x_t and x_{t-k} under the condition of knowing the sequence values. For the AR(p), the cut-off characteristic of the sequence $\{\varphi_{kk}\}$ is

$$\varphi_{kk} \neq 0 \quad k \leq p$$
$$\varphi_{kk} = 0 \quad k > p$$

For he AR(p) model, the following equation always exists:

$$\varphi_{kj} = \begin{cases} \varphi_j & 1 \leq j \leq p, \\ 0 & p+1 \leq j \leq k, \end{cases} \quad k \geq p \tag{11.16}$$

The main characteristic of AR(p) is $\varphi_{kk} = 0$ when $k > p$. Namely, the partial correlation function of AR(p) is that φ_{kk} is p step truncation, which is the unique characteristics of AR(p). Apparently, the characteristics of ACF and PCF provide the basis for model identification.

The main characteristics of the stationary time series model are listed in Table 11.1.

(3) Parameter estimation

The main content of modeling is the parameter estimation and order determination. Since the structure, statistical properties, and required prediction accuracy are different, the parameter estimation methods of models are different. The commonly used parameter estimation methods include the moment estimation method, least square method, and the maximum likelihood method. Especially, the first two methods are the most frequently used. The moment estimation method does not require meeting certain optimization constraints and is called the coarse estimation. Comparatively, the least square method, to meet the constraint condition of "minimization the sum of squared residuals," has higher precision and is called accurate estimates. When N is bigger, the precisions of different accurate estimation

Table 11.1 Characteristics of AR, MA, and ARMA model

Category	Model		
	AR(p)	MA(q)	ARMA(p, q)
Model equations	$\varphi(B)x_t = \mu_t$	$x_t = \theta(B)\mu_t$	$\varphi(B)x_t = \theta(B)\mu_t$
Model described by x_t	$\mu_t = \varphi(B)x_t$	$\mu_t = \theta - 1(B)x_t$	$\mu_t = \theta - 1(B)\varphi(B)x_t$
Model described by μ_t	$x_t = \varphi - 1(B)\mu_t$	$x_t = \theta(B)\mu_t$	$x_t = \varphi - 1(B)\theta(B)\mu_t$
Stationary conditions	The root module of $\varphi(B) = 0$ is greater than 1	Always smooth	The root module of $\varphi(B) = 0$ is greater than 1
Reversible conditions	Always reversible	The root module of $\theta(B) = 0$ is greater than 1	The root module of $\theta(B) = 0$ is greater than 1
ACF	Trailing	Truncation	Truncation
PCF	Trailing	Truncation	Truncation

methods were similar. The least square method is simple, easy to understand, and therefore, it is used in the parameter estimation.

1. Least squares estimate of AR(p) model parameter

For the AR(p) model

$$x_t = \varphi_1 x_{t-1} + \varphi_2 x_{t-2} + \cdots + \varphi_p x_{t-p} + \mu_t, \quad t = p+1,\, p+2,\ldots,N$$

it can be rewritten in the matrix form as follows:

$$Y_N = X_N \Phi_N + A_N \tag{11.17}$$

where

$$Y_N = (x_{p+1}, x_{p+2}, \ldots, x_N)^T, \Phi_N = (\varphi_1, \varphi_2, \ldots, \varphi_p)^T, A_N = (\mu_{p+1}, \mu_{p+2}, \ldots, \mu_N)^T$$

$$X_N = \begin{bmatrix} x_1 & x_2 & \cdots & x_{N-p} \\ x_2 & x_3 & \cdots & x_{N-(p-1)} \\ \vdots & \vdots & & \vdots \\ x_p & x_{p+1} & \cdots & x_{N-1} \end{bmatrix}$$

The residual is

$$A_N = Y_N - X_N \Phi_N \tag{11.18}$$

Based on the known data, the least square method minimizes the summation of residual by choosing the estimation value of the unknown parameter $\hat{\Phi}_N$, namely

$$J = \sum_{t=p+1}^{N} \mu_t^2 = A_N^T A_N = \min$$

Let

$$\frac{\partial J}{\partial \Phi_N}\bigg|_{\Phi_N = \hat{\Phi}_N} = -2X_N^T Y_N + 2X_N^T X_N \hat{\Phi}_N = 0$$

the least square estimation of $\hat{\Phi}_N$ is given by

$$\hat{\Phi}_N = (X_N^T X_N)^{-1} X_N^T Y_N \tag{11.19}$$

As long as $X_N^T X_N$ is a full rank matrix, only one least square solution exists.

2. Least squares estimation of MA(q) and ARMA(p, q) model parameters

Generally, the parameter estimation methods of MA(q) and ARMA(p, q) models are similar, and only the least squares estimation method of ARMA(p, q) model parameters is introduced here.

Consider

$$x_t = \varphi_1 x_{t-1} + \varphi_2 x_{t-2} + \cdots + \varphi_p x_{t-p} + \mu_t - \theta_1 \mu_{t-1} - \theta_2 \mu_{t-2} - \cdots - \mu_q x_{t-q} \tag{11.20}$$

Calculate the residual of Eq. (11.20) for the given sequence of $\{x_t\}$, and $x_t = \mu_t = 0$ ($t \leq 0$),

$$\begin{cases} \mu_1 = x_1 \\ \mu_2 = x_2 - \varphi_1 x_1 + \theta_1 x_1 \\ \mu_3 = x_3 - \varphi_1 x_2 - \varphi_2 x_1 + \theta_1 (x_2 - \varphi_1 x_1 + \theta_1 x_1) + \theta_2 x_1 \\ \cdots \end{cases} \tag{11.21}$$

Import Eq. (11.21) into Eq. (11.8); then Eq. (11.8) can be expressed as

$$x_t = f_t(X, \beta) + \mu_t$$

where $\boldsymbol{\beta} = (\varphi_1, \ldots, \varphi_p, \theta_1, \ldots, \theta_q)^T$, $\boldsymbol{X} = (x_1, x_2, \ldots, x_N)^T$, $f_t(X, \beta)$ is a nonlinear function of $\boldsymbol{\beta}$, and μ_t is the model residual.

Define the summation of squared residual as

$$J(\boldsymbol{\beta}) = \sum_{t=1}^{N} \hat{\mu}_t^2 = \sum_{t=1}^{N} (x_t - f_t(\boldsymbol{X}, \boldsymbol{\beta}))^2 \tag{11.22}$$

$\hat{\beta}$, minimizing $J(\beta)$, is the least square estimation of β. Because $J(\beta)$ is not an explicit function of β, corresponding parameter estimations are usually using iterative method, such as Gauss–Newton method, the steepest descent method, and damping Gauss–Newton method [159].

(4) Model validation

The model validation is to evaluate the reliability of the model and the ultimate criterion of the model validation is to check whether the residual μ_t is the white noise model or not. If the residual sequence is not white noise, useful information may exist and the model needs to be improved. Generally, the randomness of the residual sequence is validated by the ACF and the χ^2 distribution. The initial assumption of the test is the residual sequences which are independent from each other.

If the residual of a sample value $\hat{\mu}_1, \hat{\mu}_2, \ldots, \hat{\mu}_N$ has been obtained, the ACF can be calculated by

$$\hat{r}_k(\mu) = \frac{\sum_{t=1}^{N-k} \hat{\mu}_t \hat{\mu}_{t+k}}{\sum_{t=1}^{N} \hat{\mu}_t^2} \quad k = 1, 2, \ldots, N-1 \tag{11.23}$$

Construct a statistic index by

$$Q_m = N \sum_{k=1}^{m} \hat{r}_k^2(\mu) \tag{11.24}$$

In this instance, the test on the white noise of residual sequence has transferred to be the test on the χ^2 distribution of Q_m with the freedom of m. The detailed steps are listed as follows:

1. Determine the significance level of α (commonly $\alpha = 0.05$ or $\alpha = 0.01$) and obtain the value of $\chi^2_{m,\alpha}$ by checking the statistic table (Table 11.2).
2. Calculate Q_m by Eq. (11.24). If $Q_m \leq \chi^2_{m,\alpha}$, then Q_m obey the χ^2 distribution with the freedom degree of m, the sequence μ_t is the white noise, and the model is suitable. Comparatively, if $Q_m > \chi^2_{m,\alpha}$, then the model needs to be improved. Generally, m is smaller than $N/4$. When N is more than several hundred, m can be set to be 20 or 30.

Table 11.2 χ^2 distribution of different α and m

α	m										
	20	21	22	23	24	25	26	27	28	29	30
0.05	31.4	32.7	33.9	35.2	36.4	37.7	38.9	40.1	41.3	42.6	43.8
0.01	37.6	38.9	40.3	41.6	43.0	44.3	45.6	47.0	48.3	49.6	50.9

11.1.2 Application and Analysis

In this section, the ARMA model is applied to the prediction of oxidant pressure po before the injector in the LRE, which was used to judge whether any fault will happen or not. A model was established at first to accurately describe the variation of po and the corresponding one-step prediction. The data, normalized in the test, were collected on the LRE as same as Ref. [161] and the 20 samples located in the time of 1.7–2.0 s were denoted by $x_1, x_2, \ldots x_{20}$. The first 12 samples were used for the training data and the latter 8 samples were used to validate the model. An ARMA(3, 2) model was established by Matlab [162]. The result is illustrated in Fig. 11.1 and Table 11.3. It can be seen that the prediction trend of ARMA model is consistent with collected samples. However, the error, almost 4.830 %, is a litter big, which may be caused by the natural shortage of the ARMA model. In this test, the samples may not meet the assumptions that the data should be stationary, obey the normal distribution, and can be linear described by the known data, and the ARMA model cannot obtain a good prediction result. However, the ARMA model has a certain application and reference value for the real-time prediction of LRE.

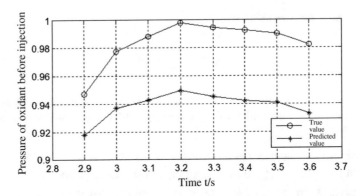

Fig. 11.1 Comparison of the prediction curve obtained by ARMA model and the actual curve

Table 11.3 Prediction result of the ARMA model	Time (s)	Actual value	Prediction value	Relative error (%)
	2.9	0.9475	0.9180	3.11
	3.0	0.9775	0.9368	4.16
	3.1	0.9885	0.9427	4.63
	3.2	0.9985	0.9497	4.89
	3.3	0.9948	0.9453	4.98
	3.4	0.9928	0.9426	5.06
	3.5	0.9905	0.9407	5.03
	3.6	0.9825	0.9333	5.01

11.2   Fault Prediction Method Based on Gray Model

Generally, things in the objective world are related to and influenced by each other and all things are constructed to be a system. The characteristics of the system are analyzed in order to figure out the running mechanism of internal system. For a class of systems, whose action principle is not clear and the internal factor is difficult to identify, people cannot establish the objective physical prototyping and is hard to understand exactly the behavior characteristics of the system. The quantitative description is difficult. A system, whose inner characteristic is partly known, is called gray system, while that whose inner characteristic is fully unknown is called black system. The gray model can pry into the essence of the objection by weakening the randomness of the data and establishing a group of continuous dynamic differential equations.

11.2.1   Introduction of Gray Model

Theory of gray system, developed by Deng in 1982, is focused on the uncertain system with poor information and small samples. The valuable information can be extracted to achieve a correct description of system behavior, evolution rule, and effective control. At present, gray theory has been basically established an emerging subject structure system [163].

11.2.2   Basic Mechanism of Gray Model

There are several axioms in the gray model theory [164].

Axiom 1 Principle of difference information Different things contain different information.

Axiom 2 Principle of nonuniqueness solution The solution of the system with incomplete and uncertain information is not unique.

Axiom 3 Principle of minimum information Minimum information, the basic criterion of the gray theory, is intended to be adequately used to solve the problem.

Axiom 4 Principle for the root of cognizance Information is the root of the cognizance.

Axiom 5 Principle for the precedence of new Information The new information is more important than the old information for the cognizance.

Axiom 6 Principle for the immortalisation of gray property The gray property of the system absolutely exists, while the white property is relative. The incompletion and uncertainty of information is universal and the completion information is relative and provisional. When the old uncertainty disappears, the new uncertainty comes soon.

These axioms are the idea, method, and technology foundations of the gray system theory. Each axiom has rich philosophical connotation.

11.2.3 Gray Prediction Method and Its Application in Fault Prediction of LRE

As the core of gray system theory, gray modeling is the bridge of integrating theory and the practice. In the gray theory, the system behavior is influenced by the gray information and the output statistics data of system is an integrated result of these gray factors. Covered by the disorderly appearance, the inherent law of system is hided in the output statistic data of system. The inherent law of system can be effectively mined by the gray modeling, which provides a new way for the prediction and control of system. At the initial of gray theory, the models of GM(n, h), where n denotes the order and h denotes the number of variables, are proposed for the prediction, decision, and control of the gray system. At present, the commonly used gray models include the models of GM (1, 1), GM (1, N), and GM (2, 1), and the model of GM (1, 1) is the most widely used in the applications of engineering, economy, ecology, and environment areas.

(1) Prediction model of GM (1, 1) [165]

Suppose that the original data is denoted by $X^{(0)} = \{x^{(0)}(1), x^{(0)}(2), \ldots, x^{(0)}(n)\}$, where $x^{(0)}(k) \geq 0$, $k = 1, 2, \ldots, n$. The general steps for the model of GM (1, 1) by $X^{(0)}$ are listed as follows:

1. Construct the accumulation generation sequence (1-AGO)

Accumulate the original data sequence to form a new sequence of $X^{(1)} = \{x^{(1)}(1), x^{(1)}(2), \ldots, x^{(1)}(n)\}$, where $x^{(1)}(k) = \sum_{i=1}^{k} x^{(0)}(i)$, $k = 1, 2, \ldots, n$.

2. Construct the background values and solve the parameter column of $[a, b]^T$

The background value sequence $z^{(1)} = \{z^{(1)}(2), z^{(2)}(3), \ldots, z^{(1)}(n)\}$, where

$$z^{(1)}(k) = \alpha x^{(1)}(k-1) + (1-\alpha)x^{(1)}(k), \quad k = 2, 3, \ldots, n \qquad (11.25)$$

Generally, let $\alpha = 0.5$ and the white differential equation of GM (1, 1) model is given by

$$\frac{dX^{(1)}}{dt} + aX^{(1)} = b \qquad (11.26)$$

Transform to be the discrete form,

$$x^{(0)}(k) + az^{(1)}(k) = b; \quad k = 2, 3, \ldots, n \tag{11.27}$$

where a and b are the model parameters. The parameter a is the development coefficient and can reflect the growth speed of the sequence $X(0)$. The parameter b is the gray action.

Calculate the parameter $\hat{a}$ by the least square method in Eq. (11.26)

$$\hat{a} = [a, b]^T = (B^T B)^{-1} B^T Y_N \tag{11.28}$$

where

$$B = \begin{bmatrix} -z^{(1)}(2) & (z^{(1)}(2))^2 \\ -z^{(1)}(3) & (z^{(1)}(3))^2 \\ \vdots & \vdots \\ -z^{(1)}(n) & (z^{(1)}(n))^2 \end{bmatrix}, Y_N = \begin{bmatrix} x^{(0)}(2) \\ x^{(0)}(3) \\ \vdots \\ x^{(0)}(n) \end{bmatrix}$$

3. Establish the prediction equation of GM (1, 1) model

Let $x^{(1)}(0) = x^{(0)}(0)$, solve Eq. (11.25), and the time response of the GM (1, 1) model can be given by

$$\hat{x}^{(1)}(k+1) = [x^{(0)}(1) - b/a]e^{-ak} + b/a; \quad k = 0, 1, \ldots \tag{11.29}$$

subtracting the sequence of $\hat{X}^{(1)}$ to obtain the prediction equation of $X^{(0)}$:

$$\hat{x}^{(0)}(k+1) = x^{(1)}(k+1) - x^{(1)}(k) = (1 - e^a)[x^{(0)}(1) - b/a]e^{-ak}, \quad k = 0, 1, \ldots \tag{11.30}$$

It can be seen that GM (1, 1) model establishes the differential equation describing a generalized energy system environment relatively unchanged based on the accumulation of temporal sequence and obtains a prediction model.

(2) Application scope of GM (1, 1) model

With the development of investigation, it is shown that the training error of GM (1, 1) model is related to the development coefficient. With the increase of the development coefficient, the training error increases quickly. When the development coefficient is lower than 0.3, the accuracy of training is bigger than 98 %. When the development coefficient is less than or equal to 0.5, the simulation accuracy is above 95 %. When the development coefficient is greater than 1, the simulation precision is less than 70 %, and when the development coefficient is greater than 1.5, the simulation precision is less than 50 %. In this instance, the following conclusions can be obtained [166]:

1. When $-a \leq 0.3$, the GM (1, 1) model can be used for the prediction with the medium and long interval.
2. When $0.3 <- a \leq 0.5$, the GM (1, 1) model can be used for the short-term forecasting.
3. When $0.5 <- a \leq 0.8$, the GM (1, 1) model should be used very carefully in the application of short-term forecasting.
4. When $0.8 <- a \leq 1$, the GM (1, 1) model with the residual error correction should be adopted.
5. When $-a > 1$, the GM (1, 1) model is not suitable for the prediction.

(3) **Validation of the** GM (1, 1) model

The validation of the GM (1, 1) model includes the residual validation, validation of post-residual, validation of correlation degree, and the validation of level windage. The residual validation is performed point by point, the validation of correlation degree is to test the similarity of the model and the given function, and the validation of post-residual is to test the statistic distribution of the residual.

1. Residual validation

Suppose that the kth value of original sequence $X^{(0)}$ is denoted by $x^{(0)}(k)$, while the estimation value of the gray model is denoted by $\hat{x}^{(0)}(k)$. The residual $q(k)$, relative error $\varepsilon(k)$, average relative error $\varepsilon(\text{avg})$, and accuracy $p0$ are defined as follows;

$$q(k) = x^{(0)}(k) - \hat{x}^{(0)}(k) \tag{11.31}$$

$$\varepsilon(k) = \frac{q(k)}{x^{(0)}(k)} \times 100\% = \frac{x^{(0)}(k) - \hat{x}^{(0)}(k)}{x^{(0)}(k)} \times 100\% \tag{11.32}$$

$$\varepsilon(\text{avg}) = \frac{1}{n-1} \sum_{k=2}^{n} |\varepsilon(k)| \tag{11.33}$$

$$p^0 = (1 - \varepsilon(\text{avg})) \times 100\% \tag{11.34}$$

Generally, $\varepsilon(k) < 10\%$ and $p0 \geq 80\%$, even $\varepsilon(k) < 20\%$ and $p0 \geq 90\%$.

2. Validation of post-residual

① Calculate the mean and variance of original sequence:

$$\bar{X}^{(0)} = \frac{1}{n} \sum_{k=1}^{n} x^{(0)}(k), \quad S_1^2 = \frac{1}{n} \sum_{k=1}^{n} (x^{(0)}(k) - \bar{X}^{(0)})^2. \tag{11.35}$$

② Calculate the mean and variance of residual sequence $q(0)$:

$$\bar{q}^{(0)} = \frac{1}{n}\sum_{k=1}^{n} q^{(0)}(k), \quad S_2^2 = \frac{1}{n}\sum_{k=1}^{n} (q^{(0)}(k) - \bar{q}^{(0)})^2 \tag{11.36}$$

③ Calculate the post-residual ratio C and the small error probability P:

$$C = \frac{S_2}{S_1}, \quad P = P\{|q^{(0)}(k) - \bar{q}^{(0)}| < 0.6745S_1\}. \tag{11.37}$$

For a given $C0 > 0$, when $C < C0$, the model is called by a qualified posterior difference model. For a given $P0 > 0$, when $P > P0$, the model is called by a qualified model of small error probability.

3. Validation of correlation

First, calculate the absolute error by referring to the sequence $\hat{X}^{(0)}$:

$$\Delta(k) = |x^{(0)}(k) - \hat{x}^{(0)}(k)|, \quad \Delta(k) = \max_k \Delta(k), \quad \Delta(k) = \min_k \Delta(k). \tag{11.38}$$

Then, calculate the correlation coefficient $\xi(k)$ and the correlation value r:

$$\xi(k) = \frac{\Delta_{\min} + \rho\Delta_{\max}}{\Delta(k) + \rho\Delta_{\max}}, \quad r = \sum_{k=1}^{n} w_k \cdot \xi(k) \tag{11.39}$$

Usually, let $\rho = 0.5, w_k = \frac{1}{n}, k = 1, 2, \ldots, n$.

4. Validation of level windage

Calculate the sequence level ratio $\sigma^{(0)}(k)$ and the model level ratio $\hat{\sigma}^{(0)}(k)$ as:

$$\sigma^{(0)}(k) = \frac{x^{(0)}(k-1)}{x^{(0)}(k)} \tag{11.40}$$

$$\hat{\sigma}^{(0)}(k) = \frac{\hat{x}^{(0)}(k-1)}{\hat{x}^{(0)}(k)} = \frac{1 + 0.5a}{1 - 0.5a} \tag{11.41}$$

Then the level windage is given by

$$\begin{aligned}
\rho(k) &= \frac{\hat{\sigma}^{(0)}(k) - \sigma^{(0)}(k)}{\hat{\sigma}^{(0)}(k)} \times 100\% \\
&= 1 - \left(\frac{1 - 0.5a}{1 + 0.5a}\right)\frac{x^{(0)}(k-1)}{x^{(0)}(k)} \times 100\%
\end{aligned} \tag{11.42}$$

Table 11.4 Classification of the model validation

Precision standard	$\varepsilon(k)$ (%)	$p0$ (%)	C	P	r	$\rho(k)$ (%)
Level 1 (GOOD)	1	99	0.35	0.95	0.90	1
Level 2 (QUANTIFIED)	5	95	0.50	0.80	0.80	5
Level 3 (RELUCTANT)	10	90	0.65	0.70	0.70	10
Level 4 (UNQUANTIFIED)	20	80	0.80	0.60	0.60	20

When $\rho(k) < \varepsilon$, a given threshold, the GM (1, 1) model of $X(0)$ has the exponential coincidence rate of ε.

For the validation of the model, the smaller of $\varepsilon(avg)$, $\varepsilon(k)$, and P, the more accurate of the model. Comparatively, the bigger of C, r, and $\rho(k)$, the more accurate of the model. When the values of $\varepsilon(k)$, $p0$, C, r, and $\rho(k)$ are given, the accurate degree of the model is determined. Generally, the precision accuracy of the model can be divided to be the degree of GOOD, QUANTIFIED, RELUCTANT, and UNQUANTIFIED based on the value of $\varepsilon(k)$, $p0$, C, r, and $\rho(k)$. Table 11.4 lists the reference value of these indexes. Generally, the validation of relative error is the most widely used. If the validations of residual, correlation value, and post-residual are passed, the model can be used for the prediction. Otherwise, the residual correction method should be used.

(4) Improved GM (1, 1) model

1. GM (1, 1) model of metabolic type

Generally, GM (1, 1) model, a continuous function, is obtained by all data before the real-time $t = n$. According to the gray system theory, GM (1, 1) forecast model can extend to any time in the future from the initial moment. However, in terms of the generalized energy system, many interference factors will continuously invade the system with the time passing by and the data before several steps of the real moment is the most useful for the prediction. The earlier the moment, the smaller the meaning of the data is. In order to consider the incoming data, the new data must be added in the original sequence to establish a GM (1, 1), a novel model. With the time passing by, the older data is gradually removed and the length of the sequency is maintained by the addition of the new data. Corresponding sequency is called equivalent dimension new information sequence [167]. By inheriting the advantages of conventional GM (1, 1) model, the improved model can consider the new disturbance of the system. According to the available sources of new interest, the improved model can be divided into two models: the complimentary model of equivalent dimension gray number and the model of metabolic type.

The modeling mechanism of equal dimension new information model and gray model are the same, but the filling vacancies of the former are the actual data of the actual system. The latter is the predicted value, using the gray number, and thus they form their respective characteristics. The former uses the new information to adjust the model. The latter uses gray number constraint gray plane size, through the timely supplement and using gray information to improve the white grade of gray level and narrow the predictive value of gray interval.

2. GOM model (gray optimization model)

① The basic idea of GOM model [168]:

First, the traditional GM (1, 1) model is used to get the model parameters a, b. Then, the $x^{(1)}$ is transformed to improve the precision. It is obvious that there is a certain relationship between the translation value and the model precision. Finally, an optimal model is established to obtain the optimal translation value of c. In the GM (1, 1) model, $x^{(1)}(k)$ was replaced by $x_1^{(1)}(k) = x^{(1)}(k) + c$. It is reflected as

$$z^{(1)}(k) \rightarrow z_1^{(1)}(k), \ a \rightarrow a_1, b \rightarrow b_1, \ \widehat{x}^{(1)}(k) \rightarrow \widehat{x}_1^{(1)}(k), \ \widehat{x}^{(0)}(k) \rightarrow \widehat{x}_1^{(0)}(k)$$

$$(11.43)$$

② Building GOM model

Step 1–2 Building by the traditional GM (1, 1) modeling Step 1–2.
Step 3 Establish the prediction formula
The prediction formula of $x_1^{(1)}$ is

$$\widehat{x}_1^{(1)}(k+1) = \left(x^{(0)}(1) - \frac{a_1}{b_1}\right)e^{-ak} + \frac{a_1}{b_1} \quad k = 0, 1, 2, \ldots \quad (11.44)$$

Among them, $a_1 = a$, $b_1 = b + ca$. Its proof is slightly more detailed. Through reduction, $x_1^{(0)}(0)$ prediction formula is

$$\widehat{x}_1^{(0)}(k+1) = \widehat{x}^{(0)}(k+1) + cb_{k+1} \quad (11.45)$$

Among them,

$$b_{k+1} = (e^a - 1)e^{-ak}, \quad c = \frac{e^a + 1}{1 - e^{-2(n-1)a}} \sum_{k=1}^{n-1} [x^{(0)}(k+1) - \widehat{x}^{(0)}(k+1)]e^{-ak}$$

GM (1, 1) model in the $x^{(1)}(k)$ for $x^{(1)}(k) + c$ obtained by the new model called GOM model, in which a and $q^{(0)}(k + 1)$ were the original GM (1, 1) model in the development of coefficients and residuals.

Theorem 4.1 $\widehat{x}_1^{(0)}(k+1) = \widehat{x}^{(0)}(k+1) + cb_{k+1}$, among $b_{k+1} = (e^a - 1)e^{-ak}$
Proved by whitening response of GM (1, 1) model, we can get

$$\widehat{x}_1^{(0)}(k+1) = \widehat{x}_1^{(1)}(k+1) - \widehat{x}_1^{(1)}(k)$$

$$= (1 - e^a)\left(x^{(0)}(1) - \frac{b_1}{a_1}\right)e^{-ak}$$

$$= (1 - e^a)\left(x^{(0)}(1) - \frac{b}{a}\right)e^{-ak} + c(e^a - 1)e^{-ak} \qquad (11.46)$$

$$= (1 - e^a)\left(x^{(0)}(1) - \frac{b}{a} + c\right)e^{-ak}$$

$$= \widehat{x}^{(0)}(k+1) + cb_{k+1}$$

Theorem 4.1 The optimal solution for the GOM model is

$$c = \frac{e^a + 1}{1 - e^{-2(n-1)a}} \sum_{k=1}^{n-1} q^{(0)}(k+1)e^{-ak} \qquad (11.47)$$

Prove set

$$q^{(0)}(k+1) = x^{(0)}(k+1) - \widehat{x}^{(0)}(k+1)$$
$$q_1^{(0)}(k+1) = x^{(0)}(k+1) - \widehat{x}_1^{(0)}(k+1) \quad (k = 1, 2, \ldots, n-1) \qquad (11.48)$$

as the residual error of model. There is

$$q_1^{(0)}(k+1) = x^{(0)}(k+1) - \widehat{x}_1^{(0)}(k+1)$$
$$= x^{(0)}(k+1) - (\widehat{x}^{(0)}(k+1) + cb_{k+1})$$
$$= x^{(0)}(k+1) - \widehat{x}^{(0)}(k+1) - cb_{k+1} \qquad (11.49)$$
$$= q^{(0)}(k+1) - cb_{k+1}$$

The above expressions reflect the direct relationship between the residuals and the translation value c. In order to determine the minimal translation value of c, we can use the following objective function:

$$\min F(c) = \min \sum_{k=1}^{n-1} [q_1^{(0)}(k+1)]^2$$
$$= \min \sum_{k=1}^{n-1} [q^{(0)}(k+1) - cb_{k+1}]^2 \qquad (11.50)$$

$F(c)$ on c derivative make it zero, and we will get

$$\frac{dF}{dc} = 2 \sum_{k=1}^{n-1} [q^{(0)}(k+1) - cb_{k+1}](-b_{k+1}) = 0 \qquad (11.51)$$

The solution is

$$c = \frac{\sum_{k=1}^{n-1} q^{(0)}(k+1)b_{k+1}}{\sum_{k=1}^{n-1} b^2_{k+1}} \qquad (11.52)$$

Also because

$$\sum_{k=1}^{n-1} b^2_{k+1} = \sum_{k=1}^{n-1} [(e^a - 1)e^{-ak}]^2 = (e^a - 1)^2 \sum_{k=1}^{n-1} e^{-2ak}$$

$$= (e^a - 1)^2 \frac{e^{-2a}(1 - e^{-2(n-1)a})}{1 - e^{-2a}} = \frac{e^a - 1}{e^a + 1}[1 - e^{-2(n-1)a}] \qquad (11.53)$$

Then

$$c = \frac{e^a + 1}{1 - e^{-2(n-1)a}} \sum_{k=1}^{n-1} q^{(0)}(k+1)e^{-ak}$$

$$= \frac{e^a + 1}{1 - e^{-2(n-1)a}} \sum_{k=1}^{n-1} [x^{(0)}(k+1) - \widehat{x}^{(0)}(k+1)]e^{-ak} \qquad (11.54)$$

(5) Case of liquid rocket engine fault prediction based on gray prediction method

Application of the above GM (1, 1) and its improved model for the prediction of liquid rocket engine combustion agent before *pf* is given. In order to predict whether the liquid rocket engine is in the presence of the fault trend, the predictive control of the pressure *pf* of the liquid rocket engine is predicted by the prediction of the pressure of the fuel injection front. At first, an appropriate model is established to accurately describe the change of pressure *pf* and single-step prediction. The samples used are from the test data of a certain type of liquid rocket engine [161]. All of these data are normalized, as shown in Table 11.5. Then divide the normalization of the data packet into groups. The 20 data in 1.74–3.64 s were recorded as $x_1, x_2, ..., x_{20}$. With the above 20 data, the former 12 data was used as the original sample point, and the later 8 data was used as the test data. Table 11.6 shows the comparison of prediction results for several GM (1, 1) models.

To further test whether the model to predict the effect is good or bad, according to the prediction effect evaluation principles and practices, we adopt the following

Table 11.5 Modeling validation sample

t(s)	1.74	1.84	1.94	2.04	2.14	2.24	2.34	2.44	2.54	2.64
pf	0.4285	0.4543	0.4729	0.4891	0.7089	0.7501	0.7545	0.8483	0.8865	0.9067
t(s)	2.74	2.84	2.94	3.04	3.14	3.24	3.34	3.44	3.54	3.64
pf	0.9233	0.9521	0.9627	0.9838	0.9916	0.9996	0.9922	0.9996	0.9889	0.9840

Table 11.6 Prediction results of GM (1, 1) and its improved model

Time (s)	True value	Predicted value of GM (1, 1) model	Predicted value of the metabolic GM (1, 1) model	Predicted value of the optimized GM (1, 1) model
2.94	0.9627	0.9312	0.9365	0.9536
3.04	0.9838	0.9420	0.9688	0.9738
3.14	0.9916	0.9625	0.9713	0.9822
3.24	0.9996	0.9706	0.9862	0.9897
3.34	0.9922	0.9723	0.9844	0.9868
3.44	0.9996	0.9775	0.9836	0.9875
3.54	0.9889	0.9734	0.9851	0.9824
3.64	0.9840	0.9716	0.9794	0.9773

Table 11.7 Comparative analysis of fitting and forecasting errors

Model	MAPE (%)	MSPE	SSE
Modeling (fitting)			
GM (1, 1)	1.87	0.0068	0.0028
GM (1, 1) Metabolic GM (1, 1)	1.24	0.0049	0.0015
Optimized GM (1, 1)	0.66	0.0027	0.0004
Prediction (Testing)			
GM (1, 1)	2.64	0.0099	0.0057
GM (1, 1) Metabolic GM (1, 1)	1.38	0.0056	0.0019
Optimized GMA (1, 1)	0.84	0.0031	0.0005

three fitting error indexes: mean absolute percentage error (MAPE), relative root mean square error (MSPE), the sum squared error (SSE) as a criterion, and the fitting and prediction effects were a full range of comprehensive evaluation (Table 11.7). Through the effect of the prediction model comparison, we can see the following: in the prediction stage, the MAPE of the GM (1, 1) and the actual value is 2.64 %, MSPE is 0.0099, SSE was 0.0057; the MAPE of metabolic GM (1, 1) and the actual value is 1.38 %, MSPE 0.0056, SSE is 0.0019; and the MAPE of optimized GM (1, 1) model premeasured value and the actual value is 0.84 %, MSPE 0.0031, SSE is 0.0005. It can be seen that the prediction accuracy of the metabolic GM (1, 1) and the optimized GM (1, 1) model is significantly better than the GM (1, 1) model.

The advantage of the gray model is that it does not require a large number of samples and not need to consider whether the data is subject to a certain probability distribution. Using the function of gray generation to weaken the randomness of the data, we can dig out the potential law. Continuous dynamic differential equation is established using discrete data sequence, which can achieve the purpose of the essence of the development of things. The above results show that results of three kinds of gray models are desirable. GM (1, 1) model of the gray optimization prediction has the highest precision and metabolism GM (1, 1) model prediction accuracy is better than GM (1, 1) model. It has some applicability and feasibility in engine fault prediction.

11.3   Rocket Engine Fault Prediction Method Based on Neural Network

Because of the variation of parameters in the working process of the liquid rocket engine, it is very difficult to establish an analytical model which is more accurate and can satisfy the online real-time fault detection. The fault detection and diagnosis method of liquid rocket engine in our country is relatively simple, and only the "red line" method is adopted for some important parameters. United Technologies Research Center (UTRC) for starting and stopping is through an adaptive learning network as classifier achieved. This network adopts the RESID (recursive structure identification) algorithm [169]. In the process of starting and shutdown, the data of the engine sensor is not stable, and the traditional method can be used in principle (autoregressive analysis, nonlinear model, and so on). However, due to the complexity of the engine structure, the computational workload is large. It does not meet the requirements of real-time monitoring and diagnosis. Therefore, it is an important content of the research of engine condition monitoring and fault diagnosis to establish the real-time engine model using advanced and practical engine modeling method. Neural network has strong learning ability, nonlinear approximation ability, and so on. In some areas, such as pattern recognition, nonlinear identification, adaptive control, intelligent detection, fault diagnosis, and so on, neural network has been a lot of successful applications.

11.3.1   Multistep Prediction Method for Dynamic Parameters of Rocket Engine Based on BP Network

Because the neural network has a good ability to approximate the nonlinear mapping, the neural network can be used to predict the nonlinear dynamic. Neural network-based prediction method has been widely concerned and achieved many successful applications [170]. The measurement parameters in the rocket engine are mostly related to. In the parameter prediction, the introduction of the relevant parameters can make up the deficiency of the forecast information and improve the prediction accuracy. Using neural network to predict the parameters can avoid the shortcomings of the traditional prediction methods, improve the accuracy of the prediction and fast, and be easy to real-time processing.

(1) Nonlinear time series analysis of correlation parameter prediction

In general, the dynamic process of the prediction problem has multiple inputs and multiple outputs, and the input and output are functions of time, which have certain correlation. Set the relevant parameters for $x_j(t)$ ($t = 1, 2, \ldots, k$), k for the parameters of the total number. The values of the time $t - 1, t - 2, \ldots, t - n$ were $x_j(t - 1), x_j(t - 2), \ldots, x_j(t - n)$. Set $y_i(t)$ ($i = 1, 2, \ldots, m$) is the value of the predictive parameter in the t time. m is the total number of prediction parameters. The values at the time $t - 1, t - 2, \ldots, t - n$ were $y_i(t - 1), y_i(t - 2), \ldots, y_i(t - n)$. If each parameter has the same number of prediction steps l, $\hat{y}_i(t + 1), \hat{y}_i(t + 2), \ldots, \hat{y}_i(t + l)$ and $\bar{y}_i(t + 1), \bar{y}_i(t + 2), \ldots, \bar{y}_i(t + l)$ are the predicted and true value of predictive parameters y_i at the time $t + 1, t + 2, \ldots, t + l$. Set

$$y_i^{(i)} = [y_i(t + 1), y_i(t + 2), \ldots, y_i(t + l)], \quad y = [y_1^{(1)}, y_2^{(2)}, \ldots, y_m^{(m)}]$$
$$\hat{y}_i^{(i)} = [\hat{y}_i(t + 1), \hat{y}_i(t + 2), \ldots, \hat{y}_i(t + l)], \quad \hat{y} = [\hat{y}_1^{(1)}, \hat{y}_2^{(2)}, \ldots, \hat{y}_m^{(m)}]^T \quad (i = 1, 2, \ldots, m)$$

and set

$$x_j^{(j)} = [x_j(t - 1), x_j(t - 2), \ldots, x_j(t - n)], \quad x' = [x_1^{(1)}, x_2^{(2)}, \ldots, x_k^{(k)}]^T$$
$$(j = 1, 2, \ldots, k)$$

then

$$\hat{y} = f(y, x') \tag{11.55}$$

In the formula, $f()$ is the prediction operator. If making $x = [y \quad x']$, then the formula is simplified as

$$\hat{y} = f(x) \tag{11.56}$$

From this, we can know that the mapping from x to y is a kind of space–time map, which has the corresponding relations in space and time. This kind of space–time mapping is both dynamic and nonlinear, and it is not suitable to use the traditional linear prediction model. Because the neural network has many excellent properties, it could solve the problem of this kind of space and time mapping.

(2) The real-time forecasting model of BP network

Real-time prediction model of BP network is shown in Fig. 11.2. Neural network forecasting is different from the traditional forecasting methods. Neural network

Fig. 11.2 Real-time
prediction model of BP

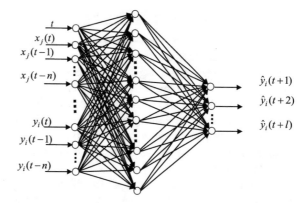

prediction is a process of thinking. It predicts the intrinsic link between the object
and the past information by predicting the object and related parameters. The
establishment process of the prediction model is the network adaptive training
learning process using the network structure parameters and the weight of the
prediction model to replace the structural parameters, historical knowledge, and
real-time knowledge. The prediction is just a simple forward calculation of the
network, and can be a single-step and multistep prediction. In general, the adaptability, fault tolerance, precision, and anti-interference ability of neural network are
very strong, so that the optimal solution of multiparameter and multistep prediction
of relevant parameters can be realized.

The most prominent advantage of multistep prediction can be achieved at different time points on the monitoring parameters for multiple predictions. It can not
only issue multiple fault alarm before the arrival of the fault which is conducive to
the identification of failure modes and to take effective measures to eliminate faults,
but by the information theory, it can also be known that the original information can
be fully utilized in the multistep prediction. The prediction results contain large
amount of information, which can overcome the error caused by chance and
randomness.

In general, the accuracy of the predicted values at different points is not the
same. These different predictive values for the early detection of failure are useful.
But the final conclusion of the fault is determined by their comprehensive value,
and the comprehensive value is given by the following formula:

$$\hat{y}_i^s(t) = \varphi_1 \hat{y}_i^1(t) + \varphi_2 \hat{y}_i^2(t) + \cdots + \varphi_l \hat{y}_i^l(t) \quad (i = 1, 2, \ldots, m) \tag{11.57}$$

In the formula $\varphi_j (j = 1, 2, \ldots, l)$ is the comprehensive coefficient of predictive
value $\hat{y}_i^j(t)$. It reflects the degree size of credibility of $\hat{y}_i(t)$ at the time of t-j. It is
need that $\sum_{j=1}^{l} \varphi j \le 1$.

(3) An example of liquid rocket engine fault prediction based on BP neural network

It can be seen that the thrust of the engine is an important indicator from the former. The thrust of the engine is a function of the oxidant flow, the flow of the combustion agent, and the pressure of the combustion chamber in a given condition of the engine system and environmental conditions. So we could make multistep prediction of the thrust F and give the results of the neural network real-time model by the relevant parameters $\dot{m}_o$, $\dot{m}_f$, and p_c. The trainings of the network learning samples and test samples are from test data, which is shown in Tables 11.8 and 11.9, respectively.

When learning and training, input the unit $\dot{m}_o(t-i)$, $\dot{m}_f(t-i)$, $P_c(t-i)$, $F(t-i)$ $i = 0, 1, 2, 3, 4; i \leq t$ and t, a total of 21, to predict the four steps of thrust. That is F $(t+l)$ (l = 1, 2, 3, 4), so there are four nodes of the output layer. The number of units in the single hidden layer is 30, and the network structure is 30-4-21. The other basic network parameters of BP network are as follows: $a = 0.7$, $\eta = 0.9$, and the maximum error of system is 0.00001. At the end of network training, the number of iterations is 17,055 times. Using the improved BP network the network parameters are as follows: $\mu = 10$, $\eta = 0.2$, and the maximum error of system is 0.00001. At the end of network training, the number of iterations is 10,300 times.

The predicted values and test values of the network are shown in Table 11.10, and the comparisons between the predicted curves and the actual curves are shown in Tables 11.3, 11.4, 11.5, and 11.6. The comparison of the predicted comprehensive values with the test values is shown in Table 11.11. The comparison between the actual curve and the prediction curve is shown in Fig. 11.7.

As shown in Figs. 11.3, 11.4, 11.5, 11.6, 11.7 and Tables 11.8, 11.9, 11.10, and 11.11, the proposed neural network model can solve the problem of real-time

Table 11.8 Forecast network modeling data sample

Time series $t(s)$	Combustion flux $(\dot{m}_f)$	Oxidant flux $(\dot{m}_o)$	Combustion chamber pressure (p_c)	Thrust (F)
0	0.3179	0.4011	0.0098	0.0000
0.2	0.5780	0.5339	0.2238	0.1418
0.4	0.6012	0.5658	0.6170	0.4113
0.6	0.6763	0.6374	0.7832	0.5674
0.8	0.8555	0.8238	0.8951	0.7730
1.0	0.9306	0.9149	0.9580	0.8865
1.2	0.9665	0.9593	0.9874	0.9362
1.4	0.9827	0.9810	0.9944	0.9574
1.6	0.9442	0.9919	0.9972	0.9787
1.8	0.9977	0.9973	1.0000	0.9929
2.0	1.0000	1.0000	1.0000	1.0000
2.2	1.0000	1.0000	1.0000	1.0000
2.4	1.0000	1.0000	1.0000	1.0000

Table 11.9 Forecast network text data sample

Time series t (s)	Combustion flux ($\dot{m}_f$)	Oxidant flux ($\dot{m}_o$)	Combustion chamber pressure (p_c)	Thrust (F)
0.1	0.3931	0.4661	0.0559	0.0496
0.3	0.6532	0.5935	0.4643	0.3262
0.5	0.6012	0.5572	0.7049	0.4681
0.7	0.7746	0.6721	0.8447	0.6596
0.9	0.9017	0.8753	0.9301	0.8440
1.1	0.9526	0.9420	0.9790	0.9163
1.3	0.9769	0.9702	0.9930	0.9504
1.5	0.9884	0.9864	0.9951	0.9716
1.7	0.9965	0.9957	0.9986	0.9858
1.9	0.9988	0.9995	1.0000	0.9986
2.1	1.0000	1.0000	1.0000	1.0000
2.3	1.0000	1.0000	1.0000	1.0000

prediction in the engine, and the prediction accuracy is high. Multiparameter multistep prediction using the relate parameters can not only overcome the difficulties brought about by the lack of information, but also to get multiple alarms of the fault in advance, avoiding traditional forecasting method of complex analysis and calculation, improving the prediction precision. The accuracy of multistep prediction is not greatly reduced with the increase of the number of steps, which overcomes the shortcomings of the traditional prediction methods and has a wide range of application prospects.

It can be seen from the above table and figure that the time interval between the prediction modeling data and the test sample is too large, the model is not precise enough, and there is a certain error in prediction accuracy. In addition, the start time of the rocket engine starting point is calculated from the start of the power of the power starter in fact. But prior to this, the filling process has been completed and the oxidant flow and combustion agent flow is not zero. In the prediction modeling data, they are regarded as zero treatment, resulting in a certain error in the prediction model. Especially in the starting process, the error is even greater. Therefore, in the application of modeling data should be refined to improve the accuracy of prediction.

11.3.2   The Prediction Method for Dynamic of Rocket Engine Based on RBF Network

Time series prediction is the main method for the performance prediction of nonlinear dynamic systems. The good function approximation ability of neural network is the basis of its application in time series prediction. According to the

Table 11.10 Comparison of prediction results and test values

Time series t(s)	Thrust (F)											
	Predictive value with one step			Predictive value with two step			Predictive value with three step			Predictive value with four step		
	Text value	BP network	Improved network	Text value	BP network	Improved network	Text value	BP network	Improved network	Text value	BP network	Improved network
0.1	0.3262	0.1783	0.1852	0.4681	0.4525	0.4770	0.6596	0.6026	0.6179	0.8440	0.7921	0.8020
0.3	0.4681	0.4958	0.4932	0.6596	0.6241	0.6333	0.8440	0.8141	0.8183	0.9163	0.9026	0.9068
0.5	0.6569	0.6173	0.6440	0.8440	0.8099	0.8365	0.9163	0.9087	0.9195	0.9504	0.9477	0.9518
0.7	0.8440	0.7865	0.8183	0.9163	0.8940	0.9142	0.9504	0.9419	0.9487	0.9716	0.9628	0.9652
0.9	0.9163	0.9093	0.9146	0.9504	0.9434	0.9449	0.9716	0.9672	0.9670	0.9858	0.9861	0.9868
1.1	0.9504	0.9510	0.9447	0.9716	0.9720	0.9677	0.9858	0.9890	0.9877	0.9986	0.9959	0.9964
1.3	0.9716	0.9703	0.9696	0.9858	0.9868	0.9861	0.9986	0.9954	0.9956	1.0000	0.9980	0.9988
1.5	0.9858	0.9881	0.9885	0.9986	0.9943	0.9938	1.0000	0.9979	0.9981	1.0000	0.9991	0.9996
1.7	0.9986	0.9949	0.9941	1.0000	0.9973	0.9968	1.0000	0.9989	0.9991	1.0000	0.9995	0.9998
1.9	1.0000	0.9971	0.9962	1.0000	0.9983	0.9979	1.0000	0.9993	0.9994	1.0000	0.9997	0.9999
2.1	1.0000	0.9981	0.9974	1.0000	0.9988	0.9985	1.0000	0.9995	0.9996	1.0000	0.9997	0.9999
2.3	1.0000	0.9988	0.9981	1.0000	0.9992	0.9989	1.0000	0.9996	0.9997	1.0000	0.9998	0.9999

Table 11.11 The comparison between the actual vales and the text vales	Time series (s)	Thrust (F)		
		Text value	BP network	Improved network
	0.1	0.3262	0.178300	0.185200
	0.3	0.4681	0.474150	0.485100
	0.5	0.6596	0.616400	0.632510
	0.7	0.8440	0.799600	0.820300
	0.9	0.9163	0.903920	0.912100
	1.1	0.9504	0.946570	0.944500
	1.3	0.9716	0.969440	0.967870
	1.5	0.9858	0.987690	0.987640
	1.7	0.9986	0.994920	0.994450
	1.9	1.0000	0.997410	0.996900
	2.1	1.0000	0.998420	0.998020
	2.3	1.0000	0.998970	0.998600

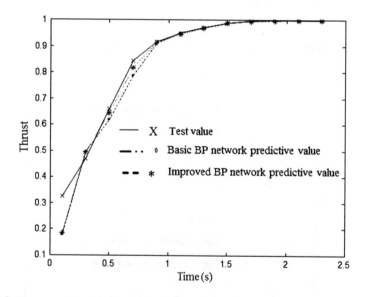

Fig. 11.3 The comparison of one-step prediction curve and test curve

approximation property of the radial basis function, we can use the radial basis function network to approximate the function relation f in the formula $x_{n+1} = f(x_1, x_2, \ldots, x_n) + e_{n+1} = \hat{x}_{n+1} + e_{n+1} \; \hat{x}_{n+1} = f(x_1, x_2, \ldots, x_n)$. At this time, the predictive value is the only one of the output nodes, and the input and output relations are satisfied:

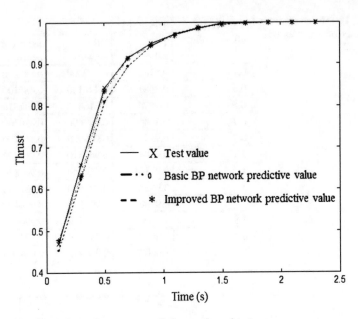

Fig. 11.4 The comparison of two-step prediction curve and test curve

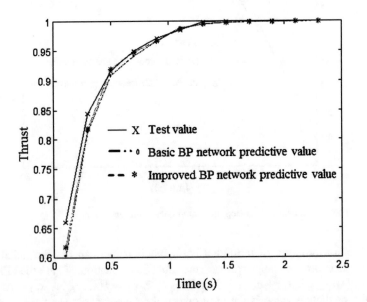

Fig. 11.5 The comparison of three-step prediction curve and test curve

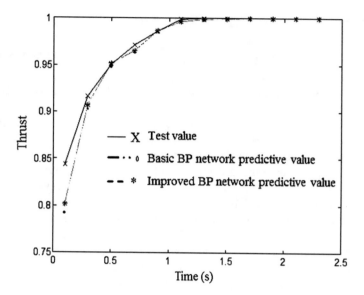

Fig. 11.6 The comparison of four-step prediction curve and test curve

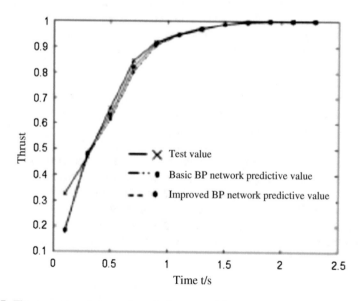

Fig. 11.7 The comparison between the actual curve and the text curve

$$E = \min_{f} \sum_{m=1}^{n+1} |x_{m+n-1} - f(x_{m+n}, \ldots, x_{m+1})|$$

The basic steps of modeling the RBF network time series prediction are as follows:

(1) Sample selection: The normalized time series of sample data is divided into training and test section;
(2) Network structure design: Determine the network's dimension of the input vector and the output vector;
(3) Network topology design: Select the activation function of the output layer of the network (using the Gauss basis function);
(4) Network initialization: Setting the initial weights of the network and the value of the target error. Among them, the initial value of the network can choose the random number in $(-0.1, 0.1)$. The target error value is selected according to the accuracy of prediction;
(5) Network training: Training network using clustering algorithm to determine the center and width parameters of Gauss basis function and the recursive least square method to determine the connection weights, fitting learning time series until the network converges to a certain standard;
(6) Network prediction: Using the test data to test the training of the network model to predict the future time series.

Whether the liquid rocket engine has a tendency to malfunction was judged through the prediction of the pressure p before the oxidant injection to control the liquid rocket engine. At first, an appropriate model is established to accurately describe the change of pressure p and single-step prediction. The samples used are from the test data of a certain type of liquid rocket engine [171]. All of these data are processed by the normalization.

Divide the normalized data into groups. The 20 data between 1.70 s and 3.60 s are recorded as $x_1, x_2, \ldots, x_{20}$. Six data form a sample. The former five values are used as input data, and the last one value is the desired output, as shown in Table 11.12. The former seven groups of the samples with 15 groups are used as the training sample for modeling, and eight groups of samples used to predict. By numerical simulation, analyze and compare the prediction performance of the two neural network prediction models of BP and RBF, and the Matlab 6.5 is used to program the simulation. The RBF network input layer is 5 nodes and the output layer is 1 node. The hidden layer node number is automatically selected according to the control error. The input of BP network is 5 nodes, the hidden layer is 8 nodes,

Table 11.12 Time series prediction samples

Input data	Expected output
$x_1, x_2, \ldots, x_5$	x_6
$x_2, x_3, \ldots, x_6$	x_7
$\vdots$	$\vdots$
$x_{15}, x_{16}, \ldots, x_{19}$	x_{20}

and the output layer is 1 node. That is, the network topology is 5-8-1, and the prediction results are better. The mean square error of the two models is controlled in 0.001.

The comparison curves of the true values and forecast results obtained by different networks are shown in Figs. 11.8 and 11.9. The time consumed by the BP network from the definition to the completion of the network is 1.086 s, while that consumed by the RBF network is 0.0625 s. Compared to BP network, the determination of the hidden layer node number manually can be eliminated in the modeling of RBF network. Choosing the average relative error, mean square error and sum squared error as the evaluation indexes, the comparison results of these two models are as shown in Table 11.13. From Figs. 11.8, 11.9 and Table 11.13 we can see that the three error indicators of RBF network are much smaller than those

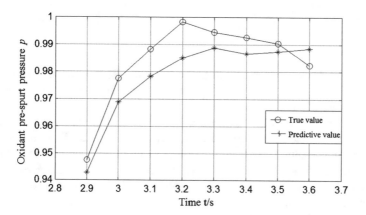

Fig. 11.8 The comparison between the predicted curve and the actual curve of the BP network model

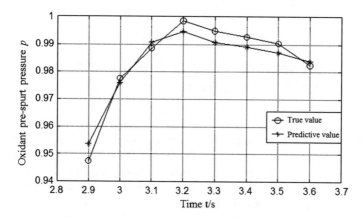

Fig. 11.9 The comparison between the predicted curve and the actual curve of the RBF network model

Table 11.13 The comparison of prediction results of network model

Time (s)	True value	Predictive value of BP network	Predictive value of RBF network
2.9	0.9475	0.9428	0.9538
3.0	0.9775	0.9688	0.9760
3.1	0.9885	0.9785	0.9908
3.2	0.9985	0.9853	0.9945
3.3	0.9948	0.9890	0.9906
3.4	0.9928	0.9868	0.9891
3.5	0.9905	0.9877	0.9869
3.6	0.9825	0.9886	0.9840
Average relative error (%)		0.730	0.340
Mean square error		0.00006	0.000014
Sum square error		0.00048	0.000112

of the BP network. The prediction accuracy of the RBF network is significantly better than that of the BP network. Moreover the training speed of RBF network is also improved a lot.

By numerical simulation examples, the prediction performances of the two neural network time series forecasting models of BP and RBF are analyzed and compared. The results show that the learning speed of RBF network is higher than that of BP network. The time required to complete the BP network from the definition to the training time is 1.086 s, and the RBF network only need 0.0625 s. And the RBF network structure can be determined adaptively. The three error indicators of RBF network model prediction are smaller than the error indicators of the BP network model. Therefore, the convergence speed and prediction accuracy of RBF network model are significantly improved than that of BP network model, which has a good application value for real-time parameter prediction of liquid rocket engine.

11.3.3   The Prediction Method for Dynamic of Rocket Engine Based on Elman Network

At present, most of the applications are static feedforward BP neural network and RBF neural network. In fact using the static feedforward neural network to identify the dynamic system is the problem that the dynamic time modeling becomes a static space modeling problem, which will lead to many problems. Unlike the static feedforward neural network, the dynamic recurrent network has the function of mapping dynamic characteristics thought storing in the internal state, so that the system has the ability to adapt to the time-varying characteristics, more suitable for the identification of nonlinear dynamic system, and represents the development

direction of neural network modeling, identification, and control, and Elman artificial neural network is a typical dynamic neural network.

   Elman artificial neural network was proposed by [172] in 1990, Elman. The
model is increased in the feedforward network with the hidden layer in an under-
taking layer, as a step delay operator, to memory, so that the system has the ability
to adapt to the time-varying characteristics, and can directly reflect the characteristics of the dynamic process. It has an obvious advantage in solving the time series
problem by the effective storage of the historical data of the system. This feature is
gradually being more attention, and thus plays a greater role in all areas. On the
basis of theoretically studying the Elman neural network and related parameters of
multiparameters and multistep prediction, using Elman neural network prediction
model to research the multiparameters and multistep fault prediction of liquid
rocket engine and applied to the practical engineering was established.

(1)   Artificial neural network of Elman type

1.   Structure and algorithm of Elman artificial neural network

   As shown in Fig. 11.10, Elman regression neural network is generally divided
into four layers: the input layer, the middle layer (hidden layer), the following layer,
and the output layer (the number of hidden layer units and the number of units to
undertake the same). The connection of the input layer, the hidden layer, and the
output layer is similar to that of the feed forward network. The unit of the input
layer only plays the role of signal transmission, and the output layer unit has a linear
weighting function. The transfer function of the hidden layer unit can be either
linear or nonlinear function, and usually the excitation function is taken as the
sigmoid nonlinear function. The following layer is also called the upper and lower
layer or the state layer, which is used to memory the output value of the first
moment of the hidden layer unit, and can be considered as a step delay operator. At

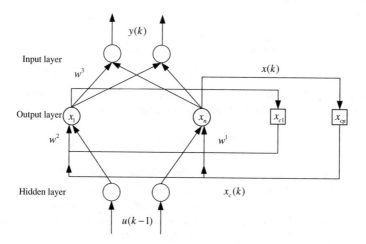

Fig. 11.10 Structure of Elman neural network

the same time, a self-feedback unit is also introduced in the input unit, which can solve the problem of higher order system identification. Here the feedforward connection part can be connected to the right of the modification, and the recursive part is fixed and cannot be modified to learn. So the Elman network is a part of the recursive network.

The input of the network at the k time includes not only the current input $u(k-1)$ but also the output value $x_c(k)$ of the first time of the hidden layer unit. At the beginning of the training, the output value of the hidden layer can be chosen as the half of its maximum value. Set the network input for the $u(k-1)$ and the output for $y(k)$. If the output of the hidden layer is denoted as $x(k)$, the nonlinear equation is expressed in the network:

$$x(k) = f(w^1 x_c(k) + w^2(u(k-1))) \tag{11.58}$$

$$y(k) = g(w^3 x(k)) \tag{11.59}$$

$$x_c(k) = x(k-1) \tag{11.60}$$

in which w_1, w_2, and w_3 represent the matrixes of the layer to the hidden layer, the input layer to the hidden layer, and the hidden layer to the output layer of the connection, respectively; $f(\cdot)$ and $g(\cdot)$ represent the excitation function of the output unit and the hidden layer, respectively. The result of an output is not only related to the current input, but also to the input of all the time before the current time. In terms of the system, the state of each time is not required to be used as the input of the network. This is the Elman network than the static neural network processing of the superiority of the dynamic system.

Elman is a dynamic recursive process. The algorithm is also called as back propagation algorithm, which is described as follows:

Elman artificial neural network learning index function using error square sum function:

$$E(w) = \sum_{k=1}^{n} [y_k(w) - \bar{y}_k(k)]^2 \tag{11.61}$$

Like BP algorithm, the calculation process of Elman artificial neural network uses the error back calculation method to correct the weight:

$$x_c(k) = x(k-1) = f(w^1_{k-1} x_c(k-1) + w^2_{k-1} u(k-2)) \tag{11.62}$$

And because of $x_c(k-1) = x(k-2)$, the formula can continue to expand. So the value of $x_c(k)$ depends on $w^1_{k-1}, w^1_{k-2}, \ldots$ of the historical period. That is, the value of $x_c(k)$ is a dynamic recursive process. Therefore, the backward propagation algorithm used in Elman artificial neural network training is called the dynamic back propagation algorithm. Specific calculation methods are mentioned as follows:

Consider the general objective function as

$$E = \sum_{p=1}^{n} E_p$$

where n is the number of output units.

$$E_p = \frac{1}{2}[y_d(k) - y(k)]^T[y_d(k) - y(k)] \tag{11.63}$$

Among them, $y_d(k)$ is the expected output of k time; and $y(k)$ is the actual output of k time.

The connection weight matrix of hidden layer to output layer w^3 is

$$\frac{\partial E_p}{\partial w_{ij}^3} = -[y_{d,i}(k) - y_i(k)]\frac{\partial y_i(k)}{\partial w_{ij}^3} = -[y_{d,i}(k) - y_i(k)]g_i'(\cdot)x_j(k) \tag{11.64}$$

To make $\delta_i^o = [y_{d,i}(k) - y_i(k)]g_i'(\cdot)$,

$$\frac{\partial E_p}{\partial w_{ij}^3} = -\delta_i^o x_j(k), \quad i = 1, 2, \ldots, n, j = 1, 2, \ldots, m \tag{11.65}$$

The connection weight matrix of the input layer to the hidden layer w^2 is

$$\frac{\partial E_p}{\partial w_{jq}} = \frac{\partial E_p}{\partial x_j(k)}\frac{\partial x_j(k)}{\partial w_{jq}^2} = \sum_{i=1}^{n}(-\delta_i^o w_{ij}^3)f_j^1(\cdot)u_q(k-1) \tag{11.66}$$

To make $\delta_j^h = \sum_{i=1}^{n}(\delta_i^o w_{ij}^3)f_j'(\cdot)$,

$$\frac{\partial E_p}{\partial w_{jq}^2} = -\delta_j^h u_q(k-1), \quad j = 1, 2, \ldots, m; q = 1, 2, \ldots, r \tag{11.67}$$

Similarly, the matrix w^1 from the following layer to hidden layer connection weight is

$$\frac{\partial E_p}{\partial w_{jl}^1} = -\sum_{i=1}^{n}\left(\delta_i^o w_{ij}^3 \frac{\partial x_j(k)}{\partial w_{jl}^1}\right), \quad j = 1, 2, \ldots, m; l = 1, 2, \ldots, m \tag{11.68}$$

The learning process of the whole network is shown in Fig. 11.11.

Fig. 11.11 Training flow chart of Elman neural network model

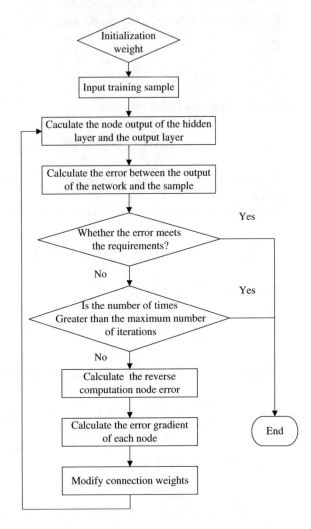

2. Characteristics of Elman artificial neural network

The characteristics of the Elman-type neural network are the output of the hidden layer by layer to undertake the delay and storage and from input to hidden layer. This way is sensitive for the history data of the state. The addition of internal feedback network increases the ability of the network to process the dynamic information, which can achieve the goal of dynamic modeling. In addition, the dynamic nature of the Elman network is provided only by the internal connection, without the use of the state as an input or training signal. This is also the superiority of the Elman network relative to the static feedforward network. Literature [173] is also shown that Elman artificial neural network has the characteristics of good dynamic characteristics, fast approaching speed, accurate and reliable prediction, and so on, to solve the prediction problem, relative to the BP network. The Elman dynamic network has more superior than the BP network. As a result of the above advantages, Elman artificial neural network has a good application prospect, and it is worth doing further research work.

(2) Fault prediction of liquid rocket engine based on Elman neural network

In fact, the fault prediction problem is a parameter prediction problem, which can be judged by the change trend of the parameters. Actually, the parameter change of the forecast is not only influenced by the information of itself, but also related to the variation of the relevant parameters:

$$\hat{y} = f(x) \tag{11.69}$$

From this, we can know that the mapping from x to y is a kind of space–time map, which has the corresponding relations in space and time. This space–time map is both dynamic and nonlinear, and it is not suitable to use the traditional linear prediction model. Elman neural network can better solve the problem of this kind of space–time mapping.

1. Data preprocessing and model number, node number selection

The transfer function of artificial neural network is different and has different output ranges. The general output ranges were $(0, 1)$ and $(-1, 1)$. Therefore, normalization of data processing can effectively prevent artificial neural network models appear abnormal, ensuring good convergence effect.

Elman-type network includes four types of nodes. The number of input and output nodes depends on the amount of available data and the purpose of establishing the model. The number of layer nodes is determined by the number of hidden layer nodes. So choosing the number of layers and nodes of the neural network model is actually the choice of the hidden layer and the number of nodes.

In general, the simulation of the complex nonlinear model needs to increase the number of layers and nodes in the hidden layer, but it must take a greater amount of calculation and computation time. In addition, a number of layers and the number of nodes will also complicate the simple problem, leading to the occurrence of learning. At present, there is no reliable theory to explain the optimal combination between the number of the hidden layer and the number of nodes and the number of input and output nodes. Usually, the method is based on the complexity of the problem and the establishment of different models to assess the effect and determine the optimal number of layers and nodes.

Although the relationship between the structures of various models and the training time or the degree of convergence cannot be quantified, the following conclusions can be drawn:

① The number of hidden nodes more computing time is longer, but there is no linear relationship between the two;

② Under the condition of the same node, the number of hidden layer will increase the computation time;

③ The increase of the node and the number of layers is not necessarily to improve the training effect, and sometimes the problem is complicated.

2. Engineering forecast example

The thrust of liquid rocket engine is an important parameter. According to it, changes can predict whether the engine will be a failure. In the case of a given engine system and environmental conditions, the thrust of the engine F related to the oxidant flow $\dot{m}_o$, combustion agent flow $\dot{m}_f$, and combustion chamber pressure p_c summed up as a function $F = f(\dot{m}_o, \dot{m}_f, p_c, t)$. Therefore, available parameters $\dot{m}_o$, $\dot{m}_f$, p_c, and t can predict the F. Especially in the engine start-up stage, the relationship between the parameters and the thrust is a relationship of multivariable, strong coupling, and serious nonlinear, and this relationship is dynamic. Therefore, using Elman neural network to establish the thrust prediction model can predict the liquid rocket engine monitoring. The training samples and test samples of the network model are derived from the test data, which are normalized to make the data in the [0, 1] range. The input structure of this network is very standardized, and the convergence speed of the network will be greatly accelerated. Due to the limitation of the test conditions, a total of 25 samples were selected, and the time interval was 0.1 s. After training, the model is established, and the 10 thrust values from 1.3 s to 3 are predicted.

When learning and training, the input unit is $\dot{m}_o(t - i)$, $\dot{m}_f(t - i)$, $P_c(t - i)$, $F(t - i)$ $i = 0, 1, 2, 3, 4; i \leq t$, and t. If using the BP network, the first is to show the time signal into space, and then give the static feedforward network as a kind of input, so as to realize the modeling and prediction of time series. However, this method increases the number of input nodes, resulting in the expansion of the network structure, slowing the training, and decreasing the accuracy. Because the Elman network is self-feedback through the hidden layer, the system dynamics is directly included in the network structure. Therefore, you can learn nonlinear dynamic system without need to learn the number of the system. It has the function of dynamic memory, and do not have to use more system state as input, thus reducing the number of input layer unit.

By reference [174] we can know that if using the BP network, then the value of i must be taken 0, 1, 2, ..., at the same time, and the number of input units is 9, 13, 17, ... For example, if i takes 0, 1, 2, then the parameter inputs are (I) oxidizer flow rate $\dot{m}_o(t), \dot{m}_o(t - 1), \dot{m}_o(t - 2)$; (II) combustion agent flow $\dot{m}_f(t), \dot{m}_f(t - 1)$, $\dot{m}_f(t - 2)$; (III) combustion chamber pressure $p_c(t), p_c(t - 1), p_c(t - 2)$; (IV) thrust F $(t), F(t - 1), F(t - 2)$; and (V) time t. So the number of input units is 13. Similarly, i takes 0, 1, 2, 3, and the number of input units is 17. So the structure of the network is more complex. Here using the Elman network it can only take $i = 0$. So the input unit is only 5, and the number of input nodes is 5. Two comparison of models structure is shown in Fig. 11.12. The thrust of the 3-step prediction is $F(t + l)$ $(l = 1,$ 2, 3). At this point, the output layer is 3 nodes. Select the single hidden layer

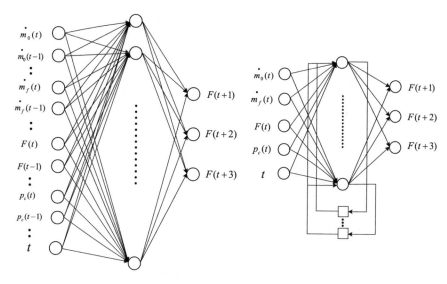

Fig. 11.12 The comparison of two kinds of network model structure

network model to avoid the problem which is too complicated. The number of hidden layer nodes is 10, 16, and 22, respectively. At this time the error is minimum value when the number of points is 16. The convergence speed of the network is relatively fast. So the number of hidden layer nodes is chosen 16.

Elman network is built using Matlab 6.5 neural network toolbox programming. Use the Tansig function as the transfer function in the input layer to the hidden layer, and the transfer function of the hidden layer to the output layer which is a pure linear function Purelin. The learning rate is 0.01, and the network control error is 0.0001. After the 1916 step training, the network can achieve the control error of the system. The error variation of training is shown in Fig. 11.13. The comparison of each step prediction curve with the actual curve is shown in Figs. 11.14, 11.15, and 11.16.

3. Result analysis

It can be seen that the forecast of the thrust change and the actual change curve trend is basically consistent and the accuracy is relatively high from the prediction of the curve and the actual curve chart. The average relative error of the first-step prediction is 0.41 %, the average relative error of the second-step prediction is 0.60 %, and the average relative error is 0.62 %. The results show that the overall prediction effect is better, and the accuracy of one-step prediction is higher than the precision of multistep prediction. The convergence rate of the network model is also the dozens of times faster than that of the traditional BP network, and it can achieve the modeling accuracy more quickly. At the same time, because the network structure is simple and the number of the input layer cells is decreased, the training

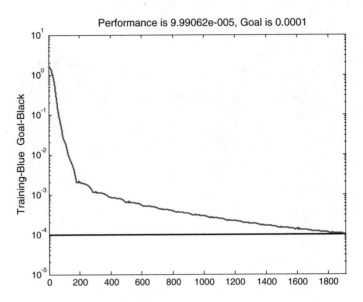

Fig. 11.13 Error variation curve of training process

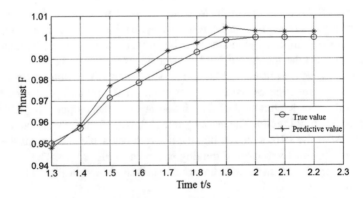

Fig. 11.14 The comparison between the first-step prediction curve and the actual curve

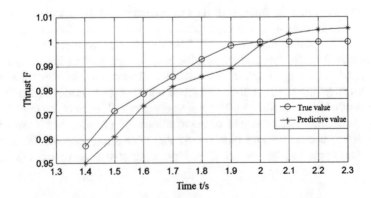

Fig. 11.15 The comparison between the second-step prediction curve and the actual curve

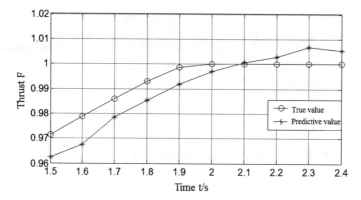

Fig. 11.16 The comparison between the third-step prediction curve and the actual curve

speed of the network is improved. But because the sample is too small and the data for modeling is not precise enough, the prediction accuracy was affected.

Real-time prediction and trend analysis of parameters are a very important part of engine fault diagnosis of monitoring. Compared with the feedforward neural network, using Elman neural network for thrust engine for real-time prediction has no number limit for the memory capacity and flexibility, which can better reflect the dynamic temporal features in input patterns. Results show that its application in nonlinear parameter prediction is effective, and has the advantages of good dynamic characteristics, fast convergence speed, high precision, and good timing prediction ability. It shows that Elman artificial neural network is suitable for prediction modeling of nonlinear dynamic systems, and it has a broad application prospect.

The application of artificial neural network in fault prediction of rocket motor is studied in order to establish a real-time accurate system model and advanced fault prediction algorithm. Because of the shortcomings of the traditional BP neural network whose convergence speed is slow and easy to fall into local minimum and so on, we put forward the improved algorithm, based on RBF neural network for time series prediction method, and give the process of RBF neural network for time series prediction modeling. The performances of two kinds of neural network time series forecasting models of BP and RBF are analyzed and compared by numerical simulation examples. The results show that the learning speed of RBF network is higher than that of BP network. BP network from the definition to the completion of the training required time is 1.086 s, and RBF network only needs 0.0625 s and network structure can be determined adaptively; RBF network model for the prediction of the three error index values is smaller than the BP neural network model of error indicators. Therefore, comparing with BP network model, the convergence speed and prediction accuracy of RBF network model are significantly improved, which has a good application value for real-time parameter prediction of liquid rocket engine.

In order to solve the problem of dynamic modeling, the fault prediction method based on Elman neural network is studied. The Elman neural network is used to predict the thrust of multiparameters in the process of rocket engine. It achieves better forecasting results. Engineering application examples show that the model structure is better than the traditional BP network and has greatly simplified, good dynamic characteristics, fast convergence speed, high precision, and the ability of good timing prediction, which provides a feasible approach to rocket engine fault prediction.

11.4   Rocket Engine Fault Prediction Based on SVM Method

The intelligent diagnosis and prediction method based on neural network has the ability of approaching the complex nonlinear system. This method requires a large number of typical fault data samples and a priori knowledge. In practice, it is difficult to find a large number of samples, and the structure and type of neural network are too dependent on the experience. The neural network model is based on the minimum principle of empirical risk, because its generalization ability cannot be guaranteed in theory, which leads to the practical application of neural network forecasting model which is more difficult, and its development in intelligent diagnosis and trend prediction.

The support vector machine found by V N Vapnik is based on statistical learning theory and structural risk on the basis of the minimum. In order to obtain the best generalization ability, it seeks the best compromise between the complexity of the model and the learning ability according to the limited sample information, which has a great advantage. Now it has been widely used in time series analysis and prediction. References [175, 176] show that the support vector machine has a good application prospect in time series forecasting and stochastic process approximation.

When using neural network to predict the model, due to the defects of the theory, it cannot guarantee the network's generalization ability. At the same time, there are no specific guidelines for the design of its structure. Usually, it only accords to observation and experience. In this case, the statistical learning theory based on the finite sample theory is conceived. The support vector machine algorithm produced by this theory also showed a good performance.

Compared with the neural network, the support vector machine method has a more mature theoretical basis. Although the small sample statistical learning theory needs to be further improved and developed, the support vector machine method has a number of unique advantages. It has great performance in small sample learning, nonlinear, high-dimensional problems, and generalization ability.

11.4.1   The Regression Estimation Based on Support Vector Machine

The original intention of the design support vector machine is to deal with the pattern recognition and classification problem. However, it is found that the support vector machine can not only solve the classification problem, but also can be used to solve the problem of time series prediction and regression. When the support vector machine is applied to deal with the problem of function approximation and regression estimation, we find that the two are essentially similar, and the mathematical programming problem is the same. So they are all called as vector regression support in reference (support vector regression, SVR).

(1)   Linear regression estimation based on support vector machine

Assuming a given l sample data (x_i, y_i), $i = 1, 2, ..., l$, where $x_i \in R^n$ is n-dimensional input samples, and $y_i \in R$ is the output samples. The function regression problem is to find a function f. After training it, through f we can get the corresponding y for the samples beyond the x.

When Vapnik put forward support vector machine regression in epsilon-SVR algorithm, he introduced the insensitive loss function. The ε-SVR controls the algorithm to achieve the accuracy of approximately by determining the ε in advance. The ε-insensitive loss function is defined as follows:

$$L(y, f(x)) = |y - f(x)|_\varepsilon = \begin{cases} 0, & \text{if } |y - f(x)| \leq \varepsilon \\ |y - f(x)| - \varepsilon, & \text{if } |y - f(x)| > \varepsilon \end{cases} \quad (11.70)$$

f is a true value function in the X domain which describes such an insensitive model. That is, if the forecast value and the actual value are less than ε, the loss is equal to 0. The purpose of the loss function is that it can use sparse data points to the following to find the estimated function.

If it is linear, we find a kind of regression estimation function in the linear function set, as shown in the following:

$$f(x) = (\omega \cdot x) + b, \quad \omega, x \in R^n, b \in R \quad (11.71)$$

$$R_{reg}[f] = \frac{1}{2}\|\omega\|^2 + CR_{emp}[f] \quad (11.72)$$

The $R_{reg}[f]$ is a regular risk; $\frac{1}{2}\|\omega\|^2$ is a regularization; C is the prespecified constant, and $Remp[f]$ is the training error which could be measured by available ε-insensitive loss function. When $y - f(x)| \leq \varepsilon$ and $Remp[f] = 0$, solution type is equivalent to solving-type optimization problem:

$$\text{min} \quad \Phi(\omega) = \tfrac{1}{2}\|\omega\|^2 = \tfrac{1}{2}(\omega \cdot \omega)$$
$$\text{s.t.} \quad \begin{cases} y_i - (\omega \cdot x_i) - b \leq \varepsilon, & i = 1, 2 \ldots, l \\ (\omega \cdot x_i) + b - y_i \leq \varepsilon, & i = 1, 2 \ldots, l \end{cases} \quad (11.73)$$

When training data contains noise, the relaxation variables are introduced to ensure that the regression function does not over fit the training samples. The optimization problem can be transformed into the minimum value problem of Function (11.74):

$$\text{min} \quad \Phi(\omega, \xi, \xi^*) = \tfrac{1}{2}(\omega \cdot \omega) + C\left(\sum_{i=1}^{l} \xi_i + \sum_{i=1}^{l} \xi_i^*\right)$$
$$\text{s.t.} \quad \begin{cases} y_i - (\omega \cdot x_i) - b \leq \varepsilon + \xi_i, & i = 1, 2, \ldots, l \\ (\omega \cdot x_i) + b - y_i \leq \varepsilon + \xi_i^*, & i = 1, 2, \ldots, l \\ \xi_i \geq 0, & i = 1, 2, \ldots, l \\ \xi_i^* \geq 0, & i = 1, 2, \ldots, l \end{cases} \quad (11.74)$$

where $(\omega \cdot \omega) = \|\omega\|^2$ reflects the complexity of the model, making the function more flat, enhancing the generalization ability; $\left(\sum_{i=1}^{l} \xi_i + \sum_{i=1}^{l} \xi_i^*\right)$ reflects the training error; $C > 0$ (given constant) was known as the regularization parameter, also known as the penalty function and regulation of the model complexity; and the training error is to make a compromise of the two. Therefore, this method is also called the penalty function method. The greater the C, the higher the fitting degree of the data; ε is used to control the pipe size regression approximation error, so as to control the support vector number and generalization ability. ε is positive, need to be set in advance, which is mainly used to control the accuracy of the algorithm to achieve. The greater the ε, the support vector is less. But the accuracy is not high.

This is a convex two optimization problem. In order to get the solution, the Lagrange function is introduced:

$$\Phi(\omega, \xi, \xi^*; \alpha, \alpha^*, \gamma, \gamma^*) = \frac{1}{2}\|\omega\|^2 + C\left(\sum_{i=1}^{l} \xi_i + \sum_{i=1}^{l} \xi_i^*\right) - \sum_{i=1}^{l} \alpha_i[(\omega \cdot x_i) + b - y_i + \varepsilon + \xi_i]$$
$$- \sum_{i=1}^{l} \alpha_i^*[y_i - (\omega \cdot x_i) - b + \varepsilon + \xi_i^*] - \sum_{i=1}^{l} (\gamma_i \xi_i + \gamma_i^* \xi_i^*)$$

$$(11.75)$$

Among them, according to Wolfe duality theorem, the original problem can be transformed into the following dual optimization problem:

$$\max \quad W(\alpha, \alpha^*) = -\varepsilon \sum_{i=1}^{l} \left(\alpha_i + \alpha_i^* \right) + \sum_{i=1}^{l} y_i \left(\alpha_i - \alpha_i^* \right) - \frac{1}{2} \sum_{i,j=1}^{l} \left(\alpha_i - \alpha_i^* \right) \left(\alpha_j - \alpha_j^* \right) \left(x_i \cdot x_j \right)$$

$$\text{s.t.} \quad \begin{cases} \displaystyle\sum_{i=1}^{l} \alpha_i = \sum_{i=1}^{l} \alpha_i^* \\ 0 \le \alpha_i, \alpha_i^* \le C, \quad i = 1, 2, \dots, l \end{cases}$$

$$(11.76)$$

The optimal solution $\alpha_i, \alpha_i^*, i = 1, 2, \dots, l$ can be obtained. By the relationship between ω, b, and α_i, α_i^* t, we could get the optimal estimation function:

$$\omega = \sum_{i=1}^{l} \left(\alpha_i - \alpha_i^* \right) x_i$$

$$b = y_i - (\omega \cdot x_i) - \varepsilon, \ \alpha_i \in (0, C) \text{ or}$$

$$b = y_i - (\omega \cdot x_i) + \varepsilon, \ \alpha_i^* \in (0, C)$$

$$(11.77)$$

From the above formula we can know that the value of the optimal estimate ω is only related to the sample vector x_i which corresponds to $\alpha_i \ne \alpha_i^*, i = 1, 2, \dots, l$. Such a sample vector x_i is called support vector. Support vector is only a small part of the total sample, so the training process can be simplified using the support vector, and the estimated function has good generalization ability.

Therefore, the support vector machine is used to select a representative feature to the quantum set (i.e., support vector) for regression estimation in the sample. Only one part of the sample data is involved in the regression to estimate the relation between the input and output functions. In the case of fewer samples, support vector machines can show its superiority.

(2) Nonlinear regression estimation based on support vector machine

If it is in the nonlinear case, the original problem can be mapped to a high-dimensional feature space by nonlinear transformation to solve the problem. In high-dimensional feature space, linear problem in inner product operation can be used to replace the nuclear function, i.e.,

$$K(x_i, x_j) = \phi(x_i) \cdot \phi(x_j) \qquad (11.78)$$

Kernel function can be used to achieve the function of the original space without to know the specific form of ϕ, so that you do not have to know the specific nonlinear transform ϕ case, only through computing the inner product in the original low-dimensional space to obtain the inner product of the high dimension space, avoiding the curse of dimensionality. So the regression equation for the nonlinear problem is

$$f(x) = \sum_{i=1}^{l} (\alpha_i - \alpha_i^*) K(x_i, x) + b \tag{11.79}$$

Among them, $K(\cdot, \cdot)$ is a kernel function. And the kernel function $K(\cdot, \cdot)$ is an arbitrary symmetric function which satisfies the Mercer condition. There are three kinds of the kernel function of regression support vector machine, namely, polynomial kernel function, radial basis kernel function, and Sigmoid kernel function. Similar to the linear case, α_i, α_i^* is the solution of the following problem:

$$\max \quad W(\alpha, \alpha^*) = -\varepsilon \sum_{i=1}^{l} (\alpha_i + \alpha_i^*) + \sum_{i=1}^{l} y_i(\alpha_i - \alpha_i^*) - \frac{1}{2} \sum_{i,j=1}^{l} (\alpha_i - \alpha_i^*)(\alpha_j - \alpha_j^*) K(x_i, x_j)$$

$$\text{s.t.} \quad \begin{cases} \sum_{i=1}^{l} \alpha_i = \sum_{i=1}^{l} \alpha_i^* \\ 0 \leq \alpha_i, \alpha_i^* \leq C, \quad i = 1, 2, \ldots, l \end{cases}$$

$$\tag{11.80}$$

Support vector for $\alpha_i - \alpha_i^* \neq 0, i = 1, 2, \ldots, l$ corresponding to the sample point. In the case of the above, linear and nonlinear optimization problems belong to the two optimization problems, which can be solved by the Matlab optimization toolbox.

(3) Least squares support vector machine regression estimation (LS-SVR)

Least squares support vector machine is proposed by Suykens et al. [178]. It uses the objective function and the constraint condition of the standard support vector machine to make the appropriate improvement, and uses the solution of the linear equation group to replace the solution standard support vector machine two times programming problem. The linear equations can be solved by the least square method. Least squares support vector machine (LS-SVM) thus named.

Standard support vector regression machine uses inequality constraints to solve the two optimization methods. Because in the optimization process the matrix size is directly proportional to the number of training samples of directly, using the inner product algorithm to solve the general size of the optimization problem is feasible. But if the large-scale optimization problem is too complex, it needs to be decomposed, cut, and other processing. Least squares support vector machine has improved the objective function using the equality to replace the inequality in the standard algorithm, and adopting the conjugate gradient iteration algorithm. The complexity of the optimization problem is reduced, and the large-scale optimization problem is solved effectively.

Two times optimization algorithm with inequality constraints is usually time-consuming and difficult to deal with real-time data. In the optimization objective, the two norms of the error are selected as the loss function by Suykens, and the inequality constraints are replaced by equality constraints. Because LS-SVR only solution of linear equations. It has the advantages of fast solving speed.

Compared with standard regression support vector machine is more suitable for online prediction.

Least squares support vector machine regression algorithm is as follows:

$$\begin{aligned} \min \quad & \Phi(\omega) = \tfrac{1}{2}(\omega \cdot \omega) + \tfrac{1}{2}C\varepsilon^2 \\ \text{s.t.} \quad & y_i - (\omega \cdot x_i) - b = \varepsilon, \quad i = 1, 2, \ldots, l \end{aligned} \qquad (11.81)$$

The corresponding Lagrange function is

$$L(\omega, \xi, \xi^*; \alpha, \alpha^*) = \frac{1}{2}(\omega \cdot \omega) + \frac{1}{2}C\varepsilon^2 - \sum_{i=1}^{l} \alpha_i [y_i - (\omega \cdot x_i) - b - \varepsilon] \qquad (11.82)$$

Optimize the former formula and solution the saddle point. On ω, b, αi, ε, we can obtain by the minimum necessary condition:

$$\begin{cases} \frac{\partial L}{\partial \omega} = 0 \Rightarrow & \omega = \sum_{i=1}^{l} \alpha_i \\ \frac{\partial L}{\partial b} = 0 \Rightarrow & \sum_{i=1}^{l} \alpha_i = 0 \\ \frac{\partial L}{\partial \alpha_i} = 0 \Rightarrow & y_i - (\omega \cdot x_i) - b = \varepsilon, \quad i = 1, 2, \ldots, l \\ \frac{\partial L}{\partial \varepsilon} = 0 \Rightarrow & C\varepsilon = \sum_{i=1}^{l} \alpha_i, \quad i = 1, 2, \ldots, l \end{cases} \qquad (11.83)$$

For $i = 1, 2, \ldots$ in the formula, eliminate ω, ξi, ξ_i^*. The formula will be written in the matrixform as

$$\begin{bmatrix} 0 & I' \\ I & XX' + \frac{E}{C} \end{bmatrix} \begin{bmatrix} b \\ \alpha \end{bmatrix} = \begin{bmatrix} 0 \\ Y \end{bmatrix} \qquad (11.84)$$

Among them, $X = [x_1, x_2, \ldots, x_l]$, $Y = [y_1, y_2, \ldots, y_l]$, $I = \underbrace{[1, 1, \ldots, 1]}_{l}^{T}$,

$\alpha = [\alpha_1, \alpha_2, \ldots, \alpha_l]'$, where E is a $l \times l$ unit array, B is the scalar, C and ε are the constants, C is the regularization parameter, also known as the penalty coefficient, and ε is the insensitive coefficient.

For the nonlinear case, there is a similar formula below:

$$\begin{bmatrix} 0 & I' \\ I & H + \frac{E}{C} \end{bmatrix} \begin{bmatrix} b \\ \alpha \end{bmatrix} = \begin{bmatrix} 0 \\ Y \end{bmatrix} \qquad (11.85)$$

Among $H(i, j) = K(x_i, x_j)$. $K(\cdot, \cdot)$ is suitable for the selection of the kernel function.

The least square regression estimation function is obtained by solving Eq. (11.93):

$$f(x) = \sum_{i=1}^{l} \alpha_i K(x_i, x) + b \tag{11.86}$$

Among

$$b = \frac{I'(H + E/C)^{-1}Y}{I'(H + E/C)^{-1}I} \tag{11.87}$$

$$\alpha = (H + E/C)^{-1}(Y - Ib) \tag{11.88}$$

By the derivation process, we can know that the optimization problem can be transformed into a linear equation, which greatly reduces the complexity of the algorithm. In addition, the use of radial basis function of LS-SVM is only required to determine two parameters C and σ. The search space of parameters is reduced from 3D to 2D by standard SVM, which greatly accelerates the modeling speed.

11.4.2 Prediction Process and Evaluation Criteria Based on Support Vector Machine

(1) Prediction process

The prediction process based on support vector machine is as follows [179]:

1. Enter the historical sample data and preprocess. Form the training and test samples;
2. Select the Kernel function and the model parameters of support vector machine;
3. Calculate the model initialization, given the α_i α_i^* and b the random initial value; use training samples to establish the objective function, forming two optimization problems, the solution to get α_i, α_i^*, and b;
4. Substitute the parameters into function estimation, and calculate the value of prediction for a certain time in the future.

(2) Model evaluation criteria

After using support vector machine regression estimation algorithm to train the samples, the prediction model is established. The performance of the model can be tested by the test samples. Set the number of data points N, y_i is the predicted data value, and $\hat{y}_i$ is the actual data value. Several commonly used predictive performance evaluation indicators are as follows:

1. Root mean square relative error

$$E_{MSE} = \sqrt{\frac{1}{N} \sum_{i=1}^{N} \left(\frac{y_i - \hat{y}_i}{y_i} \right)^2} \tag{11.89}$$

2. Average relative error

$$E_{MAPE} = \frac{1}{N} \sum_{i=1}^{N} \left| \frac{y_i - \hat{y}_i}{y_i} \right| \tag{11.90}$$

3. Root mean square absolute error

$$E_{NMSE} = \sqrt{\frac{1}{N} \sum_{i=1}^{N} (y_i - \hat{y}_i)^2} \tag{11.91}$$

4. Average absolute error

$$E_{MAE} = \frac{1}{N} \sum_{i=1}^{N} |y_i - \hat{y}_i| \tag{11.92}$$

11.4.3   An Example of Liquid Rocket Engine Fault Prediction Based on SVM Method

The thrust of rocket engine is an important index. According to its changes in advance we can determine the development trend of its performance, and early detect the potential failure of the engine to avoid accidents. Under the given conditions of the engine system and environmental conditions, the thrust of the engine F is related to the flow of the oxidant $\dot{m}_o$, the flow of the combustion $\dot{m}_f$, and the pressure of the combustion chamber p_c and t. It can be summed up as a function. So we can use the parameters $\dot{m}_o$, $\dot{m}_f$, p_c, and t to predict the F. There is a high degree of complexity and nonlinearity between the parameters and the thrust force, but the support vector regression model has a good effect on solving this kind nonlinear model which is difficult to establish its accurate mathematical model.

(1) Data preparation

Model training learning samples and test samples were from test data, and have been normalized process, which can accelerate the speed of learning. The link in the data does not reduce, and the data structure of the time series generally have a

Table 11.14 Sample data for support vector machine model

Time series t(s)	Combustion flux $(\dot{m}_f)$	Oxidant flux $(\dot{m}_o)$	Combustion chamber pressure (p_c)	Thrust (F)
0	0.3179	0.4011	0.0098	0.0000
0.1	0.3931	0.4661	0.0559	0.0496
0.2	0.5780	0.5339	0.2238	0.1418
0.3	0.6532	0.5935	0.4643	0.3262
0.4	0.6012	0.5658	0.6170	0.4113
0.5	0.6012	0.5572	0.7049	0.4681
0.6	0.6763	0.6374	0.7832	0.5674
0.7	0.7746	0.6721	0.8447	0.6596
0.8	0.8555	0.8238	0.8951	0.7730
0.9	0.9017	0.8753	0.9301	0.8440
1.0	0.9306	0.9149	0.9580	0.8865
1.1	0.9526	0.9420	0.9790	0.9163
1.2	0.9665	0.9593	0.9874	0.9362
1.3	0.9769	0.9702	0.9930	0.9504
1.4	0.9827	0.9810	0.9944	0.9574
1.5	0.9884	0.9864	0.9951	0.9716
1.6	0.9442	0.9919	0.9972	0.9787
1.7	0.9965	0.9957	0.9986	0.9858
1.8	0.9977	0.9973	1.0000	0.9929
1.9	0.9988	0.9995	1.0000	0.9986
2.0	1.0000	1.0000	1.0000	1.0000
2.1	1.0000	1.0000	1.0000	1.0000
2.2	1.0000	1.0000	1.0000	1.0000
2.3	1.0000	1.0000	1.0000	1.0000
2.4	1.0000	1.0000	1.0000	1.0000

strong nonlinear. Due to the limitation of the test conditions, the samples have 25 groups. The time interval is 0.1 s. The former 13 groups of data are for modeling with the training sample. The latter 12 groups of data are used as the test samples to test the effect of the forecast and to verify ability or generalization of the model, as shown in Table 11.14. Support vector machine is a kind of machine learning method for small sample data, and the nonlinear time series can achieve a good learning effect.

Build the prediction model to predict thrust by the standard support vector machine and least squares support vector machine. $\dot{m}_o$, $\dot{m}_f$, p_c, and t are used as the input of support vector machine regression prediction model, and the F is used as the output value of the model, that is, multiparameter prediction of the thrust F. The study shows that the multiparameter prediction can make full use of the internal connection of the relevant parameters of the liquid rocket engine to make up the deficiency of the forecast information, and improve the accuracy of the prediction.

(2) The selection of kernel function and parameters

1. Selection of kernel function

As mentioned before, the performance of the support vector regression model is not related to the type of Kernel function. Kernel parameter and penalty factor C are the main factors that affect the performance of support vector regression model. From the analysis that choosing the appropriate kernel function is good for reducing the amount of computation, when the dimension of the feature space is very high, the amount of calculation will be greatly increased, even cannot be correct results in some cases for the polynomial kernel function. Radial basis functions do not have this problem. In addition, the selection of the radial basis function is implicit. Each support vector generates a local radial basis function with its center, which can be used to find the global radial basis function parameters. Therefore, the radial basis kernel function is selected to build the prediction model.

2. Parameter selection

The radial basis kernel function is $K(x, x_i) = \exp\left(-\|x - x_i\|^2 / 2\sigma^2\right)$, where σ is the width of the radial basis function. There is no unified method to select reasonable σ, C, and ε. Cross-validation method is generally adopted. For the convenience of comparison of prediction effect of a different support vector machine, we set the penalty coefficient C and width σ the same values here. The simulation test shows that when $\sigma^2 = 20$, $\varepsilon = 0.001$, $C = 100$, two models have good effect.

(3) The analysis of the results

Average relative error and training time are introduced as evaluation indexes, and compare the prediction performance of these two methods. The average relative error is as follows:

$$E_{\text{MAPE}} = \frac{1}{N} \sum_{i=1}^{N} \left| \frac{y_i - \hat{y}_i}{y_i} \right| \times 100\% \tag{11.93}$$

Among them, the number of data points is N, y_i is the predicted data value, and $\hat{y}_i$ is the actual data value.

By Matlab 6.5 optimization toolbox programming, through a selection of more parameters testing in the training model, the selection of parameter is as follows: $\sigma^2 = 20$, $\varepsilon = 0.001$, $C = 100$. By training the data of the 13 groups of samples, the accuracy of the calculation results is higher, which can better simulate the nonlinear variation of the liquid rocket engine's thrust, as shown in Fig. 11.17.

Figures 11.18 and 11.19 show two prediction thrust curves of two different methods based on the time series. The curve with "o" points represents the actual measured value. The curve with "*" points represents the prediction results of the standard support unit and the least square support vector machine. It can be seen from the figure that the predicted results are very good approximation to the actual change curve of the thrust of the engine. Predicting the later stage of thrust change

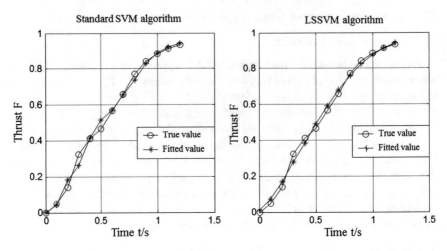

Fig. 11.17 The actual value and the fitting value of the study sample

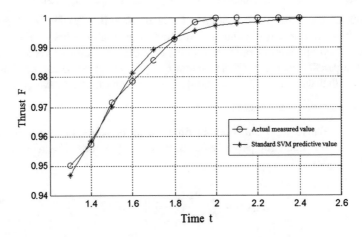

Fig. 11.18 The results based on of standard SVM prediction

tends to be stable. Meet the characteristics of the engine in normal working condition. It shows that the model has a good tendency forecast performance.

According to the average relative error and the training time as evaluation indexes, the two kinds of prediction results are compared quantitatively, as shown in Table 11.15.

By the above comparison, we can see that the effect of two different support vector machine regression algorithms for liquid rocket engine thrust prediction is more ideal, which reflects that the support vector machine (SVM) is a good learning performance and stronger generalization ability. Because the least square algorithm has been improved the objective function, the prediction accuracy of the standard

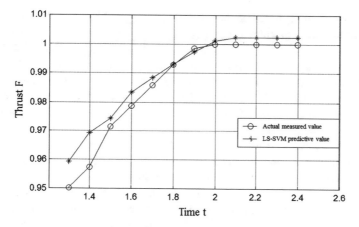

Fig. 11.19 The results based on of LS-SVM prediction

Table 11.15 Performance comparison of prediction methods

Group number project	True value	Standard SVM predictive value	LS-SVM predictive value
1	0.9504	0.9470	0.9595
2	0.9574	0.9588	0.9694
3	0.9716	0.9701	0.9745
4	0.9787	0.9816	0.9835
5	0.9858	0.9894	0.9885
6	0.9929	0.9935	0.9931
7	0.9986	0.9959	0.9976
8	1.0000	0.9974	1.0013
9	1.0000	0.9983	1.0024
10	1.0000	0.9987	1.0024
11	1.0000	0.9995	1.0024
12	1.0000	0.9999	1.0024
Average relative error (%)		0.19	0.37
Training time (s)		0.0313	0.0006

support vector machine algorithm is slightly higher than that of the least square algorithm. But the prediction accuracy of the two algorithms is very good, and it can be accepted. More importantly, the least square algorithm in the training speed is far superior than the standard algorithm. It effectively solves the problem of restricting the development of support vector machine training speed. In practice, when the scale of the problem is larger, the learning speed of the LS-SVM method is more obvious. This method also provides the possibility for real-time liquid rocket engine fault prediction.

SVM algorithm is based on the principle of structural risk minimization; find the best support vector machine suitable for the available data, avoid the phenomenon of over learning and under study, and have good generalization ability. At the same time, the SVM algorithm provides a strong theoretical basis for solving the problem of machine learning in small samples. The time series forecasting method based on support vector machine is studied, and its application to fault prediction of liquid rocket engine is simulated. The results showed that the two methods have better prediction accuracy and simpler model. The average relative errors are 0.19 and 0.37 %. The standard SVM prediction accuracy is relatively high, and the learning speed of LS-SVM method has been increased than the SVM method, whose training time from 0.0313 s is reduced to 0.0006 s. SVM has the characteristics of identifying complex nonlinear systems, and the ability to predict the high accuracy of small sample data. It determines the great potential of this method in the fault prediction research of liquid rocket engine. This algorithm is expected to be further applied in the fault prediction of liquid rocket engine.

Appendix A
Steady State Fault Model of I-Level

$$x(1) - x(3) - x(4) - x(7) - a_0 = 0 \tag{A.1}$$

$$x(2) - x(14) - x(5) - a_1 = 0 \tag{A.2}$$

$$x(14) - x(6) - x(15) = 0 \tag{A.3}$$

$$\frac{10^3 x(1)x(9)}{D_1 R_1} + \frac{10^3 x(2)x(10)}{D_2 R_2} + D_5 - 5x(23)x(17)t_1(1 - D_5^{0.2}) = 0 \tag{A.4}$$

$$10^3 D_{15} x(8) - a_2 D_{16}[x(3) + x(14)] = 0 \tag{A.5}$$

$$x(9) + \frac{a_3 x^2(1)}{R_1} + a_4 x(1)x(13) - a_5 R_1 x^2(13) = 0 \tag{A.6}$$

$$x(10) + \frac{a_6 x^2(2)}{R_2} + a_7 x(2)x(13) - a_8 R_2 x^2(13) = 0 \tag{A.7}$$

$$x(9) - x(11) - \frac{a_9}{R_1}[x(3) + x(7)]^2 - D_5 + D_8 = 0 \tag{A.8}$$

$$x(10) - x(12) - \frac{a_{10} x^2(2)}{R_2} + D_9 = 0 \tag{A.9}$$

$$x(12) - x(8) - \frac{a_{11}}{R_2} x^2(14) - \frac{a_{12}}{R_2} x^2(6) - D_{17} - D_{18} + P_1^H = 0 \tag{A.10}$$

$$x(11) - x(8) - \frac{a_{13} x^2(3)}{R_1} = 0 \tag{A.11}$$

$$\frac{a_{14} x^2(15)}{R_2} - \frac{a_{15} x^2(6)}{R_2} + P_2^H = 0 \tag{A.12}$$

$$D_{12} x(7) - x(8) - \frac{a_{16} x^2(3)}{R_1} + D_{10} + P_3^H = 0 \tag{A.13}$$

© Springer-Verlag Berlin Heidelberg and National Defense Industry Press 2016
W. Zhang, *Failure Characteristics Analysis and Fault Diagnosis*
for Liquid Rocket Engines, DOI 10.1007/978-3-662-49254-3

$$x(9) - \left(\frac{a_{17}}{R_1} + D_{11}\right)x^2(4) + D_8 - D_{10} = 0 \tag{A.14}$$

$$x(12) - \left(\frac{a_{18}}{R_2} + D_{14}\right)[x(5) + a_{19}]^2 - D_{13} = 0 \tag{A.15}$$

$$x(1) + x(2) - x(16) = 0 \tag{A.16}$$

$$\frac{x(4) + x(5)}{1 + D_6/D_7} - x(17) - a_{20} = 0 \tag{A.17}$$

$$\frac{a_{21}x(4)}{x(5)} - x(18) = 0 \tag{A.18}$$

$$\frac{x(1)}{x(2)} - x(19) = 0 \tag{A.19}$$

$$\frac{10^3 x(9)x(1)}{D_1 R_1} - x(20) = 0 \tag{A.20}$$

$$\frac{10^3 x(10)x(2)}{D_2 R_2} - x(21) = 0 \tag{A.21}$$

$$x(20) + x(21) - x(22) = 0 \tag{A.22}$$

$$a_{22}x(18) + a_{23} - x(23) = 0 \tag{A.23}$$

$$a_{24}[x(1) + x(2) - a_{25}] - 10^3 A_1 P_0 - x(24) = 0 \tag{A.24}$$

$$\frac{D_6}{D_7}x(17) - x(25) = 0 \tag{A.25}$$

$$x(26) - x(28) - x(29) - x(32) - a_{26} = 0 \tag{A.26}$$

$$x(27) - x(39) - x(30) - a_{27} = 0 \tag{A.27}$$

$$x(39) - x(31) - x(40) = 0 \tag{A.28}$$

$$\frac{10^3 x(26)x(34)}{D_{17} R_1} + \frac{10^3 x(27)x(35)}{D_{18} R_2} + D_{21} - 5x(48)x(42)t_2(1 - D_{20}^{0.2}) = 0 \tag{A.29}$$

$$10^3 D_{29}x(33) - a_{28}D_{30}[x(28) + x(39)] = 0 \tag{A.30}$$

$$x(34) + \frac{a_{29}x^2(26)}{R_1} + a_{30}x(26)x(38) - a_{34}R_1x^2(38) = 0 \tag{A.31}$$

$$x(35) + \frac{a_{32}x^2(26)}{R_2} + a_{33}x(27)x(38) - a_{34}R_2x^2(38) = 0 \tag{A.32}$$

$$x(34) - x(36) - \frac{a_{35}[x(28) + x(32)]^2}{R_1} + D_{24} = 0 \tag{A.33}$$

$$x(35) - x(37) - \frac{a_{36}x^2(27)}{R_2} + D_{25} = 0 \tag{A.34}$$

$$x(37) - x(33) - \frac{a_{37}x^2(39)}{R_2} - \frac{a_{38}x^2(31)}{R_2} + P_{21}^H = 0 \tag{A.35}$$

$$x(36) - x(33) - \frac{a_{39}x^2(28)}{R_1} = 0 \tag{A.36}$$

$$\frac{a_{40}x^2(40)}{R_2} - \frac{a_{41}x^2(31)}{R_2} + P_{22}^H = 0 \tag{A.37}$$

$$D_{27}x(32) - x(33) - \frac{a_{42}x^2(28)}{R_1} + D_{10} + P_{23}^H = 0 \tag{A.38}$$

$$x(34) - \left(\frac{a_{43}}{R_1} + D_{28}\right)x^2(29) + D_{24} - D_{10} = 0 \tag{A.39}$$

$$x(37) - \left(\frac{a_{44}}{R_2} + D_{28}\right)[x^2(30) + a_{45}]^2 - D_{13} = 0 \tag{A.40}$$

$$x(26) + x(27) - x(41) = 0 \tag{A.41}$$

$$\frac{x(29) + x(30)}{1 + D_{22}/D_{23}} - x(42) - a_{46} = 0 \tag{A.42}$$

$$\frac{a_{47}x(29)}{x(30)} - x(43) = 0 \tag{A.43}$$

$$\frac{x(26)}{x(27)} - x(44) = 0 \tag{A.44}$$

$$\frac{10^3x(34)x(26)}{D_{17}R_1} - x(45) = 0 \tag{A.45}$$

$$\frac{10^3 x(35)x(27)}{D_{18}R_2} - x(46) = 0 \tag{A.46}$$

$$x(45) + x(46) + D_{21} - x(47) = 0 \tag{A.47}$$

$$a_{48}x(43) + a_{49} - x(48) = 0 \tag{A.48}$$

$$a_{50}[x(26) + x(27) - a_{51}] - 10^3 A_2 P_0 - x(49) = 0 \tag{A.49}$$

$$\frac{D_{22}}{D_{23}}x(42) - x(50) = 0 \tag{A.50}$$

$$x(51) - x(53) - x(54) - x(57) - a_{52} = 0 \tag{A.51}$$

$$x(52) - x(64) - x(55) - a_{53} = 0 \tag{A.52}$$

$$x(64) - x(56) - x(65) = 0 \tag{A.53}$$

$$\frac{10^3 x(51)x(59)}{D_{31}R_1} + \frac{10^3 x(52)x(60)}{D_{32}R_2} + D_{35} - 5x(73)x(67)t_3(1 - D_{34}^{0.2}) = 0 \tag{A.54}$$

$$10^3 D_{43}x(58) - a_{54}D_{44}[x(53) + x(64)] = 0 \tag{A.55}$$

$$x(59) + \frac{a_{55}x^2(51)}{R_1} + a_{56}x(51)x(63) - a_{57}R_1 x^2(63) = 0 \tag{A.56}$$

$$x(60) + \frac{a_{58}x^2(52)}{R_2} + a_{59}x(52)x(63) - a_{60}R_2 x^2(63) = 0 \tag{A.57}$$

$$x(59) - x(61) - \frac{a_{61}[x(53) + x(57)]^2}{R_1} + D_{38} = 0 \tag{A.58}$$

$$x(60) - x(62) - \frac{a_{62}x^2(52)}{R_2} + D_{39} = 0 \tag{A.59}$$

$$x(62) - x(58) - \frac{a_{63}x^2(64)}{R_2} - \frac{a_{64}x^2(56)}{R_2} + P_{31}^H = 0 \tag{A.60}$$

$$x(61) - x(58) - \frac{a_{68}x^2(53)}{R_1} = 0 \tag{A.61}$$

$$\frac{a_{66}x^2(65)}{R_2} - \frac{a_{67}x^2(56)}{R_2} + R_{32}^H = 0 \tag{A.62}$$

$$D_{41}x(57) - x(58) - \frac{a_{68}x^2(53)}{R_1} + D_{10} + P_{33}^H = 0 \quad (A.63)$$

$$x(59) - \left(\frac{a_{69}}{R_1} + D_{40}\right)x(54) + D_{38} - D_{10} = 0 \quad (A.64)$$

$$x(62) - \left(\frac{a_{70}}{R_2} + D_{42}\right)[x(55) + a_{71}]^2 - D_{13} = 0 \quad (A.65)$$

$$x(51) + x(52) - x(66) = 0 \quad (A.66)$$

$$\frac{x(54) + x(55)}{1 + D_{36}/D_{37}} - x(67) - a_{72} = 0 \quad (A.67)$$

$$\frac{a_{73}x(54)}{x(55)} - x(68) = 0 \quad (A.68)$$

$$\frac{x(51)}{x(52)} - x(69) = 0 \quad (A.69)$$

$$\frac{10^3 x(59)x(51)}{D_{31}R_1} - x(70) = 0 \quad (A.70)$$

$$\frac{10^3 x(60)x(52)}{D_{32}R_2} - x(71) = 0 \quad (A.71)$$

$$x(70) + x(71) + D_{35} - x(72) = 0 \quad (A.72)$$

$$a_{74}x(68) + a_{75} - x(73) = 0 \quad (A.73)$$

$$a_{76}[x(51) + x(52) - a_{77}] - 10^3 A_2 P_0 - x(74) = 0 \quad (A.74)$$

$$\frac{D_{36}}{D_{37}}x(67) - x(75) = 0 \quad (A.75)$$

$$x(76) - x(78) - x(79) - x(82) - a_{78} = 0 \quad (A.76)$$

$$x(77) - x(89) - x(80) - a_{79} = 0 \quad (A.77)$$

$$x(89) - x(81) - x(90) = 0 \quad (A.78)$$

$$\frac{10^3 x(76)x(84)}{D_{45}R_1} + \frac{10^3 x(77)x(85)}{D_{46}R_2} + D_{49} - 5x(98)x(92)t_4\left(1 - D_{48}^{0.2}\right) = 0 \quad (A.79)$$

$$10^3 D_{57} x(83) - a_{80} D_{58} [x(78) + x(89)] = 0 \qquad \text{(A.80)}$$

$$x(84) + \frac{a_{81} x^2(76)}{R_1} + a_{82} x(76) x(88) - a_{83} R_1 x^2(88) = 0 \qquad \text{(A.81)}$$

$$x(85) + \frac{a_{84} x^2(77)}{R_2} + a_{85} x(77) x(88) - a_{86} R_2 x^2(88) = 0 \qquad \text{(A.82)}$$

$$x(84) - x(86) - \frac{a_{87} [x(78) + x(82)]^2}{R_1} + D_{52} = 0 \qquad \text{(A.83)}$$

$$x(85) - x(87) - \frac{a_{88} x^2(77)}{R_2} + D_{53} = 0 \qquad \text{(A.84)}$$

$$x(87) - x(83) - \frac{a_{89} x^2(89)}{R_2} - \frac{a_{90} x^2(81)}{R_2} + P_{41}^H = 0 \qquad \text{(A.85)}$$

$$x(86) - x(83) - \frac{a_{91} x^2(78)}{R_1} = 0 \qquad \text{(A.86)}$$

$$\frac{a_{92} x^2(90)}{R_2} - \frac{a_{93} x^2(81)}{R_2} + P_{42}^H = 0 \qquad \text{(A.87)}$$

$$D_{55} x(82) - x(83) - \frac{a_{94} x(78)}{R_1} + D_{10} + P_{43}^H = 0 \qquad \text{(A.88)}$$

$$x(84) - \left(\frac{a_{95}}{R_1} + D_{54}\right) x^2(79) + D_{52} - D_{10} = 0 \qquad \text{(A.89)}$$

$$x(87) - \left(\frac{a_{96}}{R_2} + D_{56}\right) [x(80) + a_{97}]^2 - D_{13} = 0 \qquad \text{(A.90)}$$

$$x(87) - \left(\frac{a_{96}}{R_2} + D_{56}\right) [x(80) + a_{97}]^2 - D_{13} = 0 \qquad \text{(A.91)}$$

$$\frac{x(79) + x(80)}{1 + D_{50}/D_{51}} - x(92) - a_{98} = 0 \qquad \text{(A.92)}$$

$$\frac{a_{99} x(79)}{x(80)} - x(93) = 0 \qquad \text{(A.93)}$$

$$\frac{x(76)}{x(77)} - x(94) = 0 \qquad \text{(A.94)}$$

$$\frac{10^3 x(84) x(76)}{D_{45} R_1} - x(95) = 0 \tag{A.95}$$

$$\frac{10^3 x(85) x(77)}{D_{46} R_2} - x(96) = 0 \tag{A.96}$$

$$x(95) + x(96) + D_{49} - x(97) = 0 \tag{A.97}$$

$$a_{100} x(93) + a_{101} - x(98) = 0 \tag{A.98}$$

$$a_{102}[x(76) + x(77) - a_{103}] - 10^3 A_4 P_0 - x(99) = 0 \tag{A.99}$$

$$\frac{D_{50}}{D_{51}} x(92) - x(100) = 0 \tag{A.100}$$

$$x(101) - a_{104} x(7) = 0 \tag{A.101}$$

$$x(102) - a_{104} - a_{106}[x(101) - a_{107}] = 0 \tag{A.102}$$

$$x(103) - D_{59} - x(102)(D_{60} - D_{59}) = 0 \tag{A.103}$$

$$x(104) - a_{108} x(32) = 0 \tag{A.104}$$

$$x(105) - a_{109} - a_{110}[x(104) - a_{111}] = 0 \tag{A.105}$$

$$x(106) - D_{61} - x(105)(D_{62} - D_{61}) = 0 \tag{A.106}$$

$$x(107) - a_{112} x(57) = 0 \tag{A.107}$$

$$x(108) - a_{113} - a_{114}[x(107) - a_{115}] = 0 \tag{A.108}$$

$$x(109) - D_{63} - x(108)(D_{64} - D_{63}) = 0 \tag{A.109}$$

$$x(110) - a_{116} x(82) = 0 \tag{A.110}$$

$$x(111) - a_{117} - a_{118}[x(110) - a_{119}] = 0 \tag{A.111}$$

$$x(112) - D_{65} - x(111)(D_{66} - D_{65}) = 0 \tag{A.112}$$

$$x(113) - \frac{x(103) + x(106) + x(109) + x(112)}{4} = 0 \tag{A.113}$$

$$x(114) - (1 - a_{120}) x(113) - a_{120} D_{67} = 0 \tag{A.114}$$

$$x(115) - [x(1) + x(26) + x(51) + x(76)] = 0 \tag{A.115}$$

$$x(116) - [x(7) + x(32) + x(57) + x(82)] - a_{121} = 0 \tag{A.116}$$

$$x(117) - \frac{a_{122}}{x(115)} [a_{123}x(116)x(113) - D_{68}] = 0 \tag{A.117}$$

$$x(118) - a_{124}x(25) = 0 \tag{A.118}$$

$$x(119) - a_{125} - a_{126}[x(118) - a_{127}] = 0 \tag{A.119}$$

$$x(120) - D_{69} - x(119)(D_{70} - D_{69}) = 0 \tag{A.120}$$

$$x(121) - a_{128}x(50) = 0 \tag{A.121}$$

$$x(122) - a_{129} - a_{130}[x(121) - a_{131}] = 0 \tag{A.122}$$

$$x(123) - D_{17} - x(122)(D_{72} - D_{71}) = 0 \tag{A.123}$$

$$x(124) - a_{132}x(75) = 0 \tag{A.124}$$

$$x(125) - a_{133} - a_{134}[x(124) - a_{135}] = 0 \tag{A.125}$$

$$x(126) - D_{73} - x(125)(D_{74} - D_{73}) = 0 \tag{A.126}$$

$$x(127) - a_{136}x(100) = 0 \tag{A.127}$$

$$x(128) - a_{137} - a_{138}[x(127) - a_{139}] = 0 \tag{A.128}$$

$$x(129) - D_{75} - x(128)(D_{76} - D_{75}) = 0 \tag{A.129}$$

$$x(130) - \frac{x(120) + x(123) + x(126) + x(129)}{4} = 0 \tag{A.130}$$

$$x(131) - (1 - a_{140})x(130) - a_{140}D_{77} = 0 \tag{A.131}$$

$$x(132) - [x(2) + x(27) + x(52) + x(77)] = 0 \tag{A.132}$$

$$x(133) - [x(25) + x(50) + x(75) + x(100)] - a_{141} = 0 \tag{A.133}$$

$$x(134) - \frac{a_{134}}{x(132)} [a_{143}x(130)x(133) - D_{78}] = 0 \tag{A.134}$$

$$t_1 = 0.92(4.911\beta_1 - 7.945\beta_1^2) \tag{A.135}$$

$$\beta_1 = \mu_1/c_1 \tag{A.136}$$

$$\mu_1 = 0.02094x(13) \tag{A.137}$$

$$c_1 = 44.7216\sqrt{5x(23)(1 - D_4^{0.2})} \tag{A.138}$$

$$t_2 = 0.92(4.911\beta_2 - 7.945\beta_2^2) \tag{A.139}$$

$$\beta_2 = \mu_2/c_2 \tag{A.140}$$

$$\mu_2 = 0.02094x(38) \tag{A.141}$$

$$c_2 = 44.7216\sqrt{5x(48)(1 - D_{20}^{0.2})} \tag{A.142}$$

$$t_3 = 0.92(4.911\beta_3 - 7.945\beta_3^2) \tag{A.143}$$

$$\beta_3 = \mu_3/c_3 \tag{A.144}$$

$$\mu_3 = 0.02094x(63) \tag{A.145}$$

$$c_3 = 44.7216\sqrt{5x(73)(1 - D_{34}^{0.2})} \tag{A.146}$$

$$t_4 = 0.92(4.911\beta_4 - 7.945\beta_4^2) \tag{A.147}$$

$$\beta_4 = \mu_4/c_4 \tag{A.148}$$

$$\mu_4 = 0.02094x(88) \tag{A.149}$$

$$c_4 = 44.7216\sqrt{5x(98)(1 - D_{48}^{0.2})} \tag{A.150}$$

The equations described in the above model are not a closed circuit. Due to the influence of overload and gravity, the fluid column pressure varies heavily in the fighting program and cannot approximate to be a fixed value. In this instance, the overload coefficient values in the steady state after a certain time can be used to construct the closed-loop model equations. By the above equation,

$$x(135) - [x(117) + 5.919 - a_{144}x^2(115)] = 0 \tag{A.151}$$

$$x(136) - [x(135) - a_{145}x^2(1)] = 0 \tag{A.152}$$

$$x(137) - [x(135) - a_{146}x^2(26)] = 0 \tag{A.153}$$

$$x(138) - [x(135) - a_{147}x^2(51)] = 0 \tag{A.154}$$

$$x(139) - [x(135) - a_{148}x^2(76)] = 0 \qquad\qquad \text{(A.155)}$$

$$x(140) - [x(134) + 1.211 - a_{149}x^2(2)] = 0 \qquad\qquad \text{(A.156)}$$

$$x(141) - [x(134) + 1.211 - a_{150}x^2(27)] = 0 \qquad\qquad \text{(A.157)}$$

$$x(142) - [x(134) + 1.211 - a_{151}x^2(52)] = 0 \qquad\qquad \text{(A.158)}$$

$$x(143) - [x(134) + 1.211 - a_{152}x^2(77)] = 0 \qquad\qquad \text{(A.159)}$$

Replacing D8, D24, D38, D52, D9, D25, D39, and D53 by x(136), x(137), x (138), x(139), x(140), x(141), x(142), and x(143), we can obtain the closed-loop model of the whole system (Tables A.1, A.2 and A.3).

Table A.1 Engine parameters

No.	Name	Unit	Presentation				Initial value
			Sub engine I	Sub engine II	Sub engine III	Sub engine IV	
1	Oxidant flow	kg/s	x(1)	x(26)	x(51)	x(76)	
2	Fuel flow	kg/s	x(2)	x(27)	x(52)	x(77)	
3	Oxidant flow of combustion chamber	kg/s	x(3)	x(28)	x(53)	x(78)	
4	Oxidant flow in gas generator	kg/s	x(4)	x(29)	x(54)	x(79)	
5	Fuel flow in gas generator	kg/s	x(5)	x(30)	x(55)	x(80)	
6	Flow of cooling jacket	kg/s	x(6)	x(31)	x(56)	x(81)	
7	Flow of evaporator	kg/s	x(7)	x(32)	x(57)	x(82)	
8	Pressure of combustion chamber	MPa	x(8)	x(33)	x(58)	x(83)	
9	Pressure increment of oxidant pump	MPa	x(9)	x(34)	x(59)	x(84)	
10	Pressure increment of fuel pump	MPa	x(10)	x(35)	x(60)	x(85)	
11	Pressure of oxidant before injection	MPa	x(11)	x(36)	x(61)	x(86)	
12	Branch pressure of the fuel dominant pipe	MPa	x(12)	x(37)	x(62)	x(87)	

(continued)

Table A.1 (continued)

No.	Name	Unit	Presentation				Initial value
			Sub engine I	Sub engine II	Sub engine III	Sub engine IV	
13	Turbine pump speed	r/min	$x(13)$	$x(38)$	$x(63)$	$x(88)$	
14	Fuel flow of combustion chamber	kg/s	$x(14)$	$x(39)$	$x(64)$	$x(89)$	
15	Clapboard flow	kg/s	$x(15)$	$x(40)$	$x(65)$	$x(90)$	
16	Total flow of the engine	kg/s	$x(16)$	$x(41)$	$x(66)$	$x(91)$	
17	Gas flow of turbine	kg/s	$x(17)$	$x(42)$	$x(67)$	$x(92)$	
18	Over oxygen coefficients of deputy system		$x(18)$	$x(43)$	$x(68)$	$x(93)$	
19	Mixing ratio		$x(19)$	$x(44)$	$x(69)$	$x(94)$	
20	Power of oxidant pump power	kW	$x(20)$	$x(45)$	$x(70)$	$x(95)$	
21	Power of fuel pump	kW	$x(21)$	$x(46)$	$x(71)$	$x(96)$	
22	Power of turbine	kW	$x(22)$	$x(47)$	$x(72)$	$x(97)$	
23	Calorific value of turbine gas	kJ/kg	$x(23)$	$x(48)$	$x(73)$	$x(98)$	
24	Thrust	kN	$x(24)$	$x(49)$	$x(74)$	$x(99)$	
25	Flow of sonic nozzle	kg/s	$x(25)$	$x(50)$	$x(75)$	$x(100)$	
26	Average heat capacity of oxidant in evaporator	kcal/(s °C)	$x(101)$	$x(104)$	$x(107)$	$x(110)$	
27	Effective coefficient of the evaporator heat transfer		$x(102)$	$x(105)$	$x(108)$	$x(111)$	
28	Outlet temperature of the evaporator oxidant	K	$x(103)$	$x(106)$	$x(109)$	$x(112)$	
29	Outlet temperature of the plenum collector	K	$x(113)$				
30	Inlet temperature of oxidizer tanks	K	$x(114)$				
31	Total oxidant flow	kg/s	$x(115)$				

(continued)

Table A.1 (continued)

No.	Name	Unit	Presentation				Initial value
			Sub engine I	Sub engine II	Sub engine III	Sub engine IV	
32	Total pressurization gas flow of oxidizer tank	kg/s	$x(116)$				
33	Pressure of oxidiser tank	MPa	$x(117)$				
34	Average heat capacity of the gas within the cooling device	kcal/(s °C)	$x(118)$	$x(121)$	$x(124)$	$x(127)$	
35	Efficiency factor of cooling heat exchanger		$x(119)$	$x(122)$	$x(125)$	$x(128)$	
36	Gas temperature at the outlet of cooler	K	$x(120)$	$x(123)$	$x(126)$	$x(129)$	
37	Outlet temperature of the plenum collection	K	$x(130)$				
38	Gas tank inlet temperature	K	$x(131)$				
39	Total flow of fuel	kg/s	$x(132)$				
40	Total flow of pressurized gas in the fuel tank	kg/s	$x(133)$				
41	Fuel tank pressure	MPa	$x(134)$				
42	Pressure at the five-way point of oxidant	MPa	$x(135)$				
43	Inlet pressure of oxidant pump	MPa	$x(136)$	$x(137)$	$x(138)$	$x(139)$	
44	Inlet pressure of fuel pump	MPa	$x(140)$	$x(141)$	$x(142)$	$x(143)$	

Table A.2 Interference factors

No.	Name	Unit	Presentation				Initial value
			I	II	III	IV	
1	Pump efficiency		D1	D17	D31	D45	
2	Fuel pump efficiency		D2	D18	D32	D46	
3	Turbine efficiency		D3	D19	D33	D47	
4	Turbine fall pressure ratio		D4	D20	D34	D48	
5	Servo power	kW	D5	D21	D35	D49	
6	Sonic nozzle area	m^2	D6	D22	D36	D50	
7	Turbine nozzle area	m^2	D7	D23	D37	D51	
8	Inlet pressure of oxidant pump	MPa	D8	D24	D38	D52	
9	Inlet pressure of fuel pump	MPa	D9	D25	D39	D53	
10	Saturation vapor pressure of the oxidant	MPa	D10				
11	Cavitation coefficient of oxygen system	MPa s^2/ kg^2	D11	D26	D40	D54	
12	Cavitation coefficient of evaporator	MPa s^2/ kg^2	D12	D27	D41	D55	
13	Saturated vapor pressure of fuel	MPa	D13				
14	Cavitation coefficient of fuel system	MPa s^2/ kg^2	D14	D28	D42	D56	
15	Throat area of thrust chamber	m^2	D15	D29	D43	D57	
16	Combustion efficiency of combustion chamber		D16	D30	D44	D58	
17	Oxidant density	kg/m^3	R1				
18	Fuel density	kg/m^3	R2				
19	Nozzle outlet area	m^2	A1	A2	A3	A4	
20	Atmospheric pressure	MPa	P0				
21	Oxidant temperature at the inlet of evaporator	K	D59	D61	D63	D65	
22	Exhaust temperature of turbine	K	D60	D62	D64	D66	
23	Wall temperature of pressurized oxidant pipe	K	D67				
24	Exchanged heat between the vapor and oxidizer tank wall	kcal	D68				
25	Gas temperature at the inlet of cooler	K	D69	D71	D73	D75	
26	Coolant temperature	K	D70	D72	D74	D76	
27	Wall temperature of pressurized fuel	K	D77				
28	Exchanged heat between the vapor and fuel tank wall	kcal	D78				

Table A.3 Meanings of parameters

Meaning	Presentation				Initial value
	I	II	III	IV	
Oxidant leakage	$a0$	$a26$	$a52$	$a78$	
Fuel leakage	$a1$	$a27$	$a53$	$a79$	
$\Gamma(r)\sqrt{RT}$	$a2$	$a28$	$a54$	$a80$	
Pressure loss of oxidant pump and impact factor of inlet/outlet	$a3$	$a29$	$a55$	$a81$	
Geometric influence coefficient of oxygen pump flow channel	$a4$	$a30$	$a56$	$a82$	
Influence of wheel size on oxidant pump head	$a5$	$a31$	$a57$	$a83$	
Pressure loss of fuel pump and impact factor of inlet/outlet	$a6$	$a32$	$a58$	$a84$	
Geometric influence coefficient of fuel pump flow channel	$a7$	$a33$	$a59$	$a85$	
Influence of wheel size on fuel pump head	$a8$	$a34$	$a60$	$a86$	
Resistance coefficient between oxidant pump outlet and the muzzle	$a9$	$a35$	$a61$	$a87$	
Resistance coefficient between fuel pump outlet and the branch	$a10$	$a36$	$a62$	$a88$	
Resistance coefficient between muzzle and chamber	$a11$	$a37$	$a63$	$a89$	
Resistance coefficient after fuel branch	$a12$	$a38$	$a64$	$a90$	
Resistance coefficient between the oxidant muzzle and chamber	$a13$	$a39$	$a65$	$a91$	
Resistance coefficient between the fuel muzzle and chamber	$a14$	$a40$	$a66$	$a92$	
Resistance coefficient between the fuel muzzle and chamber	$a15$	$a41$	$a67$	$a93$	
Resistance coefficient between the fuel muzzle and chamber	$a16$	$a42$	$a68$	$a94$	
Resistance coefficient between outlet of pump and cavitation tube of oxidant in deputy system	$a17$	$a43$	$a69$	$a95$	
Resistance coefficient between outlet of pump and cavitation tube of fuel in deputy system	$a18$	$a44$	$a70$	$a96$	
Fuel leakage	$a19$	$a45$	$a71$	$a97$	
Gas leakage	$a20$	$a46$	$a72$	$a98$	
Conversion coefficient of over oxidant coefficient and mixing coefficient in deputy system	$a21$	$a47$	$a73$	$a99$	
Gas calorific value of turbine I	$a22$	$a48$	$a74$	$a100$	
Gas calorific value of turbine II	$a23$	$a49$	$a75$	$a101$	
Thrust calculation experience parameters I	$a24$	$a50$	$a76$	$a102$	
Thrust calculation experience parameters II	$a25$	$a51$	$a77$	$a103$	
Averaged specific heat of oxidant in evaporator	$a104$	$a108$	$a112$	$a116$	
Calculation parameters of heat transfer coefficient I	$a105$	$a109$	$a113$	$a117$	
Calculation parameters of heat transfer coefficient II	$a106$	$a110$	$a114$	$a118$	

(continued)

Table A.3 (continued)

Meaning	Presentation				Initial value
	I	II	III	IV	
Rated heat capacity of the evaporator	$a107$	$a111$	$a115$	$a119$	
Heat transfer calculation parameter of oxidant pressurized pipe	$a120$				
Pressurized gas leakage of oxidant	$a121$				
Pressurized calculation parameters of oxidant tank I	$a122$				
Pressurized calculation parameters of oxidant tank II	$a123$				
Mean specific heat of oxidant in evaporator	$a124$	$a128$	$a132$	$a136$	
Calculation parameters of heat transfer coefficient I	$a125$	$a129$	$a133$	$a137$	
Calculation parameters of heat transfer coefficient II	$a126$	$a130$	$a134$	$a138$	
Rated heat capacity of the evaporator	$a127$	$a131$	$a135$	$a139$	
Heat transfer calculation parameter of oxidant pressurized pipe	$a140$				
Pressurized gas leakage of oxidant	$a141$				
Pressurized calculation parameter of oxidant tank I	$a142$				
Pressurized calculation parameter of oxidant tank II	$a143$				
Resistance coefficient between the oxidant tank and five-section	$a144$				
Resistance coefficient between the pump and five-section	$a145$	$a146$	$a147$	$a148$	
Resistance coefficient between the pump and fuel tank	$a149$	$a150$	$a151$	$a152$	

Appendix B
Steady State Fault Model of II-Level

$$x(1) - x(3) - x(4) - x(7) - a_0 = 0 \tag{B.1}$$

$$x(2) - x(14) - x(5) - a_1 = 0 \tag{B.2}$$

$$x(14) - x(6) - x(15) = 0 \tag{B.3}$$

$$\frac{10^3 x(1)x(9)}{D_1 R_1} + \frac{10^3 x(2)x(10)}{D_2 R_2} - 5x(23)x(17)t_1(1 - D_4^{0.2}) = 0 \tag{B.4}$$

$$10^3 D_{15} x(8) - a_2 D_{16}[x(3) + x(14)] = 0 \tag{B.5}$$

$$x(9) + \frac{a_3 x^2(1)}{R_1} + a_4 x(1)x(13) - a_5 R_1 x^2(13) = 0 \tag{B.6}$$

$$x(10) + \frac{a_6 x^2(2)}{R_2} + a_7 x(2)x(13) - a_8 R_2 x^2(13) = 0 \tag{B.7}$$

$$x(9) - x(11) - \frac{a_9}{R_1}[x(3) + x(7)]^2 - D_5 + D_8 = 0 \tag{B.8}$$

$$x(10) - x(12) - \frac{a_{10} x^2(2)}{R_2} + D_9 = 0 \tag{B.9}$$

$$x(12) - x(8) - \frac{a_{11}}{R_2} x^2(14) - \frac{a_{12}}{R_2} x^2(6) - D_{17} - D_{18} + P_1^H = 0 \tag{B.10}$$

$$x(11) - x(8) - \frac{a_{13} x^2(3)}{R_1} = 0 \tag{B.11}$$

$$\frac{a_{14} x^2(15)}{R_2} - \frac{a_{15} x^2(6)}{R_2} + P_2^H = 0 \tag{B.12}$$

$$D_{12} x(7) - x(8) - \frac{a_{16} x^2(3)}{R_1} + D_{10} + P_3^H = 0 \tag{B.13}$$

© Springer-Verlag Berlin Heidelberg and National Defense Industry Press 2016
W. Zhang, *Failure Characteristics Analysis and Fault Diagnosis*
for Liquid Rocket Engines, DOI 10.1007/978-3-662-49254-3

$$x(9) - \left(\frac{a_{17}}{R_1} + D_{11}\right)x^2(4) + D_8 - D_{10} = 0 \tag{B.14}$$

$$x(12) - \left(\frac{a_{18}}{R_2} + D_{14}\right)[x(5) + a_{19}]^2 - D_{13} = 0 \tag{B.15}$$

$$x(1) + x(2) - x(16) = 0 \tag{B.16}$$

$$\frac{x(4) + x(5)}{1 + D_6/D_7} - x(17) - a_{20} = 0 \tag{B.17}$$

$$\frac{a_{21}x(4)}{x(5)} - x(18) = 0 \tag{B.18}$$

$$\frac{x(1)}{x(2)} - x(19) = 0 \tag{B.19}$$

$$\frac{10^3 x(9)x(1)}{D_1 R_1} - x(20) = 0 \tag{B.20}$$

$$\frac{10^3 x(10)x(2)}{D_2 R_2} - x(21) = 0 \tag{B.21}$$

$$x(20) + x(21) - x(22) = 0 \tag{B.22}$$

$$a_{22}x(18) + a_{23} - x(23) = 0 \tag{B.23}$$

$$a_{24}[x(1) + x(2) - a_{25}] - 10^3 A_1 P_0 - x(24) = 0 \tag{B.24}$$

$$\frac{D_6}{D_7}x(17) - x(25) = 0 \tag{B.25}$$

$$x(26) - a_{26}x(7) = 0 \tag{B.26}$$

$$x(27) - a_{27} - a_{28}[x(26) - a_{29}] = 0 \tag{B.27}$$

$$x(28) - D_{19} - x(27)(D_{20} - D_{19}) = 0 \tag{B.28}$$

$$x(29) - (1 - a_{30})x(28) - a_{30}D_{21} = 0 \tag{B.29}$$

$$x(30) - a_{31}x(25) = 0 \tag{B.30}$$

$$x(31) - a_{32} - a_{33}[x(30) - a_{34}] = 0 \tag{B.31}$$

$$x(32) - D_{22} - x(31)(D_{23} - D_{22}) = 0 \tag{B.32}$$

$$x(33) - (1 - a_{35})x(32) - a_{35}D_{24} = 0 \tag{B.33}$$

$$10^3 D_{25}x(34) - a_{36}D_{26}[x(35) + x(36)] = 0 \tag{B.34}$$

$$x(50) - x(37) - \frac{a_{37}x(35)^2}{R_1} = 0 \tag{B.35}$$

$$x(51) - x(34) - \frac{a_{38}x^2(36)}{R_2} + P_{11}^H = 0 \tag{B.36}$$

$$x(37) - x(34) - \frac{a_{39}x^2(35)}{R_1} = 0 \tag{B.37}$$

$$10^3 D_{27}x(38) - a_{40}D_{28}[x(39) + x(40)] = 0 \tag{B.38}$$

$$x(50) - x(41) - \frac{a_{41}x(39)^2}{R_1} = 0 \tag{B.39}$$

$$x(51) - x(38) - \frac{a_{42}x^2(40)}{R_2} + P_{12}^H = 0 \tag{B.40}$$

$$x(41) - x(38) - \frac{a_{43}x^2(39)}{R_1} = 0 \tag{B.41}$$

$$10^3 D_{29}x(42) - a_{44}D_{30}[x(43) + x(44)] = 0 \tag{B.42}$$

$$x(50) - x(45) - \frac{a_{45}x(43)^2}{R_1} = 0 \tag{B.43}$$

$$x(51) - x(42) - \frac{a_{46}x^2(44)}{R_2} + P_{13}^H = 0 \tag{B.44}$$

$$x(45) - x(42) - \frac{a_{47}x^2(43)}{R_1} = 0 \tag{B.45}$$

$$10^3 D_{31}x(46) - a_{48}D_{32}[x(47) + x(48)] = 0 \tag{B.46}$$

$$x(51) - x(49) - \frac{a_{49}x(47)^2}{R_1} = 0 \tag{B.47}$$

$$x(51) - x(46) - \frac{a_{50}x^2(48)}{R_2} + P_{14}^H = 0 \tag{B.48}$$

$$x(49) - x(46) - \frac{a_{51}x^2(47)}{R_1} = 0 \tag{B.49}$$

$$a_{52}[x(56) + x(57) - a_{53}] - 10^3(A_2 + A_3 + A_4 + A_5)P_0 - x(67) = 0 \tag{B.50}$$

$$x(50) - \left(\frac{a_{54}}{R_1} + D_{34}\right)x^2(54) - D_{33} = 0 \tag{B.51}$$

$$x(51) - \left(\frac{a_{55}}{R_2} + D_{36}\right)[x(55) + a_{56}]^2 - D_{35} = 0 \tag{B.52}$$

$$x(56) + x(57) - x(59) = 0 \tag{B.53}$$

$$x(54) + x(55) - x(60) - a_{57} = 0 \tag{B.54}$$

$$\frac{a_{58}x(54)}{x(55)} - x(61) = 0 \tag{B.55}$$

$$\frac{x(56)}{x(57)} - x(62) = 0 \tag{B.56}$$

$$\frac{10^3x(52)x(56)}{D_{37}R_1} - x(63) = 0 \tag{B.57}$$

$$\frac{10^3x(53)x(57)}{D_{38}R_2} - x(64) = 0 \tag{B.58}$$

$$x(63) + x(64) - x(65) = 0 \tag{B.59}$$

$$a_{59}x(61) + a_{60} - x(66) = 0 \tag{B.60}$$

$$x(56) - x(35) - x(39) - x(43) - x(47) - x(54) = 0 \tag{B.61}$$

$$x(57) - x(36) - x(40) - x(44) - x(48) - x(55) = 0 \tag{B.62}$$

$$\frac{10^3x(52)x(56)}{D_{37}R_1} + \frac{10^3x(53)x(57)}{D_{38}R_2} - 5x(66)x(60)t(1 - D_{39}^{0.2}) = 0 \tag{B.63}$$

$$x(52) + \frac{a_{61}x^2(56)}{R_1} + a_{62}x(56)x(58) - a_{63}R1x^2(58) = 0 \tag{B.64}$$

$$x(53) + \frac{a_{64}x^2(57)}{R_2} + a_{65}x(57)x(58) - a_{66}R_2x^2(58) = 0 \tag{B.65}$$

$$x(52) - x(50) - \frac{a_{67}x^2(56)}{R_1} + D_{40} = 0 \tag{B.66}$$

$$x(53) - x(51) - \frac{a_{68}x^2(57)}{R_2} + D_{41} = 0 \tag{B.67}$$

$$x(68) - x(1) - x(56) = 0 \tag{B.68}$$

$$x(69) - x(2) - x(57) = 0 \tag{B.69}$$

$$x(70) - \frac{a_{69}}{x(68)}[a_{70}x(28)x(7) - D_{42}] = 0 \tag{B.70}$$

$$x(71) - \frac{a_{71}}{x(69)}[a_{72}x(32)x(25) - D_{43}] = 0 \tag{B.71}$$

See Table B.1

Table B.1 Parameter list of the II-level engine

No.	Name	Note	Unit	No.	Name	Note	Unit
1	Oxidant flow	$x(1)$	kg/s	18	Over oxidant coefficient of deputy system	$x(18)$	
2	Fuel flow	$x(2)$	kg/s	19	Propellant component ratio	$x(19)$	
3	Combustor oxidant flow	$x(3)$	kg/s	20	Oxidant pump power	$x(20)$	kW
4	Oxidizer flow in gas generator	$x(4)$	kg/s	21	Fuel pump power	$x(21)$	kW
5	Fuel flow in gas generator	$x(5)$	kg/s	22	Turbine power	$x(22)$	kW
6	Cooling jacket flow	$x(6)$	kg/s	23	Calorific value of turbine gas	$x(23)$	kJ/kg
7	Evaporator flow	$x(7)$	kg/s	24	Thrust	$x(24)$	kN
8	Combustion chamber pressure	$x(8)$	MPa	25	Sonic nozzle flow	$x(25)$	kg/s
9	Increased pressure of oxidant pump	$x(9)$	MPa	26	Average heat capacity of oxidant in the evaporator	$x(26)$	kcal/(s °C)
10	Increased pressure of fuel pump	$x(10)$	MPa	27	Effective coefficient of the evaporator heat transfer	$x(27)$	
11	Oxidant pressure before combustion chamber injection	$x(11)$	MPa	28	Outlet temperature of the evaporator oxidant	$x(28)$	K

(continued)

Table B.1 (continued)

No.	Name	Note	Unit	No.	Name	Note	Unit
12	Fuel pressure at branch of dominant pipe	$x(12)$	MPa	29	Tank inlet temperature	$x(29)$	K
13	Turbine pump speed	$x(13)$	r/min	30	Average heat capacity of the gas within the cooling device	$x(30)$	kcal/(s °C)
14	Fuel flow in combustion chamber	$x(14)$	kg/s	31	Effective coefficient of the cooler	$x(31)$	
15	Separator flow	$x(15)$	kg/s	32	Outlet temperature of gas in cooler	$x(32)$	K
16	Engine total flow	$x(16)$	kg/s	33	Gas temperature at the tank inlet	$x(33)$	K
17	Turbine gas flow	$x(17)$	kg/s				

Sub-engines presentations

No.	Name		Note	Unit	No.	Name	Note	Unit
34	Combustion chamber pressure	I	$x(34)$	MPa	39	Fuel pressure at the twelve ways intersection	$x(51)$	MPa
		II	$x(38)$		40	Increased pressure of oxidant pump	$x(52)$	MPa
		III	$x(42)$		41	Increased pressure of fuel pump	$x(53)$	MPa
		IV	$x(46)$		42	Oxidant flow in gas generator	$x(54)$	kg/s
35	Oxidant flow of combustion chamber	I	$x(35)$	kg/s	43	Fuel flow in gas generator	$x(55)$	kg/s
		II	$x(39)$		44	Engine oxidant flow	$x(56)$	kg/s
		III	$x(43)$		45	Engine fuel flow	$x(57)$	kg/s
		IV	$x(47)$		46	Turbine pump speed	$x(58)$	r/min
36	Fuel flow of combustion chamber	I	$x(36)$	kg/s	47	Engine total flow	$x(59)$	kg/s
		II	$x(40)$		48	Turbine gas flow	$x(60)$	kg/s
		III	$x(44)$		49	Over oxidant coefficient of deputy system	$x(61)$	
		IV	$x(48)$		50	Propellant component ratio	$x(62)$	
37	Oxidant pressure before combustion chamber injection	I	$x(37)$	MPa	51	Oxidant pump power	$x(63)$	kW
		II	$x(41)$		52	Fuel pump power	$x(64)$	kW
		III	$x(45)$		53	Turbine power	$x(65)$	kW
		IV	$x(49)$		54	Calorific value of turbine gas	$x(66)$	kJ/kg

(continued)

Table B.1 (continued)

Sub-engines presentations

No.	Name	Note	Unit	No.	Name	Note	Unit
38	Oxidant pressure at the twelve ways intersection	$x(50)$	MPa	55	Thrust	$x(67)$	kN

The whole system

No.	Name	Note	Unit
56	Engine oxidant flow	$x(68)$	kg/s
57	Engine fuel flow	$x(69)$	kg/s
58	Oxidant tank pressure	$x(70)$	MPa
59	Fuel tank pressure	$x(71)$	MPa

Appendix C
Dynamic State Fault Model of I-Level

$$
g_1 = \frac{1}{V_c^{\mathrm{I}}} \left[RT_c^{\mathrm{I}} + \frac{\partial \left(RT_c^{\mathrm{I}} \right)}{\partial r_c^{\mathrm{I}}} \left(1 + r_c^{\mathrm{I}} \right) \right] \dot{m}_{oc}^{\mathrm{I}} + \frac{1}{V_c^{\mathrm{I}}} \left[RT_c^{\mathrm{I}} - \frac{\partial \left(RT_c^{\mathrm{I}} \right)}{\partial r_c^{\mathrm{I}}} \left(1 + r_c^{\mathrm{I}} \right) r_c^{\mathrm{I}} \right] \cdot
$$
$$
\left(\dot{m}_{f1}^{\mathrm{I}} + \dot{m}_{f3}^{\mathrm{I}} \right) - \frac{1}{V_c^{\mathrm{I}}} \sqrt{RT_c^{\mathrm{I}}} \Gamma_c^{\mathrm{I}} A_{tc}^{\mathrm{I}} p_c^{\mathrm{I}} - \frac{\mathrm{d} p_c^{\mathrm{I}}}{\mathrm{d} t} = 0
$$

(C.1)

$$
g_2 = \frac{RT_c^{\mathrm{I}}}{V_c^{\mathrm{I}} p_c^{\mathrm{I}}} \left(1 + r_c^{\mathrm{I}} \right) \left[\dot{m}_c^{\mathrm{I}} - r_c^{\mathrm{I}} \left(\dot{m}_{f1}^{\mathrm{I}} + \dot{m}_{f3}^{\mathrm{I}} \right) \right] - \frac{\mathrm{d} r_c^{\mathrm{I}}}{\mathrm{d} t} = 0
$$

(C.2)

$$
g_3 = \frac{1}{V_b^{\mathrm{I}}} \left[RT_b^{\mathrm{I}} + \frac{\partial \left(RT_b^{\mathrm{I}} \right)}{\partial r_b^{\mathrm{I}}} \left(1 + r_b^{\mathrm{I}} \right) \right] \dot{m}_{o2}^{\mathrm{I}} + \frac{1}{V_b^{\mathrm{I}}} \left[RT_b^{\mathrm{I}} - \frac{\partial \left(RT_b^{\mathrm{I}} \right)}{\partial r_b^{\mathrm{I}}} \left(1 + r_b^{\mathrm{I}} \right) r_b^{\mathrm{I}} \right] \cdot
$$
$$
\left(\dot{m}_f^{\mathrm{I}} - \dot{m}_{f1}^{\mathrm{I}} - \dot{m}_{f3}^{\mathrm{I}} \right) - \frac{1}{V_b^{\mathrm{I}}} \sqrt{RT_b^{\mathrm{I}}} \Gamma_b^{\mathrm{I}} A_{tb}^{\mathrm{I}} p_b^{\mathrm{I}} - \frac{\mathrm{d} p_b^{\mathrm{I}}}{\mathrm{d} t} = 0
$$

(C.3)

$$
g_4 = \frac{RT_b^{\mathrm{I}}}{V_b^{\mathrm{I}} p_b^{\mathrm{I}}} \left(1 + r_b^{\mathrm{I}} \right) \left[\dot{m}_{o2}^{\mathrm{I}} - r_c^{\mathrm{I}} \left(\dot{m}_f^{\mathrm{I}} - \dot{m}_{f1}^{\mathrm{I}} - \dot{m}_{f3}^{\mathrm{I}} \right) \right] - \frac{\mathrm{d} r_b^{\mathrm{I}}}{\mathrm{d} t} = 0
$$

(C.4)

$$
g_5 = \frac{900}{\pi^2 J^{\mathrm{I}} n^{\mathrm{I}}} \left(N_w^{\mathrm{I}} - N_o^{\mathrm{I}} - N_f^{\mathrm{I}} - N_u^{\mathrm{I}} \right) - \frac{\mathrm{d} n^{\mathrm{I}}}{\mathrm{d} t} = 0
$$

(C.5)

$$
g_6 = \left[\left(\xi_{o1}^{\mathrm{I}} + \xi_{o2}^{\mathrm{I}} \right) \dot{m}_o^{\mathrm{I} 2} + \xi_{o3}^{\mathrm{I}} \left(\dot{m}_o^{\mathrm{I}} - \dot{m}_{o2}^{\mathrm{I}} \right)^2 + \xi_{o4}^{\mathrm{I}} \dot{m}_{oc}^{\mathrm{I} 2} + \xi_o^{\mathrm{I}} \left(\dot{m}_o^{\mathrm{I}} + \dot{m}_o^{\mathrm{II}} \right. \right.
$$
$$
\left. \left. + \dot{m}_o^{\mathrm{III}} + \dot{m}_o^{\mathrm{IV}} \right)^2 \right] / 2\rho_o + \lambda_{o4}^{\mathrm{I}} \frac{\mathrm{d} \dot{m}_{oc}^{\mathrm{I}}}{\mathrm{d} t} - \lambda_{o3}^{\mathrm{I}} \frac{\mathrm{d} \dot{m}_{o2}^{\mathrm{I}}}{\mathrm{d} t} + \left(\sum_{i=0}^{3} \lambda_{oi}^{\mathrm{I}} \right) \frac{\mathrm{d} \dot{m}_o^{\mathrm{I}}}{\mathrm{d} t}
$$
$$
+ \lambda_{o0}^{\mathrm{I}} \frac{\mathrm{d} \dot{m}_o^{\mathrm{II}}}{\mathrm{d} t} + \lambda_{o0}^{\mathrm{I}} \frac{\mathrm{d} \dot{m}_o^{\mathrm{III}}}{\mathrm{d} t} + \lambda_{o0}^{\mathrm{I}} \frac{\mathrm{d} \dot{m}_o^{\mathrm{IV}}}{\mathrm{d} t} + p_c^{\mathrm{I}} - p_{To}^{\mathrm{I}} - \Delta p_o^{\mathrm{I}} - \sum_{i=1}^{4} p_{oi}^{HI}
$$
$$
- p_{o0} = 0
$$

(C.6)

© Springer-Verlag Berlin Heidelberg and National Defense Industry Press 2016
W. Zhang, *Failure Characteristics Analysis and Fault Diagnosis
for Liquid Rocket Engines*, DOI 10.1007/978-3-662-49254-3

$$g_7 = \left[\xi_{o3}^{I} \left(\dot{m}_o^{I} - \dot{m}_{o2}^{I} \right)^2 + \xi_{o4}^{I} \dot{m}_{oc}^{I2} - \xi_{o5}^{I} \dot{m}_{o2}^{I2} \right] / 2\rho_o + \lambda_{o3}^{I} \frac{d\dot{m}_o^{I}}{dt}$$
$$- \left(\lambda_{o3}^{I} + \lambda_{o5}^{I} \right) \frac{d\dot{m}_{o2}^{I}}{dt} + \lambda_{o4}^{I} \frac{d\dot{m}_o^{I}}{dt} - p_c^{I} - \sum_{i=3}^{4} p_{oi}^{HI} + p_{bo} = 0 \qquad (C.7)$$

$$g_8 = \left[\xi_{o4}^{I} \dot{m}_{oc}^{I2} - \xi_{o6}^{I} \left(\dot{m}_o^{I} - \dot{m}_{o2}^{I} - \dot{m}_{oc}^{I} \right)^2 \right] / 2\rho_o + \left(\lambda_{o4}^{I} + \lambda_{o6}^{I} \right) \frac{d\dot{m}_{oc}^{I}}{dt}$$
$$- \lambda_{o6}^{I} \frac{d\dot{m}_{o2}^{I}}{dt} - \lambda_{o6}^{I} \frac{d\dot{m}_o^{I}}{dt} - p_{o4}^{HI} + p_c^{I} - p_{bo} = 0 \qquad (C.8)$$

$$g_9 = \left[\left(\xi_{f1}^{I} + \xi_{f2}^{I} \right) \dot{m}_f^{I2} + \xi_{f3}^{I} \dot{m}_{f1}^{I2} + \xi_{f4}^{I} \left(\dot{m}_{f1}^{I} + \dot{m}_{f3}^{I} \right)^2 \right] / 2\rho_f$$
$$+ \left(\lambda_{f1}^{I} + \lambda_{f2}^{I} \right) \frac{d\dot{m}_f^{I}}{dt} + \left(\lambda_{f3}^{I} + \lambda_{f4}^{I} \right) \frac{d\dot{m}_{f1}^{I}}{dt} + \lambda_{f4}^{I} \frac{d\dot{m}_{f3}^{I}}{dt} \qquad (C.9)$$
$$+ p_c^{I} - p_{Tf}^{I} - \Delta p_f^{I} - \sum_{i=1}^{4} p_{fi}^{HI} = 0$$

$$g_{10} = \left[\xi_{f3}^{I} \dot{m}_{f1}^{I2} - \xi_{f5}^{I} \dot{m}_{f3}^{I2} \right] / 2\rho_f + \lambda_{f3}^{I} \frac{d\dot{m}_{f1}^{I}}{dt} - \lambda_{f5}^{I} \frac{d\dot{m}_{f3}^{I}}{dt}$$
$$+ p_{f5}^{HI} - p_{f3}^{HI} = 0 \qquad (C.10)$$

$$g_{11} = \left[\xi_{f4}^{I} \left(\dot{m}_{f1}^{I} + \dot{m}_{f3}^{I} \right)^2 + \xi_{f5}^{I} \dot{m}_{f3}^{I2} - \xi_{f6}^{I} \left(\dot{m}_f^{I} - \dot{m}_{f1}^{I} - \dot{m}_{f3}^{I} \right)^2 \right] / 2\rho_f$$
$$+ \left(\lambda_{f4}^{I} + \lambda_{f6}^{I} \right) \frac{d\dot{m}_{f1}^{I}}{dt} + \left(\lambda_{f4}^{I} + \lambda_{f6}^{I} + \lambda_{f5}^{I} \right) \frac{d\dot{m}_{f3}^{I}}{dt} - \lambda_{f6}^{I} \frac{d\dot{m}_f^{I}}{dt} \qquad (C.11)$$
$$+ p_c^{I} - \sum_{i=4}^{5} p_{fi}^{HI} - p_{bf} = 0$$

$$g_{12} = \frac{dT_{o1cz}^{I}}{dt} + \frac{L_{oz}^{I} \left(\dot{m}_o^{I} - \dot{m}_{o2}^{I} - \dot{m}_{oc}^{I} \right)}{M_{o1z}} \cdot \left(T_{o1cz}^{I} - T_{o1rz}^{I} \right) + K_{oz}^{I} \cdot$$
$$\frac{A_{oz}^{I}}{\rho_{o1z}^{I} M_{o1z}^{I} C_{o1z}^{I}} \left(T_{o2cz}^{I} - T_{o1cz}^{I} \right) = 0 \qquad (C.12)$$

$$g_{13} = \frac{dT_{f1cz}^{I}}{dt} + \frac{L_{fz}^{I} \left(\dot{m}_{o2}^{I} + \dot{m}_f^{I} - \dot{m}_{f1}^{I} - \dot{m}_{f3}^{I} \right) \cdot \frac{A_{ej}^{I}}{A_{ei}^{I} + A_{ej}^{I}}}{M_{f1z}^{I}} \cdot \left(T_{f1cz}^{I} - T_{f1rz}^{I} \right)$$
$$+ K_{fz}^{I} \cdot \frac{A_{fz}^{I}}{\rho_{f1z}^{I} M_{f1z}^{I} C_{f1z}^{I}} \left(T_{f2cz}^{I} - T_{f1cz}^{I} \right) = 0 \qquad (C.13)$$

$$g_{14} = \frac{L_{og}\left(\dot{m}_o^I - \dot{m}_{o2}^I - \dot{m}_{oc}^I + \dot{m}_o^{II} - \dot{m}_{o2}^{II} - \dot{m}_{oc}^{II} + \dot{m}_o^{III} - \dot{m}_{o2}^{III} - \dot{m}_{oc}^{III} + \dot{m}_o^{IV} - \dot{m}_{o2}^{IV} - \dot{m}_{oc}^{IV}\right)}{M_{olg}} \cdot$$

$$\left(T_{olcg} - \frac{T_{olcz}^I + T_{olcz}^{II} + T_{olcz}^{III} + T_{olcz}^{IV}}{4}\right) + K_{og} \cdot \frac{A_{og}}{\rho_{olg} M_{olg} C_{olg}}\left(T_{o2g} - T_{olcg}\right)$$

$$+ \frac{dT_{olcg}}{dt} = 0$$

$$(C.14)$$

$$g_{15} = \frac{dT_{f1cg}}{dt} + \frac{L_{fg}}{M_{f1g}} \cdot \left(T_{f1cg} - \frac{T_{f1cz}^I + T_{f1cz}^{II} + T_{f1cz}^{III} + T_{f1cz}^{IV}}{4}\right) \cdot \left[\frac{A_{ej}^I}{A_{ei}^I + A_{ej}^I} \cdot \right.$$

$$\left(\dot{m}_{o2}^I + \dot{m}_{mf}^I - \dot{m}_{mf1}^I - \dot{m}_{mf3}^I\right) + \frac{A_{ej}^{II}}{A_{ei}^{II} + A_{ej}^{II}} \cdot \left(\dot{m}_{o2}^{II} + \dot{m}_f^{II} - \dot{m}_{f1}^{II} - \dot{m}_{f3}^{II}\right)$$

$$+ \frac{A_{ej}^{III}}{A_{ei}^{III} + A_{ej}^{III}} \cdot \left(\dot{m}_{o2}^{III} + \dot{m}_f^{III} - \dot{m}_{f1}^{III} - \dot{m}_{f3}^{III}\right) + \frac{A_{ej}^{IV}}{A_{ei}^{IV} + A_{ej}^{IV}} \cdot \left(\dot{m}_{o2}^{IV} + \dot{m}_f^{IV}\right.$$

$$\left.\left. - \dot{m}_{f1}^{IV} - \dot{m}_{f3}^{IV}\right)\right] + K_{fg} \cdot \frac{A_{fg}}{\rho_{f1g} M_{f1g} C_{f1g}}\left(T_{f2g} - T_{f1cg}\right) = 0$$

$$(C.15)$$

$$g_{16} = -\frac{C_{op}\overline{V}_o}{R_o}dp_o + \left(\dot{m}_o^I - \dot{m}_{o2}^I - \dot{m}_{oc}^I + \dot{m}_o^{II} - \dot{m}_{o2}^{II} - \dot{m}_{oc}^{II} + \dot{m}_o^{III} - \dot{m}_{o2}^{III} - \dot{m}_{oc}^{III}\right.$$

$$\left. + \dot{m}_o^{IV} - \dot{m}_{o2}^{IV} - \dot{m}_{oc}^{IV}\right) \cdot T_{olcg} \cdot C_{op1} + \left(q_{o1} + q_{o2}\right) - \left(A_o \cdot p_o + \frac{C_{op}\overline{V}_o}{R_o}\right)$$

$$\cdot \left(\dot{m}_o^I + \dot{m}_o^{II} + \dot{m}_o^{III} + \dot{m}_o^{IV}\right) = 0$$

$$(C.16)$$

$$g_{17} = -\frac{C_{fp}\overline{V}_f}{R_f}dp_f + \left[\left(\dot{m}_{o2}^I + \dot{m}_f^I - \dot{m}_{f1}^I - \dot{m}_{f3}^I\right)\frac{A_{ej}^I}{A_{ei}^I + A_{ej}^I} + \left(\dot{m}_{o2}^{II} + \dot{m}_f^{II} - \dot{m}_{f1}^{II} - \dot{m}_{f3}^{II}\right)\right.$$

$$\frac{A_{ej}^{II}}{A_{ei}^{II} + A_{ej}^{II}} + \left(\dot{m}_{o2}^{III} + \dot{m}_f^{III} - \dot{m}_{f1}^{III} - \dot{m}_{f3}^{III}\right)\frac{A_{ej}^{III}}{A_{ei}^{III} + A_{ej}^{III}} + \left(\dot{m}_{o2}^{IV} + \dot{m}_f^{IV} - \dot{m}_{f1}^{IV} - \dot{m}_{f3}^{IV}\right)$$

$$\left.\frac{A_{ej}^{IV}}{A_{ei}^{IV} + A_{ej}^{IV}}\right] \cdot T_{f1} \cdot C_{fp1} + \left(q_{f1} + q_{f2}\right) - \left(A_f \cdot p_f + \frac{C_{fp}\overline{V}_f}{R_f}\right) \cdot \left(\dot{m}_f^I + \dot{m}_f^{II} + \dot{m}_f^{III} + \dot{m}_f^{IV}\right)$$

$$= 0$$

$$(C.17)$$

Appendix D
Dynamic State Fault Model of II-Level

$$g^{\mathrm{I}} = \frac{1}{V_c^{\mathrm{I}}} \left[RT_c^{\mathrm{I}} + \frac{\partial(RT_c^{\mathrm{I}})}{\partial r_c^{\mathrm{I}}} (1 + r_c^{\mathrm{I}}) \right] \dot{m}_{o1}^{\mathrm{I}} + \frac{1}{V_c^{\mathrm{I}}} \left[RT_c^{\mathrm{I}} - \frac{\partial(RT_c^{\mathrm{I}})}{\partial r_c^{\mathrm{I}}} (1 + r_c^{\mathrm{I}}) r_c^{\mathrm{I}} \right] \cdot \\ \left(\dot{m}_{f1}^{\mathrm{I}} + \dot{m}_{f3}^{\mathrm{I}} \right) - \frac{1}{V_c^{\mathrm{I}}} \sqrt{RT_c^{\mathrm{I}}} \Gamma_c^{\mathrm{I}} A_{tc}^{\mathrm{I}} p_c^{\mathrm{I}} - \frac{\mathrm{d}p_c^{\mathrm{I}}}{\mathrm{d}t} = 0 \tag{D.1}$$

$$g^2 = \frac{RT_c^{\mathrm{I}}}{V_c^{\mathrm{I}} p_c^{\mathrm{I}}} (1 + r_c^{\mathrm{I}}) \left[\dot{m}_{o1}^{\mathrm{I}} - r_c^{\mathrm{I}} \left(\dot{m}_{f1}^{\mathrm{I}} + \dot{m}_{f3}^{\mathrm{I}} \right) \right] - \frac{\mathrm{d}r_c^{\mathrm{I}}}{\mathrm{d}t} = 0 \tag{D.2}$$

$$g^3 = \frac{1}{V_b^{\mathrm{I}}} \left[RT_b^{\mathrm{I}} + \frac{\partial(RT_b^{\mathrm{I}})}{\partial r_b^{\mathrm{I}}} (1 + r_b^{\mathrm{I}}) \right] \dot{m}_{o2}^{\mathrm{I}} + \frac{1}{V_b^{\mathrm{I}}} \left[RT_b^{\mathrm{I}} - \frac{\partial(RT_b^{\mathrm{I}})}{\partial r_b^{\mathrm{I}}} (1 + r_b^{\mathrm{I}}) r_b^{\mathrm{I}} \right] \cdot \\ \left(\dot{m}_f^{\mathrm{I}} - \dot{m}_{f1}^{\mathrm{I}} - \dot{m}_{f3}^{\mathrm{I}} \right) - \frac{1}{V_b^{\mathrm{I}}} \sqrt{RT_b^{\mathrm{I}}} \Gamma_b^{\mathrm{I}} A_{tb}^{\mathrm{I}} p_b^{\mathrm{I}} - \frac{\mathrm{d}p_b^{\mathrm{I}}}{\mathrm{d}t} = 0 \tag{D.3}$$

$$g^4 = \frac{RT_b^{\mathrm{I}}}{V_b^{\mathrm{I}} p_b^{\mathrm{I}}} (1 + r_b^{\mathrm{I}}) \left[\dot{m}_{o2}^{\mathrm{I}} - r_b^{\mathrm{I}} (\dot{m}_f^{\mathrm{I}} - \dot{m}_{f1}^{\mathrm{I}} - \dot{m}_{f3}^{\mathrm{I}}) \right] - \frac{\mathrm{d}r_b^{\mathrm{I}}}{\mathrm{d}t} = 0 \tag{D.4}$$

$$g^5 = \frac{1}{V_c^{\mathrm{II}\,a}} \left[RT_c^{\mathrm{II}\,a} + \frac{\partial(RT_c^{\mathrm{II}\,a})}{\partial r_c^{\mathrm{II}\,a}} (1 + r_c^{\mathrm{II}\,a}) \right] \dot{m}_{o1}^{\mathrm{II}\,a} \\ + \frac{1}{V_c^{\mathrm{II}\,a}} \left[RT_c^{\mathrm{II}\,a} - \frac{\partial(RT_c^{\mathrm{II}\,a})}{\partial r_c^{\mathrm{II}\,a}} (1 + r_c^{\mathrm{II}\,a}) r_c^{\mathrm{II}\,a} \right] \cdot \dot{m}_{f1}^{\mathrm{II}\,a} \\ - \frac{1}{V_c^{\mathrm{II}\,a}} \sqrt{RT_c^{\mathrm{II}\,a}} \Gamma_c^{\mathrm{II}\,a} A_{tc}^{\mathrm{II}\,a} p_c^{\mathrm{II}\,a} - \frac{\mathrm{d}p_c^{\mathrm{II}\,a}}{\mathrm{d}t} = 0 \tag{D.5}$$

$$g^6 = \frac{1}{V_c^{\mathrm{II}\,b}} \left[RT_c^{\mathrm{II}\,b} + \frac{\partial(RT_c^{\mathrm{II}\,b})}{\partial r_c^{\mathrm{II}\,b}} (1 + r_c^{\mathrm{II}\,b}) \right] \dot{m}_{o1}^{\mathrm{II}\,b} \\ + \frac{1}{V_c^{\mathrm{II}\,a}} \left[RT_c^{\mathrm{II}\,b} - \frac{\partial(RT_c^{\mathrm{II}\,b})}{\partial r_c^{\mathrm{II}\,b}} (1 + r_c^{\mathrm{II}\,b}) r_c^{\mathrm{II}\,b} \right] \cdot \dot{m}_{f1}^{\mathrm{II}\,b} \\ - \frac{1}{V_c^{\mathrm{II}\,b}} \sqrt{RT_c^{\mathrm{II}\,b}} \Gamma_c^{\mathrm{II}\,b} A_{tc}^{\mathrm{II}\,b} p_c^{\mathrm{II}\,b} - \frac{\mathrm{d}p_c^{\mathrm{II}\,b}}{\mathrm{d}t} = 0 \tag{D.6}$$

© Springer-Verlag Berlin Heidelberg and National Defense Industry Press 2016
W. Zhang, *Failure Characteristics Analysis and Fault Diagnosis
for Liquid Rocket Engines*, DOI 10.1007/978-3-662-49254-3

$$g^7 = \frac{1}{V_c^{\mathrm{II}\,c}} \left[RT_c^{\mathrm{II}\,c} + \frac{\partial(RT_c^{\mathrm{II}\,c})}{\partial r_c^{\mathrm{II}\,c}} \left(1 + r_c^{\mathrm{II}\,c}\right) \right] \dot{m}_{o1}^{\mathrm{II}\,c}$$
$$+ \frac{1}{V_c^{\mathrm{II}\,c}} \left[RT_c^{\mathrm{II}\,c} - \frac{\partial(RT_c^{\mathrm{II}\,c})}{\partial r_c^{\mathrm{II}\,c}} \left(1 + r_c^{\mathrm{II}\,c}\right) r_c^{\mathrm{II}\,c} \right] \cdot \dot{m}_{f1}^{\mathrm{II}\,c} \qquad \text{(D.7)}$$
$$- \frac{1}{V_c^{\mathrm{II}\,c}} \sqrt{RT_c^{\mathrm{II}\,c}} \, \Gamma_c^{\mathrm{II}\,c} A_{tc}^{\mathrm{II}\,c} p_c^{\mathrm{II}\,c} - \frac{\mathrm{d}p_c^{\mathrm{II}\,c}}{\mathrm{d}t} = 0$$

$$g^8 = \frac{1}{V_c^{\mathrm{II}\,d}} \left[RT_c^{\mathrm{II}\,d} + \frac{\partial(RT_c^{\mathrm{II}\,d})}{\partial r_c^{\mathrm{II}\,d}} \left(1 + r_c^{\mathrm{II}\,d}\right) \right] \dot{m}_{o1}^{\mathrm{II}\,d}$$
$$+ \frac{1}{V_c^{\mathrm{II}\,d}} \left[RT_c^{\mathrm{II}\,d} - \frac{\partial(RT_c^{\mathrm{II}\,d})}{\partial r_c^{\mathrm{II}\,d}} \left(1 + r_c^{\mathrm{II}\,d}\right) r_c^{\mathrm{II}\,d} \right] \cdot \dot{m}_{f1}^{\mathrm{II}\,d} \qquad \text{(D.8)}$$
$$- \frac{1}{V_c^{\mathrm{II}\,d}} \sqrt{RT_c^{\mathrm{II}\,d}} \, \Gamma_c^{\mathrm{II}\,d} A_{tc}^{\mathrm{II}\,d} p_c^{\mathrm{II}\,d} - \frac{\mathrm{d}p_c^{\mathrm{II}\,d}}{\mathrm{d}t} = 0$$

$$g^9 = \frac{RT_c^{\mathrm{II}\,a}}{V_c^{\mathrm{II}\,a} p_c^{\mathrm{II}\,a}} \left(1 + r_c^{\mathrm{II}\,a}\right) \left(\dot{m}_{o1}^{\mathrm{II}\,a} - r_c^{\mathrm{II}\,a} \cdot \dot{m}_{f1}^{\mathrm{II}\,a} \right) - \frac{\mathrm{d}r_c^{\mathrm{II}\,a}}{\mathrm{d}t} = 0 \qquad \text{(D.9)}$$

$$g^{10} = \frac{RT_c^{\mathrm{II}\,b}}{V_c^{\mathrm{II}\,b} p_c^{\mathrm{II}\,b}} \left(1 + r_c^{\mathrm{II}\,b}\right) \left(\dot{m}_{o1}^{\mathrm{II}\,b} - r_c^{\mathrm{II}\,b} \cdot \dot{m}_{f1}^{\mathrm{II}\,b} \right) - \frac{\mathrm{d}r_c^{\mathrm{II}\,b}}{\mathrm{d}t} = 0 \qquad \text{(D.10)}$$

$$g^{11} = \frac{RT_c^{\mathrm{II}\,c}}{V_c^{\mathrm{II}\,c} p_c^{\mathrm{II}\,c}} \left(1 + r_c^{\mathrm{II}\,c}\right) \left(\dot{m}_{o1}^{\mathrm{II}\,c} - r_c^{\mathrm{II}\,c} \cdot \dot{m}_{f1}^{\mathrm{II}\,c} \right) - \frac{\mathrm{d}r_c^{\mathrm{II}\,c}}{\mathrm{d}t} = 0 \qquad \text{(D.11)}$$

$$g^{12} = \frac{RT_c^{\mathrm{II}\,d}}{V_c^{\mathrm{II}\,d} p_c^{\mathrm{II}\,d}} \left(1 + r_c^{\mathrm{II}\,d}\right) \left(\dot{m}_{o1}^{\mathrm{II}\,d} - r_c^{\mathrm{II}\,d} \cdot \dot{m}_{f1}^{\mathrm{II}\,d} \right) - \frac{\mathrm{d}r_c^{\mathrm{II}\,d}}{\mathrm{d}t} = 0 \qquad \text{(D.12)}$$

$$g^{13} = \frac{1}{V_b^{\mathrm{II}}} \left[RT_b^{\mathrm{II}} + \frac{\partial(RT_b^{\mathrm{II}})}{\partial r_b^{\mathrm{II}}} \left(1 + r_b^{\mathrm{II}}\right) \right] \dot{m}_{o2}^{\mathrm{II}} + \frac{1}{V_b^{\mathrm{II}}} \left[RT_b^{\mathrm{II}} - \frac{\partial(RT_b^{\mathrm{II}})}{\partial r_b^{\mathrm{II}}} \left(1 + r_b^{\mathrm{II}}\right) r_b^{\mathrm{II}} \right] \cdot$$
$$\left(\dot{m}_f^{\mathrm{II}} - \dot{m}_{f1}^{\mathrm{II}} \right) - \frac{1}{V_b^{\mathrm{II}}} \sqrt{RT_b^{\mathrm{II}}} \, \Gamma_b^{\mathrm{II}} A_{tb}^{\mathrm{II}} p_b^{\mathrm{II}} - \frac{\mathrm{d}p_b^{\mathrm{II}}}{\mathrm{d}t} = 0$$

$$\text{(D.13)}$$

$$g^{14} = \frac{RT_b^{\mathrm{II}}}{V_b^{\mathrm{II}} p_b^{\mathrm{II}}} \left(1 + r_b^{\mathrm{II}}\right) \left[\dot{m}_{o2}^{\mathrm{II}} - r_c^{\mathrm{II}} \left(\dot{m}_f^{\mathrm{II}} - \dot{m}_{f1}^{\mathrm{II}} \right) \right] - \frac{\mathrm{d}r_b^{\mathrm{II}}}{\mathrm{d}t} = 0 \qquad \text{(D.14)}$$

$$g^{15} = \frac{900}{\pi^2 J^{\mathrm{I}} n^{\mathrm{I}}} \left(N_w^{\mathrm{I}} - N_o^{\mathrm{I}} - N_f^{\mathrm{I}} \right) - \frac{\mathrm{d}n^{\mathrm{I}}}{\mathrm{d}t} = 0 \qquad \text{(D.15)}$$

$$g^{16} = \frac{900}{\pi^2 J^{II} n^{II}} \left(N_w^{II} - N_o^{II} - N_f^{II} \right) - \frac{dn^{II}}{dt} = 0 \tag{D.16}$$

$$g^{17} = \left[\xi_{o0} \left(\dot{m}_o^I + \dot{m}_o^{II} \right)^2 + \left(\xi_{o1}^I + \xi_{o2}^I \right) \dot{m}_o^{I2} + \xi_{o3}^I \left(\dot{m}_o^I - \dot{m}_{o2}^I \right)^2 + \xi_{o4}^I \dot{m}_{o1}^{I2} \right]$$
$$/2\rho_o + \lambda_{o0} \frac{d\dot{m}_o^I}{dt} + \lambda_{o0} \frac{d\dot{m}_o^{II}}{dt} + \lambda_{o1}^I \frac{d\dot{m}_o^I}{dt} + \lambda_{o2}^I \frac{d\dot{m}_o^I}{dt} + \lambda_{o3}^I \frac{d\dot{m}_o^I}{dt}$$
$$- \lambda_{o3}^I \frac{d\dot{m}_{o2}^I}{dt} + \lambda_{o4}^I \frac{d\dot{m}_{o1}^I}{dt} + p_c^I - p_{To}^I - \Delta p_o^I - \sum_{i=0}^{4} p_{oi}^{HI} = 0 \tag{D.17}$$

$$g^{18} = \left[\xi_{o3}^I \left(\dot{m}_o^I - \dot{m}_{o2}^I \right)^2 + \xi_{o4}^I \dot{m}_{o1}^{I2} - \xi_{o5}^I \dot{m}_{o2}^{I2} \right] /2\rho_o + \lambda_{o3}^I \frac{d\dot{m}_o^I}{dt}$$
$$- \lambda_{o3}^I \frac{d\dot{m}_{o2}^I}{dt} + \lambda_{o4}^I \frac{d\dot{m}_{o1}^I}{dt} - \lambda_{o5}^I \frac{d\dot{m}_{o2}^I}{dt} + p_c^I - \sum_{i=3}^{4} p_{oi}^{HI} - p_{bo} = 0 \tag{D.18}$$

$$g^{19} = \left[\xi_{o4}^I \dot{m}_{o1}^{I2} - \xi_{o6}^I \left(\dot{m}_o^I - \dot{m}_{o2}^I - \dot{m}_{o1}^I \right)^2 \right] /2\rho_o + \lambda_{o4}^I \frac{d\dot{m}_{o1}^I}{dt}$$
$$- \lambda_{o6}^I \frac{d\dot{m}_{o3}^I}{dt} - p_{o4}^{HI} + p_c^I - p_{bo} = 0 \tag{D.19}$$

$$g^{20} = \left[\xi_{o0} \left(\dot{m}_o^I + \dot{m}_o^{II} \right)^2 + \left(\xi_{o1}^{II} + \xi_{o2}^{II} \right) \dot{m}_o^{II2} + \left(\xi_{o3}^{II} + \xi_{o4}^{II} \right) \dot{m}_{o1}^{IIa2} \right]$$
$$/2\rho_o + \lambda_{o0} \frac{d\dot{m}_o^I}{dt} + \lambda_{o0} \frac{d\dot{m}_o^{II}}{dt} + \left(\lambda_1^{II} + \lambda_{o2}^{II} \right) \frac{d\dot{m}_o^{II}}{dt}$$
$$+ \left(\lambda_{o3}^{II} + \lambda_{o4}^{II} \right) \frac{d\dot{m}_{o1}^{IIa}}{dt} + p_c^{II} - p_{To}^{II} - \Delta p^{II}o - \sum_{i=0}^{4} p_{oi}^{HII} = 0 \tag{D.20}$$

$$g^{21} = \left[\xi_{o0} \left(\dot{m}_o^I + \dot{m}_o^{II} \right)^2 + \left(\xi_{o1}^{II} + \xi_{o2}^{II} \right) \dot{m}_o^{II2} + \left(\xi_{o3}^{II} + \xi_{o4}^{II} \right) \dot{m}_{o1}^{IIb2} \right]$$
$$/2\rho_o + \lambda_{o0} \frac{d\dot{m}_o^I}{dt} + \lambda_{o0} \frac{d\dot{m}_o^{II}}{dt} + \left(\lambda_{o1}^{II} + \lambda_{o2}^{II} \right) \frac{d\dot{m}_o^{II}}{dt}$$
$$+ \left(\lambda_{o3}^{II} + \lambda_{o4}^{II} \right) \frac{d\dot{m}_{o1}^{IIb}}{dt} + p_c^{II} - p_{To}^{II} - \Delta p_o^{II} - \sum_{i=0}^{4} p_{oi}^{HII} = 0 \tag{D.21}$$

$$g^{22} = \left[\xi_{o0} \left(\dot{m}_o^I + \dot{m}_o^{II} \right)^2 + \left(\xi_{o1}^{II} + \xi_{o2}^{II} \right) \dot{m}_o^{II2} + \left(\xi_{o3}^{II} + \xi_{o4}^{II} \right) \dot{m}_{o1}^{IIc2} \right]$$
$$/2\rho_o + \lambda_{o0} \frac{d\dot{m}_o^I}{dt} + \lambda_{o0} \frac{d\dot{m}_o^{II}}{dt} + \left(\lambda_{o1}^{II} + \lambda_{o2}^{II} \right) \frac{d\dot{m}_o^{II}}{dt}$$
$$+ \left(\lambda_{o3}^{II} + \lambda_{o4}^{II} \right) \frac{d\dot{m}_{o1}^{IIc}}{dt} + p_c^{II} - p_{To}^{II} - \Delta p_o^{II} - \sum_{i=0}^{4} p_{oi}^{HII} = 0 \tag{D.22}$$

$$g^{23} = \left[\xi_{o0} \left(\dot{m}_o^{\mathrm{I}} + \dot{m}_o^{\mathrm{II}} \right)^2 + \left(\xi_{o1}^{\mathrm{II}} + \xi_{o2}^{\mathrm{II}} \right) \dot{m}_o^{\mathrm{II}\,2} + \left(\xi_{o3}^{\mathrm{II}} + \xi_{o4}^{\mathrm{II}} \right) \dot{m}_{o1}^{\mathrm{II}\,\mathrm{d}\,2} \right]$$
$$/2\rho_o + \lambda_{o0} \frac{\mathrm{d}\dot{m}_o^{\mathrm{I}}}{\mathrm{d}t} + \lambda_{o0} \frac{\mathrm{d}\dot{m}_o^{\mathrm{II}}}{\mathrm{d}t} + \left(\lambda_{o1}^{\mathrm{II}} + \lambda_{o2}^{\mathrm{II}} \right) \frac{\mathrm{d}\dot{m}_o^{\mathrm{II}}}{\mathrm{d}t} \qquad (D.23)$$
$$+ \left(\lambda_{o3}^{\mathrm{II}} + \lambda_{o4}^{\mathrm{II}} \right) \frac{\mathrm{d}\dot{m}_{o1}^{\mathrm{II}\,\mathrm{d}}}{\mathrm{d}t} + p_c^{\mathrm{II}} - p_{To}^{\mathrm{II}} - \Delta p_o^{\mathrm{II}} - \sum_{i=0}^{4} p_{oi}^{H\,\mathrm{II}} = 0$$

$$g^{24} = \left[\xi_{o3}^{\mathrm{II}} \dot{m}_{o1a}^{\mathrm{II}\,2} + \xi_{o4}^{\mathrm{II}} \dot{m}_{o1a}^{\mathrm{II}\,2} - \xi_{o5}^{\mathrm{II}} \dot{m}_{o2}^{\mathrm{II}\,2} \right] /2\rho_o + \lambda_{o3}^{\mathrm{II}} \frac{\mathrm{d}\dot{m}_{o1a}^{\mathrm{II}}}{\mathrm{d}t}$$
$$+ \lambda_{o4}^{\mathrm{II}} \frac{\mathrm{d}\dot{m}_{o1a}^{\mathrm{II}}}{\mathrm{d}t} - \lambda_{o5}^{\mathrm{II}} \frac{\mathrm{d}\dot{m}_{o2}^{\mathrm{II}}}{\mathrm{d}t} + p_c^{\mathrm{II}} - \sum_{i=3}^{4} p_{oi}^{H\,\mathrm{I}} - p_{bo}^{\mathrm{II}} = 0 \qquad (D.24)$$

$$g^{25} = \left[\xi_{f0} \left(\dot{m}_f^{\mathrm{I}} + \dot{m}_f^{\mathrm{II}} \right)^2 + \left(\xi_{f1}^{\mathrm{I}} + \xi_{f2}^{\mathrm{I}} \right) \dot{m}_f^{\mathrm{I}\,2} + \xi_{f3}^{\mathrm{I}} \dot{m}_{f1}^{\mathrm{I}\,2} \right.$$
$$\left. + \xi_{f4}^{\mathrm{I}} \left(\dot{m}_{f1}^{\mathrm{I}} + \dot{m}_{f3}^{\mathrm{I}} \right)^2 \right] /2\rho_o + \lambda_{f0} \frac{\mathrm{d}\dot{m}_f^{\mathrm{I}}}{\mathrm{d}t} + \lambda_{f0} \frac{\mathrm{d}\dot{m}_f^{\mathrm{II}}}{\mathrm{d}t}$$
$$+ \left(\lambda_{f1}^{\mathrm{I}} + \lambda_{f2}^{\mathrm{I}} \right) \frac{\mathrm{d}\dot{m}_f^{\mathrm{I}}}{\mathrm{d}t} + \lambda_{f3}^{\mathrm{I}} \frac{\mathrm{d}\dot{m}_{f1}^{\mathrm{I}}}{\mathrm{d}t} + \lambda_{f4}^{\mathrm{I}} \frac{\mathrm{d}\dot{m}_{f1}^{\mathrm{I}}}{\mathrm{d}t} \qquad (D.25)$$
$$+ \lambda_{f4} \frac{\mathrm{d}\dot{m}_{f3}^{\mathrm{I}}}{\mathrm{d}t} + p_c^{\mathrm{I}} - p_{Tf}^{\mathrm{I}} - \Delta p_f^{\mathrm{I}} - \sum_{i=0}^{4} p_{fi}^{H\,\mathrm{I}} = 0$$

$$g^{26} = \left[\xi_{f3}^{\mathrm{I}} \dot{m}_{f1}^{\mathrm{I}\,2} - \xi_{f5}^{\mathrm{I}} \dot{m}_{f3}^{\mathrm{I}\,2} \right] /2\rho_f + \lambda_{f3}^{\mathrm{I}} \frac{\mathrm{d}\dot{m}_{f1}^{\mathrm{I}}}{\mathrm{d}t}$$
$$- \lambda_{f5}^{\mathrm{I}} \frac{\mathrm{d}\dot{m}_{f3}^{\mathrm{I}}}{\mathrm{d}t} + p_{f5}^{H\,\mathrm{I}} - p_{f3}^{H\,\mathrm{I}} = 0 \qquad (D.26)$$

$$g^{27} = \left[\xi_{f4}^{\mathrm{I}} \left(\dot{m}_{f1}^{\mathrm{I}} + \dot{m}_{f3}^{\mathrm{I}} \right)^2 + \xi_{f5}^{\mathrm{I}} \dot{m}_{f3}^{\mathrm{I}\,2} \right.$$
$$\left. - \xi_{f6}^{\mathrm{I}} \left(\dot{m}_f^{\mathrm{I}} - \dot{m}_{f1}^{\mathrm{I}} - \dot{m}_{f3}^{\mathrm{I}} \right)^2 \right] /2\rho_f$$
$$+ \left(\lambda_{f4}^{\mathrm{I}} + \lambda_{f6}^{\mathrm{I}} \right) \frac{\mathrm{d}\dot{m}_{f1}^{\mathrm{I}}}{\mathrm{d}t} + \left(\lambda_{f4}^{\mathrm{I}} + \lambda_{f6}^{\mathrm{I}} + \lambda_{f5}^{\mathrm{I}} \right) \frac{\mathrm{d}\dot{m}_{f3}^{\mathrm{I}}}{\mathrm{d}t} \qquad (D.27)$$
$$- \lambda_{f6}^{\mathrm{I}} \frac{\mathrm{d}\dot{m}_f^{\mathrm{I}}}{\mathrm{d}t} + p_c^{\mathrm{I}} - \sum_{i=4}^{5} p_{fi}^{H\,\mathrm{I}} - p_{bf}^{\mathrm{I}} = 0$$

$$g^{28} = \left[\xi_{f0} \left(\dot{m}_f^I + \dot{m}_f^{II} \right)^2 + \left(\xi_{f1}^{II} + \xi_{f2}^{II} \right) \dot{m}_f^{II\,2} + \left(\xi_{f3}^{II} \right.\right.$$
$$\left.\left. + \xi_{f4}^{II} \right) \dot{m}_{f1}^{II\,a\,2} \right]/2\rho_o + \lambda_{f0} \frac{d\dot{m}_f^I}{dt} + \lambda_{f0} \frac{d\dot{m}_f^{II}}{dt}$$
$$+ \left(\lambda_{f1}^{II} + \lambda_{f2}^{II} \right) \frac{d\dot{m}_f^{II}}{dt} + \left(\lambda_{f3}^{II} + \lambda_{f4}^{II} \right) \frac{d\dot{m}_{f1}^{II\,a}}{dt}$$
$$+ p_c^{II} - p_{Tf}^{II} - \Delta p_f^{II} - \sum_{i=0}^{4} p_{fi}^{H\,II} = 0 \tag{D.28}$$

$$g^{29} = \left[\xi_{f0} \left(\dot{m}_f^I + \dot{m}_f^{II} \right)^2 + \left(\xi_{f1}^{II} + \xi_{f2}^{II} \right) \dot{m}_f^{II\,2} + \left(\xi_{f3}^{II} + \xi_{f4}^{II} \right) \dot{m}_{f1}^{II\,b\,2} \right]$$
$$/2\rho o + \lambda_{f0} \frac{d\dot{m}_f^I}{dt} + \lambda_{f0} \frac{d\dot{m}_f^{II}}{dt} + \left(\lambda_{f1}^{II} + \lambda_{f2}^{II} \right) \frac{d\dot{m}_f^{II}}{dt}$$
$$+ \left(\lambda_{f3}^{II} + \lambda_{f4}^{II} \right) \frac{d\dot{m}_{f1}^{II\,b}}{dt} + p_c^{II} - p_{Tf}^{II} - \Delta p_f^{II} - \sum_{i=0}^{4} p_{fi}^{H\,II} = 0 \tag{D.29}$$

$$g^{30} = \left[\xi_{f0} \left(\dot{m}_f^I + \dot{m}_f^{II} \right)^2 + \left(\xi_{f1}^{II} + \xi_{f2}^{II} \right) \dot{m}_f^{II\,2} + \left(\xi_{f3}^{II} + \xi_{f4}^{II} \right) \dot{m}_{f1}^{II\,c\,2} \right]$$
$$/2\rho_o + \lambda_{f0} \frac{d\dot{m}_f^I}{dt} + \lambda_{f0} \frac{d\dot{m}_f^{II}}{dt} + \left(\lambda_{f1}^{II} + \lambda_{f2}^{II} \right) \frac{d\dot{m}_f^{II}}{dt}$$
$$+ \left(\lambda_{f3}^{II} + \lambda_{f4}^{II} \right) \frac{d\dot{m}_{f1}^{II\,c}}{dt} + p_c^{II} - p_{Tf}^{II} - \Delta p_f^{II} - \sum_{i=0}^{4} p_{fi}^{H\,II} = 0 \tag{D.30}$$

$$g^{31} = \left[\xi_{f0} \left(\dot{m}_f^I + \dot{m}_f^{II} \right)^2 + \left(\xi_{f1}^{II} + \xi_{f2}^{II} \right) \dot{m}_f^{II\,2} + \left(\xi_{f3}^{II} + \xi_{f4}^{II} \right) \dot{m}_{f1}^{II\,d\,2} \right]$$
$$/2\rho_o + \lambda_{f0} \frac{d\dot{m}_f^I}{dt} + \lambda_{f0} \frac{d\dot{m}_f^{II}}{dt} + \left(\lambda_{f1}^{II} + \lambda_{f2}^{II} \right) \frac{d\dot{m}_f^{II}}{dt}$$
$$+ \left(\lambda_{f3}^{II} + \lambda_{f4}^{II} \right) \frac{d\dot{m}_{f1}^{II\,d}}{dt} + p_c^{II} - p_{Tf}^{II} - \Delta p_f^{II} - \sum_{i=0}^{4} p_{fi}^{H\,II} = 0 \tag{D.31}$$

$$g^{32} = \left[\left(\xi_{f3}^{II} \dot{m}_{f1a}^{II\,2} + \xi_{f4}^{II} \dot{m}_{f1a}^{II\,2} - \xi_{f5}^{II} \dot{m}_{f2}^{II\,2} \right)/2\rho_f + \lambda_{f3}^{II} \frac{d\dot{m}_{f1a}^{II}}{dt} \right.$$
$$\left. + \lambda_{f4}^{II} \frac{d\dot{m}_{f1a}^{II}}{dt} - \lambda_{f5}^{II} \frac{d\dot{m}_{f2}^{II}}{dt} + p_c^{II} - \sum_{i=3}^{4} p_{fi}^{H\,II} - p_{bf}^{II} = 0 \right. \tag{D.32}$$

$$g^{33} = \frac{dT_{o1cz}}{dt} + \frac{L_{oz}\left(\dot{m}_o^I - \dot{m}_{o2}^I - \dot{m}_{o1}^I\right)}{M_{o1z}} \cdot (T_{o1cz} - T_{o1rz}) + K_{oz} \cdot$$
$$\frac{A_{oz}}{\rho_{o1z}M_{o1z}C_{o1z}}(T_{o2cz} - T_{o1cz}) = 0 \tag{D.33}$$

$$g^{34} = \frac{dT_{f1cz}}{dt} + \frac{L_{fz}^I\left(\dot{m}_{o2}^I + \dot{m}_f^I - \dot{m}_{f1}^I - \dot{m}_{f3}^I\right) \cdot \frac{A_{ej}^I}{A_{ei}^I + A_{ej}^I}}{M_{f1z}^I} \cdot (T_{f1cz} - T_{f1rz})$$
$$+ K_{fz} \cdot \frac{A_{fz}}{\rho_{f1z}^I M_{f1z}^I C_{f1z}^I}(T_{f2cz} - T_{f1cz}) = 0 \tag{D.34}$$

$$g^{35} = -\frac{C_{op}\overline{V}_o}{R_o}\frac{dp_o}{dt} + \left(\dot{m}_o^I - \dot{m}_{o2}^I - \dot{m}_{o1}^I\right)T_{o1cg} \cdot C_{op1} + (q_{o1} + q_{o2})$$
$$- \left(A_o \cdot po + \frac{C_{op}\overline{V}_o}{R_o}\right) \cdot \left(\dot{m}_o^I + \dot{m}_o^{II}\right) = 0 \tag{D.35}$$

$$g^{36} = -\frac{C_{fp}\overline{V}_f}{R_f}\frac{dp_f}{dt} + \left(\dot{m}_{o2}^I + \dot{m}_f^I - \dot{m}_{f1}^I - \dot{m}_{f3}^I\right)\frac{A_{ej}^I}{A_{ei}^I + A_{ej}^I} \cdot T_{f1} \cdot C_{fp1}$$
$$+ (q_{f1} + q_{f2}) - \left(A_f \cdot p_f + \frac{C_{fp}\overline{V}_f}{R_f}\right) \cdot \left(\dot{m}_f^I + \dot{m}_f^{II}\right) = 0 \tag{D.36}$$

References

1. Han H, Chen J (2008) 21 century foreign deep space exploration development plans and their progresses. Spacecr Eng 17(3):1–22
2. Wang Z (2008) Military space technology and its development. Spacecr Eng 17(1):12–17
3. Staszewski WJ, Boller C, Tomlinson GR (2004) Health monitoring of aerospace structures: smart sensor technologies and signal processing. Wiley, England
4. Shang B, Song B, Yang J (2008) New sensor thechnology in aircraft structural health monitoring. Nondestr Test 30(5):289–292
5. Newlon RO, White DJ, Dussault PL (2007) Condition monitoring of rotorcraft structures: why we need it and how to achieve it. AIAA-2007-1671
6. Jaw LC (2005) Recent advancements in aircraft engine health management (EHM) technologies and recommendations for the next step. ASME GT2005-68625
7. Komar DR, Christenson RL (1996) Reusable launch vehicle engine systems operations analysis. AIAA-96-4246, September 1996
8. Maul WA, Meyer CM (1991) Space engine safety system. AIAA-91-3604, September, 1991
9. Hawman MH (1990) Health monitoring system for the SSME-program overview. AIAA-90-1987, July 1990
10. Tulpule S, Galinaitis WS (1990) Health monitoring for the SSME—fault detection algorithms. AIAA-90-1988, July 1990
11. Jurado EP, Shade JB, Weise MA (1990) Selection of monitoring techniques for a liquid propellant rocket engine. AIAA-90-2698, July 1990
12. Bonkob EB (1977) Reliability theory basis for the rocket engine. National Defense Industry Press, Beijing, pp 374–381
13. Wu J, Zhang Y, Chen Q (1994). Steady fault simulation and analysis of liquid rocket engine. J Propuls Technol 3: 6–13
14. Huang M, Zhang Y, Feng X (1994) The failure detection system of liquid rocket engine based on neural network. J Natl Univ Def Technol 16(2):55–60
15. Xie T, Zhang Y (1997) GA-HCM hybrid clustering algorithm and its application in fault detection for liquid rocket engine system. J Propuls Technol 18(1):36–42
16. Huang M, Zhang Y, Chen Q (1997) Fault diagnosis of liquid rocket engine based on fuzzy measuring of fuzzy rule sets. J Propuls Technol 18(5):5–8
17. Wu J, Zhang Y, Chen Q (1994) Transient performance simulation of a large liquid rocket engine under fault conditions. J Aerosp Power 9(4):361–365
18. Huang W, Wang K (1996) Fault diagnosis of liquid rocket engine based on a qualitative model. J Propuls Technol 17(4):5–9
19. Zhen W, Wu J (2004) Qualitative bond-graph based fault diagnosis for liquid-propellant rocket engine. J Astronaut 25(6):604–608
20. He H, Hu X, Jiang Z et al (2008) SVM implemented in fault diagnosis of liquid rocket engine. J Rocket Propuls 34(3):7–12

© Springer-Verlag Berlin Heidelberg and National Defense Industry Press 2016
W. Zhang, *Failure Characteristics Analysis and Fault Diagnosis
for Liquid Rocket Engines*, DOI 10.1007/978-3-662-49254-3

21.   Wang Y, Hu X, Li Z et al (2006) Fault diagnosis based on rough set theory for liquid rocket engine. Missile Space Veh 2:50–54
22.   Zhao Y, Zhang Y, Li M et al (2002) Study of the fault diagnosis technology for liquid propellant rocket engine by plume UV-VIS radiation. J Astronaut 23(1):34–39
23.   Yu X, Liao M, Zhao C (2002) Condition monitoring and fault diagnosing system of liquid propellant rocket engine turbo pump. Missile Space Veh 4:54–59
24.   Zhang W, Zhang Y, Huang X (2002) Real-time intelligent diagnosis system for missile propellant filling equipment. J Propuls Technol 23(6):489–491
25.   Chen Q (2003) Some progress in researches on fault detection and diagnosis of liquid rocket engine. J Astronaut 24(1):1–11
26.   Zha C, Li M (2004) A simple discuss about condition monitoring and fault diagnosis technique and its application in modern aero engine. Aviat Maint Eng 2004(2):47–48
27.   Zhu H, Wang K, Chen Q (2000) Adaptive thresholds algorithm for fault detection of liquid rocket engine in ground test. J Propuls Technol 21(1):1–4
28.   Yang X, Zhang Z, Xiang S et al (2005) Adaptive threshold algorithm based on autoregressive model for fault detection of LRE in ground test. J Aerosp Power 20(6):1088–1092
29.   Xie G, Hu N, Hu L (2006) Improved adaptive correlation thresholds algorithm for turbopump real-time fault detection. J Propuls Technol 27(1):5–8
30.   Zhong Y, Jin T, Qing S (2005) New envelope algorithm for Hilbert-Huang transform. J Data Acquis Process 20(1):13–17
31.   Zhang W, Zhang Y (2006) Fault mechanism analysis and diagnosis technology for the missile engine. Northwestern Polytechnical University Press, Xi'an
32.   He Z (2001) Advanced engine health management applications of the SSME real-time vibration monitoring system. J Rocket Propuls 2001(6):29–33
33.   Li T, Li A, Wang T et al (2004) Study of fault diagnosis for valve train based on time sequence analysis. Chin Intern Combust Engine Eng 25(6):73–75
34.   Wei C, Qin H (2002) The application of the theory of grey system to the diagnosis of equipment breakdown. J Jishou Univ (Nat Sci Edn) 23(2):70–74
35.   Zhu D, Yu S (2002) Survey of knowledge-based fault diagnosis methods. J Anhui Univ Technol 19(3):197–205
36.   He F (2007) Study on the fault diagnosis technology of shielded pump in the rocket power system. Dissertation of the Xi'an Institute of High-Tech, Xi'an
37.   Chen Q, Liu H (2006) Intelligent fault diagnosis for liquid rocket engine. J Rocket Propuls 32(1):1–6
38.   Li G, Li X, He X (2005) Application in aero-engine fault diagnosis based on reformative ART1 neural network. Control Autom 21(25):156–158
39.   Zhang W, Zhang Y, Zhan R et al (1996) An improved neural networks algorithm for rotary machinery fault diagnosis. J Vib Eng 9(1):31–37
40.   Zhang W, Xu Z, Huang X (2003) Study on application of multi-layer structure neural network to fault diagnosis of liquid rocket engine. Mechan Sci Technol 22(2):292–294
41.   Tian L, Zhang W, Yang Z (2009) Application of elman neural network on liquid rocket engine fault prediction. J Project Rockets Missiles Guidance 19(1):191–194
42.   Gao Z, Zhang W, Ma B et al (2004) Application of fuzzy pattern recognition to fault diagnosis of liquid propellant rocket engines. J Vib Eng z1:415–418
43.   Zhang S, B Xu (2001) A study of fuzzy cluster of samples by neural network in fault diagnosis. J Beijing Inst Machin 16(2):1–4
44.   LES, Ryan RS (1992) (NASA) The role of failure/problems in engineering: a commentary of failure experienced. N92-22235, March 1992
45.   Wu J, Zhang Y, Chen Q (1996) Fault analysis for large liquid rocket engine with turbo-pump system. Missiles Space Veh 1:10–15

46. Yang E, Zhang Z, Cui D (1999) Study on fault monitoring and diagnosis techniques for thrust chamber and turbo-pump systems of liquid rocket engines. J Beijing Univ Aeronaut Astronaut 25(5):619–622

47. Zhang Y, Wu J (1994) Model based fault detection and diagnosis for the propulsion system. J Propuls Technol 5:1–8

48. Wu J, Zhang Y, Chen Q (1995) A survey on fault detection and diagnosis for liquid rocket propulsion system. J Propuls Technol 16(6):45–50

49. Wu J, Zhang Y, Chen Q (1994) Steady fault simulation and analysis of liquid rocket engine. J Propuls Technol 1994(3):6–13

50. Zhao Z, Yao H, Li W (1993) Simplified mathematical model and digital simulation of aero-engine. J Aerosp Power 8(2):138–204

51. Tan S, Liu H (1991) Steady character simulation on the large liquid rocket engine. J Rocket Propuls 4

52. Guo K (1987) Discussion on the steady simulation of liquid rocket engine. J Rocket Propuls 3:41–47

53. Guo K, Li Z (1987) Nonlinear analysis for static characteristics of liquid propellant rocket engine. J Astronaut 4:44–49

54. Guo K (1987) Liquid rocket propellant engine system analysis and evaluation. J Xi'an Inst High-Tech 6

55. Zhang Y (1998) Health monitoring technology for the liquid rocket engine. National Defense University of Technology Press, Changsha

56. Sun G (1993) Fault mode analysis of liquid rocket engine. Missile Space Veh 5:27–37

57. Zhang D (1991) Computing liquid dynamics. Zhonsan University Press, Guangzhou

58. Liu H (1991) Boundary layer theory. China Communication Press, Beijing

59. Zhu W (1993) Hydrodynamics. Higher Education Press, Beijing

60. Kong L (1990) Engineering liquid dynamics. Shandong University of Technology Press, Jinan

61. Guo X (1990) Liquid rocket engine tests. Astronautic Press, Beijing

62. Wu J (1995) Researches on fault detection and diagnosis of liquid rocket engine. Dissertation of National Defense University of Technology, Changsha

63. Yang E (1999) Research on direct/inverse problems and algorithms of condition monitoring and fault diagnosis for liquid rocket engine. Dissertation of Beijing University of Aeronautics & Astronautics, Beijing

64. Chen Q (1993) Control and dynamic characteristic theory of liquid rocket engine. National Defense University of Technology Press, Changsha

65. Mason JR, Southwich RD (1989) Large liquid rocket engine transient performance simulation system. Pratt & FR-20282-3, July 1989

66. Ruth EK, Ahn H, Baker RL, Brosmer MA. Advanced liquid rocket engine transient model. The Aerospace Corp, EI Segundo, CA. AIAA-90-2299

67. Cheng M, Zhang Y (2000) Dynamic characteristics of priming process in space craft propulsion system (I) theoretical model and simulation results. J Propuls Technol 4

68. Lu S, Zhang Y (1996) Real-time fault simulation model of liquid rocket engine. J Propuls Technol 17(5):14–17

69. Chen M, Ling Y (1992) Computing methods. Xi'an Jiaotong University Press, Xi'an

70. Yuan Y, Sun W (2001) Optimal theories and methods. Science Press, Beijing

71. Li Q, Mo Z, Qi L (1999) Numerical method of the nonlinear equations. Science Press, Beijing

72. Lin J, Han Z (1994) An improved Hopfield neural network for solving nonlinear equations. Control Desis 9(2):111–114

73. Kennedy MP, Chua IO (1988) Neural networks for nonlinear programming. IEEE Trans, CAS-35(5)

74. Zhang W, Xu Z, Huang X (2003) Simulation of steady-state fault for whole liquid propellant missile power system. J Syst Simul 15(9):1205–1207

75. Li S (2004) Ant colony algorithms with applications. Harbin Institute of Technology Press, HarBin

76. Zhan S, Xu J (2003) An ant colony algorithm which applying to the multidimensional function optimization problems. J Basic Sci Eng 11(3):223–229

77. Xiong W, Yu S, Wei P (2003) A design based in ant colony algorithm for function optimization. Microelectron Comput 20(1):23–30

78. Zhan S, Xu J, Wu J (2003) The optimal selection on the parameters of the ant colony algorithm. Bull Sci Technol 19(5):381–386

79. Zhou M, Sun S (1999) Genetic algorithms theory and applications. National Defense Industry Press, Beijing

80. Yan P, Zhang C (2005) Artificial neural networks and simulated evolution computation. Tsinghua University Press, Beijing

81. Zhang W, Xu Z, Huang X (2004) Simulation of dynamic fault for whole liquid propellant missile power system. J Syst Simul 16(5):971–973

82. Zhang W, Xu Z, Huang X (2004) Simulation of dynamic fault for whole liquid propellant missile power system. J Syst Simul 16(6):1178–1180

83. Yu H (2001) Equipment fault diagnosis engineering. China Machine Press, Beijing

84. Zhang W (1995) Neural network technology and its application in the fault diagnosis of large rotating machinery. Dissertation of Xi'an Jiaotong University, Xi'an

85. Zhang X (1995) Investigations on fuzzy neural networks theory, methods and applications. Dissertation of Xia'an Jiaotong University, Xi'an

86. Zhang Y, Zhang W, Liu G (1998) Fault detection and diagnosis of the power system based on multi-layer neural network. J Vib Eng 9:11

87. Powell MID (1985) Radial basis function for multivariable interpolation. Review IMA conference on algorithms for the approximation of functions data RMCS Shrivanham

88. Broomhead DS, Lowe D (1988) Multivariable function interpolation and adaptive network. Complex Syst 2:321–355

89. Zhou K, Kang Y (2005) Neural network model and its simulation design by Matlab. Tsinghua University Press, Beijing

90. Pal NR et al (1993) Generalized clustering networks and kohonen's self-organizing schemes. IEEE Trans NN 4:549–557

91. Xu Y, Wen X, Yi X, et al (2001) ART-2A same phasic data with different amplitudes and resolution methods. J Data Acquis Process 16(4):393–398

92. Shen A, Yu B, Guan H (1996) Research on a classifier using the ART-2 neural network. J North Jiaotong Univ 20(2):146–151

93. Xu Y, Wen X, Yi X, et al (2002) New ART-2A un-supervised clustering algorithm and its application on BIT fault diagnosis. J Vib Eng 15(2):167–172

94. Chen Z, Zhou R, Liu H et al (1996) A novel adaptive harmonic oscillation algorithm. J Softw 7(8):458–465

95. Grossman A, Morlet J (1984) Decomposition of hardy functions into square integrable wavelets of constant shape. SIAM J Math Anal 15(4):723–736

96. Meyer Y (1992) Wavelet and operators. Cambridge University Press, Cambridge, UK

97. Daubechies I (1988) Orthonormal bases of compactly supported wavelets. Comm Pure Appl Math 41:909–996

98. Mallat S (1989) multiresolution approximations and wavelet orthonormal bases of L2 (R) Tran. Am Math Soc 315:69–87

99. Mallat S (1989) A theory for multiresolution signal decomposition: the wavelet represention. IEEE Trans PAMI 11(7):674–693

100. Chui CK, Wang JZ (1991) A cardinal spline approach to wavelet. Proc Am Math Coc 113 (3):785–793

101. Cui J, Cheng Z (1995) An introduction to wavelets. Xi'an Jiaotong University Press, Xi'an

102. Coifman RR, Meyer Y, Wickerhauser MV (1992) Wavelet analysis and signal processing. In: Wavelets and their applications, pp 153–178

103. Goodman TNT, Lee SL (1994) Wavelets of multiplicity. Trans Am Math Soc 342 (1):307–324
104. Geronimo JS, Hardin DP, Massopust PR (1994) Practical function and wavelet expansions based on several scaling function. J Approx Theor 78(3):373–401
105. Steffen P, Heller PN, Gopinath RA et al (1993) Theory of regular M-band wavelet bases. IEEE Trans 41(12):3497–3511
106. Li C, Zheng N (1999) Construction theory of the compacted supported wavelet frame with high dimension. Sci China (Ser E) 29(4):321–332
107. Vetterli M, Herley C (1992) Wavelet and filters banks: theory and design. IEEE Trans 40 (9):2207–2232
108. Strang G, Nguyen T (1996) Wavelet and filters. Wellesley Cambridge Press, Wellesley
109. Sweldens W (1997) The lifting scheme: A custom-design construction of second generation wavelets. SIAM J Math Anal 29(2):511–546
110. Sweldens W, Schroder P (1996) Building your own wavelets at home. In: Wavelets in computer graphics, pp 15–87. ACM SIGGRAPH course notes
111. Sweldens W (1996) The lifting scheme: a custom-design construction of biorthogonal wavelets. Appl Comput Harmon Anal 3(2):186–200
112. Mallat S (1991) Zero-crossings of a wavelet transform. IEEE Trans IT 37(4):1019–1033, July 1991
113. Gao Q, He Z (2000) Application of harmonic wavelet and its time-frequency profile plot to diagnosis of rotating machinery. J Xi'an Jiaotong Univ 34(9):62–66
114. Jiao L, Du H, Liu F et al (2006) Immune optimization—computing, learning and cognition. Science Press, Beijing
115. Li T (2004) Computer immunology. Publishing House of Electronics Industry, Beijing
116. Hofmeyr SA, Forrest S (1999) Immunity by design: an artificial immune system. In: Proceedings of GECCO'99 Orlando, Florida, USA
117. Ballet, P, Rodin V (2000) Immune Mechanisms to regulation multi-agent systems. GECCO 2000, Las Vegas, Nevada, USA
118. Dasgupta D (1998) An artificial immune system as a multi-agent decision support system. In: Proceedings of the IEEE international conference on systems, man and cybernetics (SMC), San Diego
119. Ishida Y (1996) Active diagnosis by immunity-based agent approach. In: Proceedings of international work-shop on principle of diagnosis (DX 96), Val-Morin, Canada
120. De Castro LN, Von Zuben FJ (2001) AiNet: an artificial immune network for data analysis. Int J Comput Intell Appl (IJCIA) 1(3):231–259
121. Timmis J, Neal M (2001) A resource limited artificial immune system for data analysis. Knowl Based Syst 14(3–4):2001
122. Garrett SM (2005) How do we evaluate artificial immune systems. Evol Comput 13 (2):145–178
123. Forrest S, Perelson A, Cherukuri R (1994) Self-nonself discrimination in a computer. In: Proceedings of 1994 IEEE computer society symposium on research in security and privacy. IEEE Computer Society, Los Almitos, CA, USA, pp 202–212
124. Timmis J, Neal M, Hunt J (2000) An artificial immune system for data analysis. Biosystem 55(1–3):143–150
125. Chen J, Wang H (2002) The application of system dynamic simulation technology to the development of LOX-LH2 staged combustion rocket engine. Missiles Space Veh 5:29–35
126. Liu H, Zhang E, Dong X (1999) Start-up simulation of a liquid propellant staged combustion rocket engine J Propuls Technol 20(3):5–9
127. Yang L, Gao Y (2003) Fuzzy logic theory mechanism and applications. South China University of Technology Press, Guangzhou
128. Gao Z (2005) Application of the fuzzy theory in the fault diagnosis of liquid rocket engine. Institute of Xi'an high technology, Xi'an

129. Shi Y, Wang G (2000) Fault classification of the power boiler loss based on fuzzy pattern recognition. J Eng Therm Energy Power 15(5):526–528

130. Chen J, Fu L, Cui X (2002) Fuzzy cluster analysis based on Max-⊙ transitivity. J Chongqing Univ 25(12):98–101

131. Li X, Liu X, Wang K (2003) Multi-machine power system eigenvalue calculation based on fuzzy clustering analysis. Electron Power Autom Equip 23(1):76–78

132. Liu S, Hu X (2001) Fuzzy clustering analysis of the surface subsidence coefficients. J Nanjing Archit Civil Eng Inst (Nat Sci) 57(2):27–31

133. Huang M (1998) Study on the neural networks diagnosis of liquid rocket engine. Dissertation of National University of Defense Technology, Changsha

134. Jang JSR, Sun CT (1993) Functional equivalence between radial basis function networks and fuzzy inference systems. IEEE Trans Neural Netw 4:156–158

135. Hunt KJ (1996) Extending the functional equivalence of radial basis function network and fuzzy inference system. IEEE Trans Neural Netw 3:776–781

136. Comes C, Vapnik VN (1995) Support vector networks. Mach Learn 20:273297

137. Vapnik VN (1998) Statistical learning theory. Wiley, New York

138. Vapnik VN, Zhang X (2000) The nature of the statistical learning theory. Tsinghua University Press, Beijing

139. Zhu Y (2003) Support vector machine and its application in the fault pattern recognition of machine. Dissertation of Xi'an Jiaotong Univeristy, Xi'an

140. Zheng R (2005) Study on the support vector machine in the fault diagnosis. Dissertation of Zhongnan University, Changsha

141. Zhang X (2000) On the statistic learning and support vector machine. Acta Autom Sinica 26(1):32–42

142. Jaakkola TS, Haussler D (1999) Probabilistic kernel regression models. In: Proceeding of the conference on A.I. and statistics

143. Opper M, Winther O (2000) Gaussian process and SVM: mean field and leave-one-out. In: Smola AJ, Barlett PL et al (eds) Advances in large margin classifiers. MIT Press, Cambridge, pp 311–326

144. Chapella O, Vapnik V (1999) Model selection for support vector machines. In: Advances in NIPS'99

145. Gu Y, Shi D, Wang Y (2001) Applications of hidden Markov models in biological sequence analysis. Nat Mag 23(5):273–277

146. Rabiner LR, Juang BH (1986) An introduction to hidden Markov models. IEEE ASSP Mag 3(1):4–6

147. Huang Y (2015) Applications of hidden Markov models in bioinformatics. Electron Sci Tech 28(8):185–189

148. Luo X (2006) Research on faults diagnosis methods and system based hidden Markov models(HMM) for power transformer. Dissertation of Zhejiang Univeristy, Hangzhou

149. Xie J (1995). Application of hidden Markov model in sound processing. Huazhong University of Science and Technology (HUST) Press, Wuhan

150. Bahl LR, Brown PF et al (1986) Maximum mutual information estimation of hidden Markov parameters for speech recognition. In: Proceedings of ICASSP'86, pp 49–52

151. Atlas L, Ostendorf M, Bernard GD (2000) Hidden Markov models for monitoring machining tool-wear. IEEE international conference on acoustics, speech, and signal processing, vol 6, pp 3887–3890

152. Ocak H, Loparo KA (2001) A new bearing fault detection and diagnosis scheme based on hidden Markov modeling of vibration signals. IEEE international conference on acoustics, speech, and signal processing, vol 5, pp 3141–3144

153. Song X, Ma H, Liu J et al (2006) Feature extraction and clustering technology in the fault diagnosis based on HMM. J Vib Meas Diagn 26(2):92–96

154. Winger LL (2001) Linearly constrained generalized Lloyd algorithm for reduced codebook vector quantization. IEEE Trans Signal Process 49(7):1501–1509

155. Rabiner LR (1989) A tutorial on hidden Markov models and selected application in speech recognition. Proc IEEE 77(2):257–286

156. Liu C (2008) Study on the application of hidden Markov model in the fault diagnosis of turbo pump. Institute of high technology in Xi'an, Xi'an

157. Liu X, Liu G, Qiu J (2006) HMM-SVM based fault diagnosis model and its application. Chin J Sci Instrum 27(1):45–48

158. Hu G (2005) Strategy: prediction and decision. Tsinghua University Press, Beijing

159. Wu J, Sun D (2006) Modern data analysis. China Machine Press, Beijing

160. Li H, Hu Q (2005) Prediction and decision. Xidian University Press, Xi'an

161. Yang E (1996) Research on the condition monitoring and fault diagnosis engineering application system of YF75 engine. Dissertation of Beihang University

162. Zhang S, Lei Y (2007) Application of MATLAB in the time series analysis. Xidian University Press, Xi'an

163. Xiao X, Song Z, Li F et al. (2005) Basic theory and application of grey technology. Science Press, Beijing

164. Liu S, Dang Y, Fang Z et al. (2004) Grey system theory and applications, 3rd edn. Science Press, Beijing

165. Zhang D, Jiang S, Shi K (2002) Theoretical shortage and improvement of the grey prediction formula. Theor Pract Syst Eng 8:140–142

166. Liu S, Deng J (2000) Application rang of the GM(1,1) model. Theor Pract Syst Eng 5:121–124

167. Xiao J, Sun D, Qin Y (2005) Application of the grey model in electric power predicting and its optimization. Control Theor Appl 24(2):19–21

168. Sfetsos A, Siriopoulos C, Combinatorial time series forecasting based on clustering algorithms

169. Ruiz CA, Wauman MW, Gadinaitis WS (1992) Algorithms for real-time fault detection of the space shuttle main engine. AIAA 92-3169

170. Zhang Y, Mo J, Li R, Zhang W (1999) Fault detection and diagnosis of power system based on multiple neural network. In: Eighth international space conference of Pacific-basin societies, Xi'an, China, June 1999

171. Duan X (2003) Research on the novel algorithm of undetermined engineering investment based on RBF neural network. Dissertation of Liaoning Engineering and Technology University

172. Elman JL (1990) Finding structure in time. Cognit Sci 14:179–211

173. Yang J, Wu X, Wang N et al (2002) Research on load prediction model of air condition system based on Elman neural network. J Chongqing Univ (Nat Sci Edn) 25(8):25–27

174. Cui D, Yang E, Zhang Z, et al (1996) Neural networks approach to the multi-steps prediction of multi-parameters with correlative parameters. Acta Aeronautica et Astronautica Scinica. 17(3):310–316

175. Enrico C (2004) On support vector machines and sparse approximation for random processes Neurocomputing 56:39–60

176. Tay FEH, Cao L (2001) Application of support vector machines in financial time series forecasting. Omega 29:309–317

177. Ai N, Wu Z, Ren J (2005) Support vector machine and artificial neural network. J Shangdong Univ Technol (Sci & Tech) 19(5):45–49

178. Suykens JAK, Vandewalle J (1999) Least squares support vector machine classifiers. Neural Process Lett 9(3):293–300

179. Wang H, Zhang J, Xu X (2005) Machinery trend condition support vector machine intelligent prediction. Mach Tool Hydraul 5:170–172

Printed in the United States
By Bookmasters